Mines and Minerals of California

California Miner's Association

with an introduction by Kerby Jackson

This work contains material that was originally published in 1899 by the California Miner's Association.

Introduction

It has been many years since the California Miner's Association released it's important publication "Mines and Minerals of California". First released in 1899, this important volume has now been out of print and has been unavailable to the mining community since those days, with the exception of expensive original collector's copies and poorly produced digital editions.

It has often been said that *"gold is where you find it"*, but even beginning prospectors understand that their chances for finding something of value in the earth or in the streams of the Golden West are dramatically increased by going back to those places where gold and other minerals were once mined by our forerunners. Despite this, much of the contemporary information on local mining history that is currently available is mostly a result of mere local folklore and persistent rumors of major strikes, the details and facts of which, have long been distorted. Long gone are the old timers and with them, the days of first hand knowledge of the mines of the area and how they operated. Also long gone are most of their notes, their assay reports, their mine maps and personal scrapbooks, along with most of the surveys and reports that were performed for them by private and government geologists. Even published books such as this one are often retired to the local landfill or backyard burn pile by the descendents of those old timers and disappear at an alarming rate. Despite the fact that we live in the so-called "Information Age" where information is supposedly only the push of a button on a keyboard away, true insight into mining properties remains illusive and hard to come by, even to those of us who seek out this sort of information as if our lives depend upon it. Without this type of information readily available to the average independent miner, there is little hope that our metal mining industry will ever recover.

This important volume and others like it, are being presented in their entirety again, in the hope that the average prospector will no longer stumble through the overgrown hills and the tailing strewn creeks without being well informed enough to have a chance to succeed at his ventures.

Kerby Jackson
Josephine County, Oregon
October 2015

Introduction.

HE subject matter of this work is confined as far as possible to a plain statement of facts, regarding our great state and its mineral wealth. It has been a difficult and laborious task to gather all the data and statistics, as aside from a few counties, whose citizens were public spirited enough to look after their county interests, all matter has been prepared from the Secretary's office.

In the early stages of the labor, much assistance was promised by members of the Association and others, in the form of special articles upon subjects of great interest to the industry, as well as reports upon various counties and their resources. As the work progressed, however, it was found that many substitute articles would be necessary, as promises make poor reading matter. The Association is indebted to the contributors, whose articles upon various subjects will be found highly interesting and instructive; to the advertisers, whose solid financial support has materially assisted in defraying the cost of publication; to Mr. J. O. Denny, whose able and conscientious services have been indispensable in the editorial work, and to the publishing house, for the extra care and attention given to the general make-up.

Aside from the advertisers, the Association is under obligation to the following public spirited people, whose generous subscriptions have made it possible to issue this work, without drawing upon the general fund in the treasury:

Hon. James D. Phelan	Phoebe Hearst
Miller, Sloss & Scott	Alvinza Hayward
Dunham, Carrigan & Hayden Co.	Chas. D. Lane
Southern Pacific R. R. Co.	Edward Coleman

Pacific Telephone Co.	J. C. Pascoe
D. B. Hinckley	Tubbs Cordage Co.
Ensign, Bickford & Co.	P. Geo. Gow
Washburn & Moen Co.	Cal. Cap Co.
J. Palache	Cal. Fuse Works
W. B. Bourn	Sussman, Wormser & Co.
A. Halsey	Jas. Spires, Jr.
W. A. Farrish	Alf. von der Ropp
J. A. Rathbone	Milton Andrus
D. E. Hayes	Gutta Percha Rubber Co.
G. W. Beaver	Levi Strauss
E. J. Molero	C. W. White

The purpose of this work is to call attention to our bountiful mineral wealth and our thousands of acres of undeveloped and unprospected country, feeling that by so doing, we may be able to interest some of the people who are seeking legitimate mining investments.

Very truly,

Edward H. Benjamin, Secretary.

The Mineral Industry of California.

BY

CHARLES G. YALE.

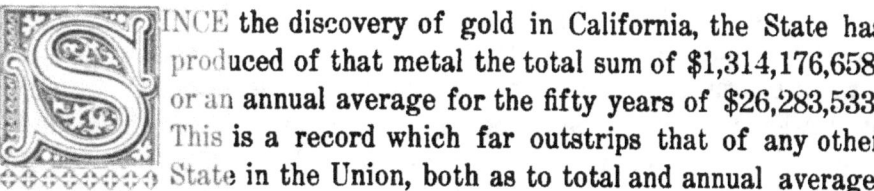

INCE the discovery of gold in California, the State has produced of that metal the total sum of $1,314,176,658, or an annual average for the fifty years of $26,283,533. This is a record which far outstrips that of any other State in the Union, both as to total and annual average, from time of first production. And this is not the end, by any means, since gold production still exceeds that of all other mineral products of the State, and California continues to yield from fifteen to seventeen millions yearly, with good prospects of a gradual increase for some time to come.

California has well earned her title of the "Golden State," as this record shows. A single section of the Union, which has been producing, on an average for fifty years, over twenty-six millions of dollars in gold annually, and still continues yielding many millions, is entitled to attract investors to its gold fields. It is true that the yearly yield is not as great now as when the famous placers of the State were opened and before they were exhausted, but, at the present time, the gold mainly comes from quartz mines and is naturally more difficult to obtain. In the year 1852, the State produced $81,294,700, and it was not until 1857 that the yield fell below fifty millions. It was nearly all placer mining that was done in those days. Since then we have been getting our gold, not only from placers, but from quartz, pocket, seam, hydraulic, drift, ocean-beach sand, dredging, wing-damming, dry-washing, and other forms of gold mining. The relative importance, at the present time, of the different forms of precious metal mining in California is shown in the record of 1898, when the quartz mines produced $12,488,321; the placers, $1,841,473;

the drift mines, $1,028,547, and the hydraulic mines, $962,192. In this record, the Chinese miners are called placer miners, and the surface placers include products of bars, river beds, gulches, flats, dredgers, etc.

It should be borne in mind, in considering the mineral product of the State, that the figures quoted relate only to gold mines, and that the mines of silver, copper, lead, borax, quicksilver, etc., etc., are not referred to, but will be, further on, in this chapter. While the combined gold and silver product of this State was, in 1898, $16,320,533, the total value of all the mineral products of the State that year, including gold and silver, was $27,289,079, so that it may be seen that the State is rich in other mineral wealth beside gold and silver.

In the earlier history of gold mining in California, the main portion of the precious metal was procured from the placers or surface washings, in the gulches, canyons, and river bars and beds, and for many years an enormous yield was maintained from these sources. Gradually, however, as was to be expected, the area available for this kind of mining was narrowed, as ground was worked out, and then attention was turned to other sources of gold supply.

Finding, as the miners of those days did, these auriferous deposits only at certain points, they very naturally looked near by for the source of the free placer gold. This led to the search for quartz veins, and also for the large bodies of auriferous gravel, which were contained in the hills of the mining counties of the State. In time, the deep gravel deposits were found and located, and the quartz ledges opened and developed. Then the character of the mining and the character of the mining population changed. It was no longer possible for a nomadic miner, with pick and pan, to gather a fortune in a few days, from a deposit which nature had concentrated for him in a few yards of earth. It became necessary to employ capital, as well as labor, in the mining operations of the State. Companies took the place of the individual miner, built ditches to bring in water to the gravel claims, and mills to crush the ore from the quartz veins. The miners gradually stopped working for themselves, and were employed by the companies for daily wages. They became more settled in their habits, gave up their nomadic instincts, and became permanent residents of the little mining camps and larger mining towns, where they were sure of steady employment in the mines or mills. Of course, there are still many prospectors in the mountain counties, as well as miners who are working their own claims, but the majority of the

A Typical Hydraulic Mine in Trinity County. An Illustration of a High Bench of the Present River System.

mining population is now engaged in work for the companies, who are operating the mines on a more extensive scale than is usually possible for individuals.

The era of speculative mining, incidental to newly settled mining regions, has also passed by in California.

EARLY MISTAKES.—In the earlier history of gold mining in this State, some very foolish and extravagant ideas prevailed, and many mistakes were made. As a result, many mines had to be abandoned, after more or less loss, and people came to look upon gold quartz mining as a risky business. It was found, ultimately, that the fault lay more with the men themselves than with the mines. Basing the entire yield of a lode on a few assays from rich specimens of ore from a "chimney" was found to be fallacious; and experience taught the lesson that a good solid ledge of average ore was far more valuable than the mere rich pocket mine bright with specimens. It was found, too, that because a mine was a gold mine, it did not follow that all sorts of extravagant expenditures could be indulged in; that its operations must be conducted on a business basis, and economy be practiced in all departments, the same as in a manufacturing or other enterprise. These things are true of other States, as well as California, and are here mentioned simply that the changes and advances may be noted, which have gradually led to the present condition of the quartz-mining industry in California.

It may be confidently stated, that the quartz industry of the State is in a better condition to-day than it has been since its inception. The appliances for saving gold which have been perfected are applied intelligently and carefully, and every effort is made to gain a large percentage of the gold that is in the ores.

What are termed low-grade ores vary in value from $3.50 to $8 per ton in gold; and high-grade ores are those yielding from $15 to $30 per ton. The average value of the ore treated at present may be stated at $10 to $12 a ton. In the low-grade ores, the gold can seldom be seen with the naked eye, but the high-grade ore often shows particles of free gold. Mineralogically, the ores worked consist generally of a quartz gangue, carrying free gold and iron pyrites. Quartz is the characteristic matrix of the veins, though other matrices occur.

The milling of gold ores in California has been greatly perfected of late years, and it is now possible to work ores profitably that twenty years since were practically worthless. This has come about by the adoption of more economical methods in both mine and mill, and greater care in the handling of the ore, as well as increased knowledge of the proper way to treat the ores in the mill. Improved appliances are also important factors. The present form of stamp-mill combines durability of construction with an automatic simplicity in the manipulation of the ores. The important object in the treatment is to secure the largest yield consistent with the least waste of the

CROPPINGS OF QUARTZ LEDGE ON THE MOTHER LODE.

metals, by the adoption of the most rapid, simple, and economical method of reduction and amalgamation. The cyanide process, chlorination process, and canvas plants, play an important part in the ultimate saving of all the gold.

Although the ordinary stamp-mill still continues to maintain its prestige for the reduction of ores in California, it is proper to state that, within the past few years, various forms of roller-mills and other forms of pulverizers have been largely put into use, especially on small mines, where the owners were unwilling, or unable, to go to the expense of erecting stamp-mills. There are several types of the pulverizers, varying more or less in principle, but forming in themselves complete ready-built mills, simply needing a foundation to be ready for work. These cost less than stamps, and have been found very effective on certain classes of ores. In fact, their use is increasing gradually, as they are well adapted for prospecting mills and mines owned by individuals or small companies. There are some persons who advocate their use in preference to the stamp, while others do not believe in anything but the stamp for crushing ore. However, numbers are made, sold, and put into use, and, in certain instances, their effectiveness cannot be doubted.

There have been many and costly failures in quartz mining, and, from the fact that all quartz veins cannot be prolific gold-bearers, there will be failures in the future; yet experience has been a good teacher, and imparted knowledge where ignorance once existed, and improvements in machinery, and better knowledge of working ores, have done much of late in giving a reliability to quartz mining that it did not possess in earlier years, when there was a general inexperience as to the characteristics of mineral-bearing veins, and a woeful lack of knowledge in properly manipulating the ores. Early quartz mining was attended with great expense, and only because of the richness of many veins was it possible to obtain a profit when there was such an absence of skill and economy. Quartz miners have learned, however, that gold-bearing veins are not rich as a rule, but are yet sufficiently reliable, in a majority of cases, to yield a profit if the same prudence and intelligence is used as is necessary in other trades and industries. This has been so well demonstrated that, of late years, quartz mining has grown into much larger dimensions than is generally supposed, not through the agency of corporations that figure in the stock markets, but mainly through local or individual enterprise, until there are now, in all parts of the mining regions, quartz mines that are being worked regularly and with a profit, but whose output is not a matter of publication any more than would be the details of other private business.

It is to be borne in mind that gold mining in this State has none of the characteristics of stock gambling in any manner. Of all the numbers of quartz, hydraulic, drift, placer and other gold mines, now being operated in California, not a single one is called at the mining stock exchanges in San Francisco. It has never been the custom to list gold mines, and the few attempts made in that direction by the brokers have not met the approval of the gold miners. Of course, shares of stock in the incorporated companies are bought and sold, but they are not dealt in at the exchanges in the usual manner of ordinary stocks. It is possibly for this reason, among others, that many suppose gold mining to be on the wane in this State, though the contrary is the case. In fact, there has not been one word said or printed of the gold mines of California, where there have been many thousands of the famous Klondike in the past few years. Nevertheless, California has each year exceeded the gold yield of the entire Klondike. The records of the United States Mint at San Francisco show that, while the gold yield of the Klondike in 1898 was 595,318 standard ounces, or $11,038,478, the gold yield of California in the same period was $15,906,478.

WIDE DISTRIBUTION.—Quartz mines are found and worked in a great many counties in California, from Siskiyou on the north to San Diego on the south. Most of these are in the mountain, foot-hill and desert regions, there being none of note in the valley counties or Coast Range. At present, most of the quartz mines are worked by companies organized for the purpose, though some are operated by individuals; most of the latter, however, being worked on a comparatively small scale. There are great numbers of "prospects," or undeveloped quartz mines, scattered throughout the mining counties, which need the assistance of capital for their full development, and to prove their worth. They are mainly owned by the prospectors who found them, but who are unable from lack of means to properly equip them with machinery and open them. The possibilities for capital in this direction are practically unlimited. Most purchasers, however, look for "going" mines, that is, those which are developed and paying. Naturally, these are scarce, since the owners of a paying mine do not care to sell except at a high price. With the owners of smaller or less developed mines, it is different, and this is the reason so many claims are for sale. It is not that the men do not themselves think the claims valuable, but because they have not the means to put the mines in shape to become productive. A very few thousands of dollars would purchase any one of hundreds of these claims, but more money must be put into development before there is possibility of profit.

The stamp-mills, machinery and other appliances used at the quartz mills of this State do not vary materially from those used elsewhere in this country, and therefore it is not necessary to enter into any detailed descriptions of them. Many of the appliances now generally used originated here, as California was a pioneer in quartz mining, and they were brought to a degree of perfection simply through the experience in their use.

FLUME TO CARRY 600 MINER'S INCHES OF WATER FROM THE STANISLAUS RIVER
TO THE 120-STAMP MILL AT THE MELONES MINE.

ABANDONED AND RE-OPENED MINES. — It is worthy of note that a great many of the mines which were abandoned after a few years' working have of late years been re-opened and worked with success and profit. This has been particularly the case with the mines on what is known as the Mother Lode, which extends through five counties of the State, viz.: El Dorado, Amador, Calaveras, Tuolumne and Mariposa. Between 1850 and 1860, numbers of quartz mines were opened in different parts of the State, by men with no experience or knowledge of the business, and many were the failures in consequence. Mine after mine was closed down and so remained for many years. In 1858 there were 280 quartz mills in California, and in 1861 the State Geological Survey found that there were only forty or fifty of these in successful operation, and several of these were leading a very precarious existence.

This shows the change which had occurred in a few years, and how many quartz operations had been failures. Mills were built before they were sure of having a mine; the percentage saved from the ore was small as compared with these days; sulphurets were not saved, and the whole business was run in a hap-hazard way. The early locations were made at any point where croppings were found, and mills built before developments warranted. Millions were squandered by men ignorant of quartz mining or milling, and upon the exhaustion of free gold surface ore, and when sulphurets began to appear, the mines closed down. About this time, in 1860, the superior attractions of Comstock silver mining in Nevada caused many to leave for that State, and the gold quartz properties of California were practically overlooked.

As stated, this general abandonment was more marked along the Mother Lode than elsewhere, since the grade of the ore was not as high as where smaller and richer ledges were found. Moreover, the Mother Lode mining was found to be very expensive, requiring large plants and considerable capital. In numerous instances, where water became troublesome, and the ore of too low grade to pay with small mills, under conditions existing at that time, the mines were left idle and practically abandoned, only work enough being done to comply with the laws.

During these earlier periods numbers of quartz mines were opened, which, under the conditions prevailing and the nature of the gold-saving appliances, could not possibly be made to pay. On some of the Mariposa estate mines, in 1863, where they worked about 65,000 tons of ore averaging from $7.77 to $8.05 per ton, the cost was $5 per ton for all expense; and in one of these mines, when they were working $29 rock, the tailings showed $16 per ton. Prof. Ashburner, of the State Geological Survey, said that seventy per cent of the gold in the quartz worked at the Benton mills was lost. In Tuolumne County, as late as 1869, according to official reports, the owners of the mines were mostly men without capital, depending entirely on the daily yield of the mines for opening and developing them, and in that year three-quarters of the mills in the county were idle. Many mines were also involved for years in litigation relating to titles of the property.

A CHANGE IN METHODS.—The experience gained from the many failures in the earlier days of quartz mining finally brought about an entire change in the methods of conducting operations. High-priced and useless officials have been done away with, and skilled men only employed in the different departments. Railroads, better wagon-roads, cheaper supplies, improved means of transportation, better machinery at less cost, highly improved reduction appliances and methods, giant powder, power drills, water and electric power, and generally improved systems and appliances, have all contributed to bring about a change for the better. Under these new conditions, a

HORSE WHIM. WOODSTOCK MINE, CHILI GULCH, CALAVERAS COUNTY.

large number of the old abandoned mines throughout the State have within the past ten or fifteen years been rehabilitated and made producers. Experience proved that the most successful mines were those which had been properly opened with sufficient capital to develop and equip them without waiting for the mine to pay for this. Numbers of individuals or small companies had opened mines on the Mother Lode, but had to give them up before pay ore was reached. The newer and larger companies have had the advantage of all the earlier work.

It is a noteworthy fact, borne out by the records of the U. S. Mint at San Francisco, that the four largest producing gold mines in the State of California have been, for the past five years, mines which were for years idle and abandoned, but eventually brought to a producing point by the introduction of capital and economical systems of mining and milling. These examples of many on a smaller scale which might be cited have encouraged others to open properties which have been idle for years, and which did not pay under the conditions referred to.

Another feature to which attention may be called, in the case of such abandoned mines which had once paid, is that the new companies, having capital at command, now sink shafts 1,200 or 1,500 feet deep direct, without stopping to drift or cross-cut until the desired depth is reached. Such a method of working was never thought of in the early days of quartz mining, where the mine was expected to pay its own way in sinking, etc. This deep sinking involves large expenditure for hoisting and pumping machinery, for which reason only large companies, with abundant capital, can undertake this class of work. Several instances might be cited of the successful and economical result of this kind of work, as compared with the desultory development accomplished by the old style of mining.

PRODUCT OF RE-OPENED MINES.—Along the Mother Lode, few of the old mines abandoned or idle were re-opened, developed, and equipped before the year 1890 by the companies now owning them. A brief note of the record of these re-opened mines, made between the years 1890 and 1898 inclusive, as to bullion production, will serve to show what success has been made in deep mining on such properties as were once given up as useless by those who could not, or did not know how, to work them profitably.

There are no available records of Mariposa County, but in El Dorado, Amador, Calaveras, and Tuolumne counties, twenty-three of these once abandoned mines have, since being re-opened and worked, yielded a total of $23,941,661, or an average of over a million dollars each in gold. Some of these mines referred to have been worked for the full period named, but several of them have only been worked two or three years since being re-opened.

ON AND OFF THE MOTHER LODE.—The conditions of mining on the Mother Lode are fully given in another chapter in this volume, so it is unnecessary to say much in this place. It is to be noted, however, that the fame of that lode is so wide that many persons imagine the whole quartz interests of the State center at that point, and that there is little of value outside of it. This is an entirely mistaken idea. Last year there was a production of gold in thirty-one counties of California, and the Mother Lode only traverses five of them. The largest gold-producing county in the State is not in the Mother Lode region at all. The only county which produced over two millions in gold last year is not on the Mother Lode. Of the seven counties which yielded over a million in gold last year, three are Mother Lode counties and four are not. Of the four largest producing mines in the State last year, two are Mother Lode mines, and the other two are in Kern and San Diego counties, away off the Mother Lode. Of twenty-nine gold mines in California which yielded over $100,000 each, in 1898, there are eight in the Mother Lode counties and twenty-one which are not. These twenty-one are in the counties of Butte, Kern, Mono, Nevada, Placer, San Bernardino, Shasta, San Diego, Sierra, and Trinity. This list includes the drift mines of Placer County. Of mines which produced over $200,000 last year, seven are on the Mother Lode and eight are not. Of eight mines which produced over $300,000 last year, five are on the Mother Lode and three are not; but the two largest producers of the State are in Mother Lode counties.

This statement is not made with any idea whatever of detracting from the well-known merits of the mines on the Mother Lode, but simply that it may be understood that the quartz mining interests of the State are not entirely confined to that section, or to the five counties through which the Mother Lode courses. The large mines on the Mother Lode have had the advantage of introduction of abundant capital for their development and have paid handsomely on the investment. There are still many other opportunities for further investment in this direction, with equal chances of success. But, at the same time, there are many other places in the State where capital may be introduced profitably in the purchase, development, and equipment of quartz properties, and where less money will be required to accomplish proportionate results. That is to say, that properties off the Mother Lode may be originally acquired at less price than upon it, and persons looking for smaller investments may well look elsewhere. Of course, it may be admitted that the larger companies or syndicates desiring extensive properties and intending to invest largely in deep mining, only possible with abundant capital, will continue to turn their attention to the quartz mines of the Mother Lode.

UNDERGROUND SCENE IN THE RED POINT DRIFT MINE.

"Prospects" and Mines.—The greatest difficulty to overcome here, as in other mineral regions, is to induce people with money to take hold of undeveloped properties. They are usually looking for "going" mines, that is, those already developed to a producing stage. But the problem of how best to turn a "prospect" into a mine can only be solved by the individual prospector, since there is no general rule to go by. There are thousands of prospects for sale, but there are few purchasers for this class of property, and, as a result, every county in the State has an abundance of little claims with which the owners can do nothing without the assistance of capital, and, therefore, nothing is known of their actual value.

Growth of the Industry.—During the era of the largest gold production in this State—say from 1850 to 1855 inclusive—the annual output of gold averaged only about $55,000,000. At that time nearly everybody was engaged in mining, while now the miners represent but a small proportion of the population. Those who work for wages do nearly as well now as they ever did, all things considered. But it now requires a larger amount of both skill and capital to accomplish much in our mines than were needed in the early days—a condition of things that puts the mere wage-earner and worker at a disadvantage.

With the above two factors supplied, the opportunities for making money in both the gold and silver mines of California are really better now than they ever were before. This the history of the business during the past decade fully demonstrates. It has, during that period, been growing steadily, and even rapidly, in all the elements of desirable investment—profit, efficiency, economy, and safety—a condition that cannot fail to draw to it capital and insure the countenance of both legislators and the people. Purged of the grosser abuses that so long militated against it, its mistakes largely corrected, and so many other gains made, there can be no doubt that mining will, from this time on, make rapid advances in this State. It would take but little time and energy now to double the present number of our miners, and reach a corresponding increment in our output of gold.

Were California even poor in the precious metals, she would yet become a great mining State. With such wealth as this, she is, in this respect, destined to be a very important factor in the financial affairs of the world. Gold mining is with us in an embryotic state. It has not yet reached even the stage of sturdy infancy. Our true golden era rests in the future, not in the past. Our El Dorado has not yet been revealed to us.   It lies buried deep in the bowels of the earth.   The placer deposits that have made for us such a name, and given to mining such impetus and eclat, were but driblets which Nature, having released from her matrices, brought within our easy reach as a means of encouraging us to further efforts, and leading us on to that greater and more enduring wealth stored away in the rocky ribs of the mountains.

TYPICAL MINER'S CABIN IN REMOTE MINING REGION, PLACER COUNTY.

In considering the bullion made in this State for the past fifteen or twenty years, it must be borne in mind that comparatively little money has, during that period, been invested in our mines, preference having been given by capitalists to real estate, timber, railroads, and various other lines of investment.   But there are, in this respect, indications of a change.   With the many improvements effected, it is found that mining is becoming a tolerably safe, as well as profitable, industry in California.   Some reverses have come, as they will come in every pursuit, yet scarcely any notable losses have here occurred in the business of late years; none, in fact, except by reason of management clearly defective, if not criminally faulty, or through investments of a purely speculative kind.

As regards the extent of our mining field, it is simply illimitable. A hundred millions of additional capital might as well be invested there as not, nor would fifty thousand men crowd it any more than twenty thousand. Of the mineral deposits that actually exist in California, not a tithe, probably, has yet been discovered, nor has a much larger proportion of those already discovered been developed to a productive condition. We have made a good beginning—hardly more.

INCREASING GOLD PRODUCT.—The gold output of the State has been slowly but gradually on the increase of late years, though the drought of two years past has affected it materially. Still, those familiar with the conditions affecting gold mining in California predict a further augmentation of the yield for some years to come. This opinion is based on the fact that, in late years, more capital has been invested in quartz and gravel mines than had previously been the case for a long time. Moreover, it has come to be recognized that deep mining in the gold leads will pay handsomely where sufficient capital is invested for proper development and plant, and many mines, once given up, have been re-opened by organized companies with means to do the work. When properly and thoroughly opened the mines, as a general rule, repay all the outlay and a handsome profit. The deep sinking now being done at many points, especially on the Mother Lode, involves large expenditure for hoisting and pumping machinery, for which reason companies only with abundant ready capital can undertake such operations. Many of these ventures will not reach a productive stage for one or two years, nor are they expected to, but, when those now being prosecuted reach this point, the gold yield of the State should naturally be materially augmented.

It may be stated, also, in further explanation of the gradual increase in gold production of California, that a great many mines have been started again in operation after a long period of idleness, through coming into possession of men or companies financially able to bring them to a producing stage. In numerous instances, where water became troublesome or the ore of too low grade to pay well under the conditions existing ten or twenty years ago, the mines were left practically abandoned and idle, only work enough being done to hold them. There are still hundreds and hundreds of such mines, which are gradually being rehabilitated by the infusion of capital in the mining counties. Moreover, the demand for gold properties, which has become manifest within the past five years, has encouraged many to open and develop small claims upon which little work had previously been done. Many of these have become producers on the resumption of operations.

"PROSPECTING."—The prevailing conditions have, moreover, given marked impetus to prospecting throughout the gold regions of the State, which has resulted in the discovery of a great many new claims, many of which are being opened and developed, and some of which have become producers on a moderate scale. Another feature to be considered is that certain agricultural or horticultural operations having become less profitable, some men formerly engaged in those operations have turned their attention to prospecting or working mining ground either overlooked or unconsidered, and others have regularly engaged in mining as a business. The merchants and business men of the cities and towns have also engaged in mining, within the past few years, to a larger extent than formerly, and have been willing to assist miners and prospectors in finding or developing claims. Investments are being made by small companies, the members of which pay for their stock in monthly installments, extending over a period of one or two years, the money collected being spent almost entirely in purchase, development, and equipment of the claims, and none of the officers, aside from the acting superintendent, being paid salaries. The mining is thus carried on in an economical manner, and men of small means are afforded an opportunity of becoming interested in mining operations. Of course, the organization of such companies involves the purchase of mining properties and the employment of men, which is a good thing for the mining regions and the mining industry.

Those who have claims, but no means to open them, are thus enabled to sell outright, or get the financial assistance necessary. In the case of large mines, there is now no difficulty whatever in procuring capital, either for purchase, equipment, or development, the demand for such being greater than the supply.

DISTRIBUTION OF GOLD.—The precious metals find a wide distribution in California, and the deposits occur under favorable conditions for cheap exploitation. The principal gold field in this State extends north and south with regularity and scarcely any interruption for a distance of nearly 700 miles, besides which there are several outlying districts producing gold and silver. Gold was mined last year in thirty-one of the fifty-seven counties of the State, and is found in several other counties where the deposits are not at present being worked.

Except in the desert regions occupying the south-eastern part of the State, and the country lying east of the Sierra Nevada, the mining districts of California are nearly all well supplied with wood and water, prime essentials for cheap and successful mining and milling. The soil and climate in most of these districts are well adapted for fruit and stock-raising, rendering the life of the miner one of comfort, and the cost of subsistence extremely moderate. In certain

RAMSEY'S CABIN ON THE TRAIL TO THE RALSTON DIVIDE, PLACER COUNTY.

counties the men work at horticulture in certain seasons and mining in others, the work being carried on oftentimes on the same "ranch."

THE HYDRAULIC MINES.—The hydraulic mines of California are noted the world over for the extensive plants connected with them, which are of special interest to engineers in view of the extensive dams and reservoirs required, the long pipe lines, the immense quantities of water used under enormous pressure, and the vast quantity of material moved under this system. To all miners the process is of great interest, since by it the cheapest method of mining for gold is accomplished. The gold may be in very small quantities in the gravel, since the cost of the hydraulic method is only from one and a half to eight cents per ton of material treated, according to conditions.

Briefly described, hydraulic mining may be said to consist in the plan of breaking down and disintegrating the auriferous gravel as it stands in place, and carrying it into the gold-saving appliances by means of water discharged through iron pipes upon such gravel, under great pressure. The most notable feature of this process, the employment of water under pressure, the breaking up of the indurated gravel with gunpowder or other explosives, and effecting its further disintegration by dropping it over precipitous falls—in short, all that is most characteristic about it is so new and wholly without precedent that it may be considered exclusively the product of California. Limited and imperfect at the start, this industry developed into one of overshadowing magnitude, the mines worked by this process having turned out, before they were prevented from running except with debris dams, more than a third of the annual gold product of the State.

ORIGIN OF HYDRAULIC MINING.—The discovery of gold in California drew to this part of the Pacific Coast a numerous, eager, and energetic population. As the most of these newly arrived immigrants repaired at once to the mines and there engaged in the business of gold gathering, the more shallow and accessible placers were worked out with such rapidity that but for the timely discovery of more permanent deposits, and the employment of improved ways and means for working them, placer mining here would have undergone speedy curtailment and perhaps suffered early extinction. These later discoveries consisted of the auriferous gravel found in the inhumed beds of the pliocene rivers, and the vast accumulations of like material lying over and adjacent to these ancient channels.

The finding of these pliocene deposits was, like the original discovery of gold, the result of chance rather than of any preconceived theory or systematic plan of exploration. In the prosecution of their labors, the pioneer miners were early led to observe that the gulches and canyons, entering the rivers immediately above the exceptionally rich bars, were apt to contain great quantities of gold dust. Following up and washing the gravel along such gulches and canyons, these miners encountered on either side thereof what appeared to be fertile streaks or "leads," to the breaking down and disintegration of which these depressions were evidently indebted for their special enrichment.   These "leads," on being traced up, were found to run under the banks of the canyons, the courses of which, before their destruction, they had crossed at various angles.   The further explorations of these strange formations not only showed them to be receptacles of great wealth, but disclosed the additional fact that they occupied deep, water-worn channels, once the beds of broad and swift-flowing streams.   To these latter, so extinguished and entombed during the long ages of the past, the miners of the period gave the striking and appropriate name, "Dead Rivers," a term by which they have continued to be designated ever since.

The bulk of this auriferous detritus, as we find it resting in the old river channels, or massed in the deep hydraulic banks, is made up, for the most part, of silicious sand, fine and coarse gravel, clay, formed from volcanic ashes or other sedimentary deposits, lava, tufa, conglomerates, fossilized wood, and boulders.   The latter occur, usually, near the bottom and are often of great size, single stones weighing as much as fifteen and twenty tons being frequently met with.   This detritus, in the form of secondary deposits, is found also gathered on the modern river bars and alluvial flats and basins, or spread out over the low prairies and rolling foot-hills, where both hydraulic and drift mining are sometimes prosecuted with good results, though not generally on a large scale.   The rocky portion of this material is composed mostly of quartz, accompanied with fragments of every rock common to the Sierra Nevada, such as slate, serpentine, granite, etc., all more or less rounded and water-worn.   There is also intermixed with it a small percentage of magnetic iron ore, in the shape of black sand.   Fossils, both animal and vegetable, and, at some points in great quantity, occur in the middle and upper strata of gravel.   These organic remains consist of the bones and teeth of various animals, the trunks and branches of trees partially or wholly silicified, lignites and carbonized wood, in all of which latter the vegetable structure is clearly apparent.   Cords of petrified wood have, in some places, been uncovered; also, Indian mortars and similar rude implements, proving the existence of man prior to the filling in of these pliocene rivers.

"THE DEEP GRAVELS."—It is unnecessary at this place to enter into a discussion of the origin of these deep gravels, but it may be said that the blue lead theory, like the glacial and marine theories before it, has now been generally discarded by the investigators of the auriferous gravels, and there has been generally accepted the "fluvi-

OPEN CUT, SHOWING EARLY WORKINGS OF THE MELONES MINE, CALAVERAS
COUNTY, FROM WHICH $3,000,000 WAS TAKEN.

atile" theory in its essence, ascribing the origin of the gravel accumulations to the depositions of ancient rivers with courses similar to the present rivers. The character of the deposits shows that there has been an intermittent action. The torrential period and the quiescent period are both shown in alternating heavy and light material of the channels. But while there is this unanimity of opinion as to the origin of the gravel accumulations, there still exists a difference of opinion regarding the geological age of the deposits.

Sometimes the contents of these old river channels and hydraulic banks are composed of gravel from bottom to top; oftener, however, they consist of layers of gravel, sand, volcanic ash or clay, and lava, occurring usually from the bottom up, in the order here mentioned. Again we find this material interstratified without much regard to regularity, the lava, where any is present, constituting always the upper and the gravel the lower stratum. These bodies of sand, like the layers of volcanic ash, are sometimes flat, but often lenticular in shape; both vary much in size, from a few feet to six or eight and, occasionally, as much as ten or twelve yards in thickness, their linear being always many times greater than their vertical extent. This sand, also a volcanic product, is sometimes so loose and dry that, if left unsupported, it crumbles under its own weight.

Where not denuded, as large portions of them have been, the contents of these pliocene channels are covered with alluvial drift and volcanic flows, either in the shape of scoria, sand, and mud, or basaltic cappings, which latter take, usually, the form of mesas or table mountains. These lava flows, which reach sometimes a thickness of more than a hundred feet, prove fatal to hydraulic washing wherever they occur in much volume, their hardness and thickness interposing an insuperable obstacle to their removal.

The lava may be a true lava, brought into its present position by a lava flow, or, as is more generally the case, the term refers to lava of a fragmental and tufaceous character. Lava boulders are more or less rounded, water-worn boulders of volcanic origin, brought into their present condition by running water. Volcanic breccia and volcanic conglomerates are likewise designated by the term "lava." The most common form of lava is the material of the "light-colored, fine-grained, homogeneous beds, resulting from the consolidation of the ashes and volcanic mud." The lava of the tufaceous variety is sometimes very compact, and has a metallic ring when struck, resembling "clink-stone" (phonolite) in this respect. In this lava, crystals of glassy feldspar often occur.

The lava is nearly always found "capping" (overlying) the gravel, but in some localities it occurs interstratified with the gravel.

Beautiful dendritic markings (called "photograph rocks") are often found covering the lava. These markings are usually mistaken for impressions of plants, but they are of inorganic origin, having been formed by percolating waters, carrying oxide of manganese and

THE BREAKFAST OF A MINE-SURVEYING PARTY ON THE MOTHER LODE.

oxide of iron, which substances have assumed the dendritic or tree-like form.

Cement is a term of variable signification. In some districts it designates volcanic material of brecciated or conglomerated character. In other localities it refers to a quartz (non-volcanic rock) conglomerate cemented by oxide of iron.

Pipe-clay is more or less indurated clay, of non-volcanic origin, forming finely laminated beds. It generally occurs interstratified with beds of sand, etc.

The width of channel worked in hydraulicking varies from 150 to 1,000 feet. The width of the deep rivers on top varies from a few hundred feet to several thousand. At Columbia Hill, Iowa Hill, and other places, the rivers are from one mile to a mile and a half wide. In some places, where the gravel deposits have a great extent, laterally the width may be due to the confluence of two or more streams or tributaries. Broad expanses of alluvions may also indicate the embouchures of the ancient rivers at these points.

The uniformity that is generally considered as a characteristic feature of the distribution of the gold in gravel channels rarely exists. On the other hand there is almost always a great fluctuation in the value of gravel taken across the stream and likewise along the course of the stream. The larger channels are usually less "spotted" as regards the value of the gravel than the smaller channels.

In drift mining this tendency is well marked in the occurrence of the "pay leads," to which the extraction of gravel is restricted. These leads are usually near the lowest depression of the channel, but often make to one or the other side, and have most sinuous courses and varying widths. The width of these pay leads is determined primarily by the cost of mining. Where low-grade gravel can be profitably worked, the width of the pay lead obviously would be greater than where a more costly system prevails.

Pay leads 100 to 150 feet are regarded as of about the average width. Where rich gravel, $5 to $8, is the minimum grade worked, such leads are often only from fifty to seventy-five feet wide. When gravel of $2 to $4 is being mined, the width of the pay lead may often reach 300 or 400 feet.

DISTRIBUTION OF THE GOLD.—In speaking of the distribution of gold in these deposits, John Hays Hammond says:

The paragenesis of the gold in the auriferous gravels of California differs from that in similar formations elsewhere, in the comparative paucity of the associated minerals.

Zircon, magnetic pyrites (chiefly in the form of sand), and garnets are the most abundant minerals accompanying the gold; but platinum, iridosmine, rutile, epidote, diamonds, chronite, topaz, cassiterite, and other minerals, also occur. The gold appears in size from minute particles (finest flour) to large nuggets. Gold from size of flax seed to melon seed is, as a general thing, considered coarse gold. Gold from top gravel is textually fine, and also of high grade, the coarser gold being nearer the bedrock. The gold usually is of a flattened character, well rounded on its edges. Occasionally the gold is rough and but little water-worn, and frequently still in its matrix. Gold found in upper gravel often has preserved its crystalline form. Rusty gold is often found in the alluvions. The rusty character of the gold is probably due to a coating of silica and sesquioxide of iron. It is not readily, indeed at times not at all, amalgamable. Such gold, when saved, is recovered entirely by reason of its specific gravity. The richest gravel, as a rule, occurs on and near the bedrock. Sometimes, however, equally rich gravel is found in the upper horizons of the channels, notably on benches occurring on the inner side of a bend in the course of a channel.

The old washings of surface ground, near Malakoff, Nevada County, from 1870 to 1874, is estimated at about 3,250,000 cubic yards, the yield of which was about two and nine-tenths cents per cubic yard. The Bloomfield Company, from November 29, 1876, to October 13, 1877, washed 1,591,730 cubic yards of top gravel, which yielded three and eight-tenths cents per cubic yard. During the same period that company washed 702,200 cubic yards of bottom gravel, which yielded thirty-two and nine-tenths cents per cubic yard. The bottom gravel extended from bedrock to a height of sixty-five feet. All the gravel above that horizon was considered as top gravel.

The depth of the top gravel varied from a few feet to over two hundred feet.

The steep bedrock is more favorable for rich gravel than where the bedrock is flat, but often the gravel is poor where the channel is narrow and steep.

Gravel that yields remunerative results by the drift process has been worked, in some instances, from sixty to one hundred and fifty feet above bedrock. The coarser gravel ("heaviest wash") is the richest. The light "wash," and material of sandy character, is usually poor in gold. The top gravel of hydraulic mines rarely pays for piping, unless worked in conjunction with the underlying richer auriferous deposits. The gold is often found in the grass roots of the vegetation covering the gravel deposits, but not in considerable quantities. The top gravel of a deposit is usually richer than that lying a foot or two below the surface, owing to the concentration of the gold superficially. In schistose rocks, as well as in decomposed granitic rocks, the gold is often found in the bedrock at a depth of from one to five feet below the surface. In drift mining, the bedrock is stripped to recover this gold. In hydraulic mining, bedrock of this character, when not too hard, is piped. When necessary, bedrock is blasted before being piped.

The mining of these deep placers is done by either drift or

hydraulic mining.   The restriction of hydraulic mining gave quite an impetus to drift mining, but there are many grand deposits that are peculiarly adapted to profitable hydraulic mining that are not susceptible of remunerative drift mining.   In some mines the gold is more or less uniformly distributed throughout the gravel, and the lowest stratum is not rich enough to pay for the drifting, while hydraulicking can be profitably conducted.

The lithological character of the material filling the old river channels will determine which of the systems is to be pursued.   Where a great depth of lava or other non-paying material is superimposed

EN ROUTE TO THE BEACH MINES, HUMBOLDT COUNTY.

upon the auriferous pay gravel, drift mining must be resorted to. The exploitation of such large quantities of barren material would not be practicable, and by the system of drift mining this is obviated. In some places there is a concentration of the gold in a stratum immediately overlying the bedrock, while the rest of the deposit carries an amount of gold too small to admit of profitably hydraulicking the entire deposit.   When such is the case, recourse is had to drift mining.   Again, lack of grade, lack of dump, or the absence of other conditions requisite for hydraulic mining, make it necessary to adopt drift mining instead of the hydraulic system.

The principal feature of expense of large hydraulic mines is the water system, as reservoirs, ditches, flumes and pipe lines have to be constructed. The water used in hydraulicking is derived from the streams fed chiefly by the rains and melting snows. Where operations are conducted upon a comparatively small scale (the companies having no storage reservoir of any considerable capacity), the hydraulicking season is about co-extensive with the rainy season, rarely being prolonged more than a month or two after the cessation of the rains. Under such conditions, five to six months is about the duration of hydraulicking.

Companies hydraulicking more extensively have storage reservoirs commensurate with the magnitude of their operations. Such companies are enabled to continue their piping with but little interruption during the year. The reservoirs constructed for mining purposes (chiefly for hydraulic mining) on Yuba, Bear, Feather and American rivers have an aggregate storage capacity of about fifty billion gallons.

The source of water supply of the North Bloomfield Gravel Mining Company, the Milton Mining and Water Company, and the Eureka Lake and Yuba Canal Company is about the headwaters of Big Canyon Creek and Middle Yuba River in Nevada and Sierra counties. The catchment area embraced in those sections represents in the aggregate 68.6 square miles. There are eleven principal reservoirs, varying in area from ten acres to four hundred and eighty-seven and one-half acres (high water area) each, and having a capacity of from $2\frac{1}{2}$ to 796.7 million cubic feet. The total area of the reservoirs of these companies is over eleven thousand six hundred acres (at high water mark), with a total capacity of over two billion one hundred and ninety-five million cubic feet.

In order to obtain efficient head or pressure, it is often necessary to bring the water from great distances. To overcome the topographical obstacles, considerable engineering skill is sometimes required. The water provided by the water companies in the central mining counties of the State was used mainly for hydraulic mining in the palmy days of that industry. The La Grange Ditch and Mining Company has twenty-five miles of ditch, costing $450,000. The Union Water Company, Calaveras County, has a plant which cost $200,000. The hundred miles of ditch of the Mokelumne and Campo Seco Company cost $500,000. The El Dorado Water and Deep Gravel Company's plant cost $600,000, and that of the California Water and Mining

Company, with two hundred and fifty miles of ditches, cost the same figure. The Park Canal and Mining Company, with two hundred and ninety miles of ditch, eight miles of pipe, etc., with 2,200 inches capacity, spent $2,000,000 on its plant. The North Bloomfield Mining Company has a plant costing $708,000. The length of ditches, including reservoirs, of this company is one hundred and fifty-seven miles. The Milton Company's system cost $391,579; Auburn and Bear River, $350,000; Amador Canal, $400,000; Dardanelles, $125,000; Gold Run, $150,000; Little York the same; Natoma, $390,000; Phœnix Ditch Company, $880,000; California, $550,000; Eureka Lake, $732,000; South Yuba, $1,100,000; Smartsville ditches, $1,000,000; Spring Valley and Cherokee, $500,000; Blue Tent, $200,000. In this list only larger ditches are named, but it will be seen that the water systems required for this class of mines is very extensive to cost such sums. Nothing is said here of cost of mines or equipment outside of water supply. Of course, to supply such systems, immense and costly reservoirs, dams, etc., must be built.

In order to utilize the pressure due to the elevated position in which the water is brought, with respect to the gravel to be washed, the water is conducted from the ditches into a tank called the "pressure box" or "bulk-head." From the pressure box or bulk-head, by means of a feed-pipe (main pipe), the water is brought to a distributer. The size of the feed-pipe is determined by the quantity of water to be used. Twenty-two-inch (in diameter) mains are generally used in the larger hydraulic workings.

The pipes are made of wrought-iron, the thickness of which increases with the diameter of pipe, and the hydrostatic pressure to which the pipe is subjected, Nos. 16, 14, 12 (Birmingham gauge). These numbers correspond to thickness of .065, .083, 1.09 inches, respectively. To prevent the pipes from corroding, they are coated with a preparation of asphalt and coal tar.

The "distributer" is a box which serves the purpose of a hydrant, and by means of valves enables the partition of the stream of water and the diversion of the branch streams into two or more pipes, whereby more than one part of the gravel bank can be simultaneously hydraulicked.

From the distributer the streams are piped to the "monitors" or "giants." These are the discharge pipes which concentrate the stream and enable its projection upon any desirable point. The giants and monitors are labor saving. Before their introduction at North

GRAVEL-WASHING PLANT AT THE HIDDEN TREASURE DRIFT MINE, PLACER COUNTY.

Bloomfield, there were ten to fifteen streams playing, while now one stream, tended by one man, does as much work.

The "nozzles" of the monitor are from four to nine inches in diameter. Two or more monitors are employed, depending upon the magnitude of the workings. The streams played upon the bank are one hundred or more feet in length. The large mines have a dozen or more monitors, but rarely have enough water to supply at once more than four or five monitors.

The disintegrating power of the jet, which at the larger mines equals one thousand to one thousand five hundred inches of water (1,500 to 1,750 cubic feet per minute), weighing sometimes upwards of one hundred thousand pounds, under pressure of one hundred and fifty feet to four hundred and fifty feet, is enormous. Yet, notwithstanding this, the gravel is sometimes so tenaciously cemented that assistance of powder becomes necessary to shatter or break up the bank preparatory to its disintegration by water.

It is desirable to get the monitor or giant as near the bank as is consistent with the safety of the miners and the machine, in order to obtain the full force of the stream. Where the banks are high—over two hundred feet or so—they are accordingly usually worked in benches of from one hundred and fifty feet to two hundred feet in depth.

Bank blasting is generally resorted to in order to loosen or disintegrate the auriferous material before it is piped. It is essential, also, that there should be "dump" room, that is a place to deposit, or get rid of, the debris or tailings resulting from the washing down of the banks. It was this feature which brought about the "war" against hydraulic mining, since the debris or tailings, when allowed to pass into the canyons and gulches, eventually went into the navigable streams and injured property along the banks.

A condition prerequisite to the successful working of the hydraulic mine is the attainment of the grade necessary for the treatment of the detrital material piped from the bank.

To get the requisite grade for the sluices and, at the same time, to obtain a suitable place for the deposition of the debris from the washing out of the gravel bank, often involves the driving of a long bedrock tunnel.

The topographical features of the environs determine the location of the tunnel. The mouth of the tunnel should be sufficiently below the bedrock of the deposit to be prepared for the contingency of a change in the grade of channel, whereby the tunnel might be placed above the level of the drainage and rendered practically useless. This difference of level also admits of the use of chimneys or shafts, connecting the surface of the bedrock of the channel with the face (interior end) of the tunnel, whereby a drop is obtained which facilitates the disintegration of the obdurate conglomerate.

The tunnels are from a few hundred to several thousand feet in length. The tunnel of the North Bloomfield Company, in Nevada County, is 7,874 feet long, and its dimensions seven by eight feet. It cost about $500,000. The tunnels are from four to eight feet wide and from five to nine feet high.

The material "piped" from the bank is carried by the stream of water through bedrock cuts and sluices to the chimneys or shafts (where such exist).

The bedrock cut, as its name implies, is a trench carried in the bedrock from the upper end of the line of sluices to the gravel bank. These cuts are about as wide as the sluices and are sometimes twenty to forty feet deep. The gravel is washed into these cuts and thus is brought to the sluices.

The gold from the auriferous material is caught in the quicksilver in the riffles of the sluices, etc., undercurrents being also used as gold-savers. The cost of mining by this method is between one and a half cents and eight cents per ton of material treated. In this description on hydraulic mining the writer has freely drawn on the article by John Hays Hammond, in the Ninth Report of the State Mineralogist of California.

The damage done by the debris, or tailings, from these hydraulic mines caused a great deal of contention and litigation between the miners and farmers of certain sections of the State. The history of this is given in another chapter in this volume. It is only to the counties in the drainage basin of the Sacramento and San Joaquin rivers that the Federal law, requiring the debris to be impounded, applies. Elsewhere in the State the mines run freely without restriction.

"HYDRAULIC GRAVEL ELEVATORS."—Before the injunctions of the courts stopped the large hydraulic mines of the State, a machine was in use for working shallow banks and low deposits in basins and creeks, called the hydraulic gravel elevator. Many valuable auriferous deposits are found in basins and flats lying along creeks, far too low to be underrun by bedrock tunnels, open cuts, or drains at any cost, and these deposits had to be worked by shovels and barrows, whims, engines, derricks, and pumps. To overcome these difficulties, and to dispense with the use of expensive pumping plants, the hydraulic elevator was devised and put in use.

It was at first looked upon with some doubt as a practical apparatus to do any work on a large scale; but, when the large mines were prevented from working by the ordinary hydraulic method, this appliance was tried in these claims, and has been found to answer the purpose of working them, though on a more limited scale than was possible by the old method.

The principle upon which the operation of these hydraulic gravel elevators is based is that of driving gravel up hill by hydraulic force. It is only necessary to give the impelling water more velocity than it has through an ordinary flume to make it acquire sufficient force to carry gravel up an inclined plane. This fact suggested the construction of a form of machine which should so direct and confine the inherent hydraulic force of a stream of water as to impel before its power masses of earth and gravel.

These machines are being worked in various parts of the State, where the vast areas of ground which have been piped out in the past are utilized as reservoirs for impounding the debris which may be hydraulicked by the elevator system. The gravel from the standing banks is washed down by giants, sluiced to the elevator and forced up, as described, to such points that it may be led to old workings or pits, and there retained and not flow into the streams, the gold being retained in the sluices and other gold-saving appliances.

Without this apparatus for elevating and throwing back the debris, many mines would be unable to impound their debris without very great expense, since old workings can be utilized as debris reservoirs. This appliance has been greatly improved of late years and is largely used in river-bed mining, the material being elevated from the bed to the sluices above without any handling, the water doing all the work and the operation being continuous.

"DRIFT MINING."—As far as the occurrence of the deposits are concerned, all that relates to the pliocene, or "dead" rivers of the State, applies to drift as well as hydraulic mining. The drift mining interests of California have been materially advanced of late years. In the first place, some of those gravel mines, formerly worked by hydraulic process, stopped washing away the entire bank of gravel, and only "drifted" out the lower or richer stratum, washing it after it was taken to the surface, or, if cemented, crushing it in stamp mills, the same as quartz. And again, capital turned its attention more to drift mining when hydraulic mining was no longer possible to be carried on. The term drifting, when applied to this class of gravel mining, relates to the mode of extraction of the auriferous gravel by means of tunnels and gangways or galleries. This system is rendered necessary in consequence of the capping of volcanic lava overlying the ancient channels in which the gold is found, and rendering hydraulic operations impossible. In hydraulic mining, the entire face and body of the bank is removed by the pipe; in drift claims, only the lower stratum of gravel lying on the bedrock is mined and washed.

These gravel deposits are worked by tunnels generally, although it is sometimes necessary to sink shafts. The underground work is very similar to that in coal mines.

The conditions of drift-mining ground may be briefly stated as follows: A "divide" or ridge between two deep river canyons, the sides cut into by small ravines tributary to the main streams, the top sometimes several miles wide, mesa-like and comparatively level, and having only one main slope in the direction of the ancient buried rivers. This top is usually composed of lava several hundred feet thick, and somewhere under this lava cap, and between walls of true country rock termed the "rims," is the channel of an ancient river, sometimes with little or no gravel in its bed, sometimes with gravel a couple of hundred feet thick; this gravel sometimes barren of gold, but more frequently found to be rich in the precious metal. Under all of the above are granites, metamorphic slates, and talcose rocks. The problem of drift mining is to find the position of these ancient river channels and open up and prospect them with the least possible expenditure, and, if a lead of pay gravel is found, to work it at the least possible cost. The gold found in these channels varies from flour-like fineness to large lumps. It is most concentrated on the gravel lying immediately on the bedrock.

In some of these drift mines very extensive tunnel work has been done. In the Red Point, on the Forest Hill Divide, Placer County, the tunnel is two and a quarter miles in length, and numerous mines have works extending over a mile underground. It is estimated that in that county alone there are still about two hundred miles of the dead river channels yet unworked. Details of the drifting operations in the county will be found in the chapter on Placer County, by I. H. Parker, in another portion of this volume, and for this reason little is here said of this important industry. Those desirous of detailed information will find in the Tenth Report of the State Mineralogist of California an excellent article, by Ross E. Browne; in the Eighth Report, one by Russell L. Dunn; and in the Ninth Report, one by John Hays Hammond, all dealing with the drift-mining interests of the State. The article by Mr. Parker in this volume is, of course, of recent date, and gives the particulars of the present conditions of the industry. Drift mining is carried on in several counties of California. It entails the application of capital, since the preliminary operations are very expensive. The future of drift mining in California is promising. The better knowledge of the character and situation of the buried auriferous channels, and the more systematic methods of exploration and mining, have taken away much of the uncertainty which caused so many failures among the earlier drift-mining operations. The channels, when opened, yield from $150 to $5,000 per running foot, counting the width mined. In some places the ground is much richer than in others.

RIVER-BED MINING.—Another form of gravel mining common in this State is river-bed mining. In the earlier days it was practiced more than at present, but is still continued in certain localities, and within the past few years has increased in favor. The object of this branch of gold mining is to recover the gravel forming the bottoms of river channels or streams, and known to be auriferous. To do this various expedients are resorted to, such as draining the channel, wholly or in part, dredging, etc. Where it is sought to drain the whole bed of a stream, the water is diverted by means of dams into a ditch or flume constructed along the bank of the stream to a point below the section to be reclaimed, and there the entire flow is returned to the channel. By this means such section can be so far freed from water that it is possible to control the seepage by pumps, wheels, etc. Where there exist natural facilities for running tunnels, the entire river-bed can in like manner be laid bare by such means. When the design is to dry and work only a strip along one side of the river-bed, this is effected by what, in mining parlance, is termed a "wing-dam," that is, a water-tight wall which starts from the bank and is carried out a short distance into and down the river, the wall being continued back to the bank. The water inside the space so enclosed is then raised with wheels or hand pumps and emptied into flumes that discharge it into the river.

While not peculiar to California, river-bed mining has been pursued here on a scale not paralleled in other countries, and the efficiency of our methods greatly surpass those employed elsewhere. Outside this State the business does not appear to have reached large proportions, nor has any great amount of gold been gathered elsewhere by this method.

Working the beds of the rivers that traverse the mining regions of California was begun here at an early day. The first crop of gold dust harvested by this method, however, was very bountiful. Like some other kinds of gold mining here, this branch of the business, after having prospered and attained large dimensions, underwent a marked decline. It has for several years past been on the increase, however, both as regards the number and magnitude of the operations. While workings of this kind are pursued to some extent throughout nearly all parts of our gold fields, the Klamath, Salmon, Scott, and Feather rivers are the localities of the largest operations.

It is rather odd, and to some extent paradoxical, that the places where most capital has been put into this branch of business have been the scenes of the most prominent failures. As far as experience goes, the men who have been the most successful have been those practical ones who did, or superintended, the work themselves and put very little money into the operation. The Chinese are very successful river miners, and number of whites are engaged in the business, especially on the Klamath River and its tributaries. But the two largest operations ever undertaken in this line were failures. On the Big Bend of the Feather River, Butte County, a couple of million dollars were expended with no commensurate results, in fact, with no results at all, though the fourteen miles of river-bed, laid bare by the tunnel operation, were supposed to be exceptionally rich. They ran a tunnel 12,000 feet long, and finding it too small to carry the volume of water of the river, enlarged it to thirteen by sixteen feet, built a dam, sluices, pumps, electric plants, etc. They got little or nothing, though the tunnel carried the water of the river at its low stages all right. The gold was not in the gravel in the quantities expected. Nearly three millions of dollars were spent by an English company, near Oroville, Butte County, where they built dams, and built a rock and concrete flume to carry the river water and leave the old bed dry. They used hydraulic elevators to lift the gravel to the sluices. The highest lift was seventy feet from bottom of the river. They only found five or six spots in the river which were not worked by the miners of early days, and from one of these were reported to have taken $76,000. But, generally speaking, the enterprise was very unprofitable, and it was finally abandoned about a year ago. But the smaller river-bed claims seem to be quite profitable where worked by small companies of miners or by individuals.

DREDGING MACHINES.—Another form of river-bed mining is that carried on by dredging machines. About all of the first attempts in this direction in California were failures, and, for twenty or thirty years, no dredgers were operated. But, within the past few years,

SUSPENSION BRIDGE ACROSS TRINITY RIVER, AT JUNCTION CITY.
Carries a Pipe-Line for 3,000 inches of Water under 500 feet Head.　Bridge is 350 feet Span.　Cables are 620 feet long, 5 inches in diameter, and weigh 11 tons each.

the success of some dredging enterprises has led others into the business, and it is now on the increase. The improved machinery is the reason of this renewal of dredging operations. Dredgers are now being operated in this State on the Feather, Yuba, American, Klamath, Sacramento, and Trinity rivers. The greatest number is at Oroville, Butte County, where there are six at work, and two more being constructed. Most of these are in the Feather River, but one is working on gravel a mile from the river, where a hole about twenty-five feet deep has been dug, and in this the dredger was built or placed, water enough flowing from the ground to float it. This machine is at work in the middle of an orchard. Some of these dredges have a capacity of 2,000 cubic yards daily, and the buckets have a capacity of three and a quarter cubic feet. If the ground is such that a reasonable full capacity can be taken up by the buckets, they can work very close to this figure, but in hard ground they handle from 800 to 1,000 yards daily. One machine on the Yuba River is said to have made a profit on ground worth only six cents per cubic yard. Endless-chain bucket or steam-shovel dredges are preferred. The cost varies from three to six cents per yard, accord-

HYDRAULICKING ON THE TRINITY RIVER, NEAR JUNCTION CITY.

ing to consistency of material and depth excavated. Details of this class of work are given in another chapter in this volume.

The ordinary forms of placer mining, such as working with long-tom, sluices, ground sluicing, etc., are practiced in many places in this State, but mainly in a small way, by Chinese or individual white miners. Dry-washing is carried on in Kern County, some of the desert counties, and along the banks of the Colorado River.

OCEAN-BEACH AURIFEROUS SANDS.—Besides those already mentioned and partially described, the gold-bearing deposits of California occur in several other forms, all designated by names more or less fit, a few being, perhaps, a little fanciful. The most of these deposits are, in fact, distinguished not so much by any inherent peculiarities as by the conditions under which they are found, and the methods and appliances adopted in working them.

The auriferous beach sands, which once afforded profitable employment to many men, have years since become so impoverished that they figure but slightly among our available mineral resources. These ocean placers have, in fact, responded feebly to the attempts made of late to work them. But, for all this, we have these deposits of low grade in infinite quantity occurring at intervals. They reach along the sea-shore for many miles, extending at several points, in the form of buried channels, some distance inland. So abundant, but now so poor, these gold-bearing sands await the coming machine that is to make their further working profitable.

Meantime, the auriferous beaches continue to be worked at a few points and in a small way. Along the sea-shore in Humboldt and Del Norte counties, formerly the chief sites of this class of mining, the residents of that section of the State gather from these sands by hand-sluicing a little gold every year. Along the shore further south some little work of this kind is also being done. Between Point Sal and Point Concepcion, off the coast of Santa Barbara, several small companies have been engaged in washing the beach sand. The most successful machine for treating the "black sands" seems to be what is known as the "Oregon Tom," which is described and illustrated in the Thirteenth Report of the State Mineralogist.

PRESENT GOLD PRODUCTION.—As to the present condition of the gold mining industry generally, in California, it may be stated that in the year 1898 there were thirty-two counties in the State which produced gold, and there was a product of silver also from twenty of these. Of the thirty-one counties, several produced over a million dollars each in precious metals, and one of these exceeded the two-million dollar mark. Nevada County, which held the leading place in 1896 and 1897, still maintains its position as the largest gold producer among the counties of the State. Amador County, which was fifth in rank in 1897, now takes second place, followed closely by Tuolumne, which is third. The fourth county in rank among those producing over a million dollars is Placer, which was third in 1897. Shasta County, which is fifth, now for the first time takes a place among the counties yielding over a million. The sixth is Kern County, which also appears on the same list for the first time. Calaveras drops back from fourth to seventh place. Trinity, which was in this list of honor in 1897, drops below the million rank in 1898.

PRODUCT OF CALIFORNIA BY COUNTIES—1898.

COUNTY	GOLD	SILVER	TOTAL
Amador	$1,806,363 00	$1,742 00	$1,808,105 00
Butte	514,508 00	9,317 00	523,825 00
Calaveras	1,019,023 00	3,432 00	1,022,455 00
Del Norte	9,057 00		9,057 00
El Dorado	501,966 00	4,174 00	506,140 00
Fresno	27,557 00		27,557 00
Humboldt	57,512 00		57,512 00
Inyo	137,107 00	73,503 00	210,610 00
Kern	1,017,930 00	6,543 00	1,024,473 00
Lassen	37,460 00	300 00	37,760 00
Los Angeles	21,300 00		21,300 00
Madera	94,884 00	50 00	94,934 00
Mariposa	336,418 00	993 00	337,411 00
Mono	446,017 00	66,667 00	512,684 00
Nevada	2,017,628 00	19,476 00	2,037,104 00
Placer	1,488,022 00	5,670 00	1,493,692 00
Plumas	369,609 00		369,609 00
Riverside	189,188 00	1,384 00	190,572 00
Sacramento	57,301 00		57,301 00
San Bernardino	261,512 00	32,000 00	293,512 00
San Diego	673,196 00	300 00	673,496 00
San Luis Obispo	1,000 00		1,000 00
Santa Barbara	1,000 00		1,000 00
Shasta	860,180 00	171,768 00	1,031,948 00
Sierra	399,063 00	519 00	399,582 00
Siskiyou	768,804 00	321 00	769,125 00
Stanislaus	19,400 00		19,400 00
Trinity	859,255 00	314 00	859,569 00
Tulare	12,400 00		12,400 00
Tuolumne	1,734,953 00	15,582 00	1,750,535 00
Yuba	166,865 00		166,865 00
	$15,906,478 00	$414,055 00	$16,320,533 00

PRODUCT OF DIFFERENT KINDS OF GOLD MINES.—The relative importance of the different kinds of gold mining carried on in the counties of California is a matter of interest as to the origin of the metal, whether from alluvial washings or rock in place. The appended table shows that not only is quartz mining by far the most important branch of the gold-mining industry, but that the importance is steadily on the increase as compared with other forms. Comparing these figures of source of gold and silver for 1898 with corresponding ones for the previous year, it is seen that quartz mining shows a gain of $1,098,191; placer mining, a loss of $538,441; hydraulic mining, a loss of $388,896; and drift mining, a loss of $174,511. It must be remembered, however, that placer, drift, and hydraulic mining suffered more, proportionately, from the very dry seasons than did quartz mining, since these forms of alluvial washing depend in a very large measure on an abundant water supply for successful work.

It is somewhat difficult to separate properly in the returns these forms of gravel mining, as many simply say in response to the questions "placer," when they may be engaged in drift or hydraulic mining. For this and the further reason that, unless otherwise specified, the Chinese are enumerated as "placer miners" (since they never work quartz), some of the gold attributed to "placers" may have come from hydraulic or drift claims. The drift and hydraulic tailings being worked by Chinese in various places are also tabled under the heading of "placer." The surface placers include bars, river-beds, ravines, gulches, flats, etc., while the hydraulic and drift mines are deep gravels, the drift mines being covered with a lava capping which prevents the banks of gravel being washed by the hydraulic process.

What gold was obtained by dredging machines is classed under "placer." Only a few of these machines were working in California during 1898, but others have been completed since the close of that year, and several more are being built. The aggregate yield from this source is not as yet very important.

The following table will show in detail, by counties, the source of gold and silver in California in 1898:

SOURCE OF GOLD AND SILVER—1898.

COUNTY	PLACER	HYDRAULIC	QUARTZ	DRIFT
Amador	$15,500	....	$1,792,605	....
Butte	197,400	$1,550	277,242	$47,633
Calaveras	33,400	26,154	957,401	5,500
Del Norte	9,057			
El Dorado	48,913	17,188	432,454	7,585
Fresno	8,000	319	19,238	
Humboldt	39,293	18,219	....	....
Inyo	31,000	....	179,610	....
Kern	33,900	....	990,033	540
Lassen	160	....	37,600	....
Los Angeles	13,000	....	8,300	
Madera	3,800	....	91,134	....
Mariposa	17,955		319,456	
Mono	2,500	....	510,184	....
Nevada	62,550	146,091	1,751,598	76 865
Placer	380,594	30,704	341,602	740,792
Plumas	126,140	79,797	141,415	22,257
Riverside	....	....	190,572	....
Sacramento	42,301	15,000	....	....
San Bernardino	5,000	....	288,512	....
San Diego	30,000	....	643,496	....
San Luis Obispo	1,000	....	....	
Santa Barbara	1,000			
Shasta	16,300	1,015	1,014,633	....
Sierra	98,279	48,610	194,230	58,463
Siskiyou	224,112	224,450	258,469	62,094
Stanislaus	19,400	....	....	....
Trinity	227,322	348,642	281,055	2,550
Tulare	....	....	12,400	....
Tuolumne	12,453	....	1,738,082	....
Yuba	141,144	4,453	17,000	4,268
	$1,841,473	$962,192	$12,488,321	$1,028,547

RATES OF MINERS' WAGES.—There is considerable variation in the rate of wages paid miners in the different camps of the counties of California. In most places they are paid by the day, but in some counties it is the custom to pay a certain sum per month, with or without board. The Chinese are paid smaller wages than the whites and always board themselves. As a general rule, however, these men work for themselves or Chinese companies, leasing or buying placer or hydraulic ground from the whites. They are able to make satisfactory wages from ground abandoned by white miners, or from tailings from large mines. The highest wages are paid to white miners in the quartz mines who work under-ground, the surface men getting less pay. As a general proposition, wages are smaller in the gravel mines than those of quartz.

In the principal quartz-mining counties, the daily wages for miners may be said to be $2.50, though the rate, according to locality, is from $2 to $3. In the principal Nevada County quartz-mining districts, the rate is $3, as it is in Mariposa. Along the Mother Lode, above Mariposa, rates vary from $2 to $2.75, a $3 rate being exceptional. There is a similar variation in the lower scales of the placer mining regions. In portions of Trinity County, wages run from $2 to $2.75 per day, and $40 per month and board. In the beach and other placer mines of Humboldt County, the prevailing rates are $1.25 to $1.75. The highest rate is in Mono County, where, at Bodie, the miners receive $4 per day.

NUMBER OF MINERS.—The annexed table shows approximately the number of men employed in the gold, silver, lead, and copper mines of California in 1898. As compared with the number in 1897, an increase of 3,060 is shown. This includes 1,200 men at work in the copper mines and smelting works at Keswick, Shasta County. The table is made from returns on the blanks sent to producers from the U. S. Mint, at San Francisco, and also answers from postmasters, express agents, bankers, gold-dust buyers, and others in the numerous camps in the different counties. The answers, in 1898, were more complete than in the previous year, which, in a measure, accounts for the apparent increase; but even then it is possible that omissions are made of many prospectors and persons mining in a small way on their own account. The list does not include miners employed in quicksilver, petroleum, borax, or any mines except gold, silver, copper and lead.

MINERS EMPLOYED IN CALIFORNIA — 1898.

Amador	1,220	Madera	148	Santa Barbara	10
Butte	887	Mariposa	814	Shasta	1,948
Calaveras	1,286	Mono	248	Sierra	545
Del Norte	45	Nevada	1,949	Siskiyou	1,429
El Dorado	1,110	Placer	1,248	Stanislaus	15
Fresno	161	Plumas	576	Trinity	1,205
Humboldt	105	Riverside	225	Tulare	50
Inyo	277	Sacramento	140	Tuolumne	1,637
Kern	1,060	San Bernardino	468	Yuba	290
Lassen	55	San Diego	590		
Los Angeles	72	San Luis Obispo	10		19,823

The silver-mining industry in California, as in other States, is now of little moment, by reason of the decreased value of the metal. The mines worked for silver alone are now all closed down. What silver is produced in the State comes from the copper smelters, mines of argentiferous galena, and associated with the gold in quartz mines.

PRODUCTS OF ECONOMIC VALUE.—In addition to the output of gold and silver, California produces of the economic minerals several millions in value every year. In fact, there are some seventy substances of economic value found here, and at present forty-two of them are being mined. In many cases where these substances are known to exist but are not utilized, the reason is either lack of capital or distance from lines of cheap transportation. Where the gold output was about $16,000,000 last year, the entire mineral product of the State was valued at upwards of $27,000,000. In 1898 the total value of metallic substances mined, including precious metals, was $20,023,638; of non-metallic substances, $2,102,072; of hydrocarbons and gases, $3,070,594; and of structural materials, $2,093,379; total, $27,289,070.

The gradual increase in the value of the mineral product of the State of late years is shown by the following from official reports of the State Mineralogist of California:

1893	$18,806,261.00	1896	$24,291,398.00
1894	20,203,294.00	1897	25,142,441.00
1895	22,844,644.00	1898	27,289,079.00

This is an increase yearly of over $1,500,000, and makes a total for the six years named of $138,577,117. These figures may surprise some who had an idea that the mining industry of California was in a state of decadence. The values include precious metals, quicksilver, borax, copper, petroleum, structural materials, and all products of the mineral world. Instead of decreasing in value, it is seen that

California is yearly increasing her output, and this increase will be very materially larger shortly as the copper and petroleum fields of the State are developed.

From returns received at the California State Mining Bureau, in answer to inquiries, a table has been compiled showing the amount and value of the mineral products of the State for the year 1898, and this will give a good idea of the diversity of the mineral industry in this State. It is as follows:

MINERAL PRODUCT OF CALIFORNIA, 1898.

Asbestos	10 tons	$200.00
Antimony	40 "	1,200.00
Asphalt	25,690 "	482,175.00
Bituminous rock	46,836 "	137,575.00
Borax	8,300 "	1,153,000.00
Cement	50,000 barrels	150,000.00
Clay, brick	100,102 M.	571,362.00
" pottery	28,947 tons	33,747.00
Coal	143,045 "	337,475.00
Copper	21,543,229 pounds	2,475,168.00
Gold		15,906,478.00
Granite	98,369 cu. feet	147,732.00
Gypsum	3,100 tons	23,600.00
Lead	655,000 pounds	23,907.00
Lime	297,860 barrels	254,010.00
Limestone	27,686 tons	24,548.00
Macadam	452,691 "	369,082.00
Magnesite	1,263 "	19,075.00
Manganese	440 "	2,102.00
Marble	8,050 cu. feet	23,594.00
Mineral paint	653 tons	9,698.00
" waters	1,429,809 gallons	213,817.00
Natural gas		74,424.00
Paving blocks	1,144 M.	21,725.00
Platinum	300 ounces	1,800.00
Petroleum	2,249,088 barrels	2,376,420.00
Pyrites	6,000 tons	30,000.00
Quicksilver	31,092 flasks	1,188,626.00
Rubble	724,674 tons	445,395.00
Salt	93,421 "	170,855.00
Sandstone	37,264 cu. feet	46,384.00
Serpentine	750 " "	3,000.00
Silver		414,055.00
Slate	400 squares	2,800.00
Soda	7,000 tons	154,000.00
Sulphur	2 "	50.00
Total		$27,289,079.00

RANK OF THE COUNTIES.—The relative rank of the counties of the State, in point of mineral production, is given in the following table. In each case the value given includes that of all mineral substances combined, produced in the respective counties for the year.

Some counties produce, in addition to gold and silver, five, six, or seven other substances, while other counties, which yield little or no gold or silver, produce, in large quantities, quicksilver, mineral oils, copper, lead, asphalt, structural materials, etc. The figures after the names of the counties indicate aggregate value of all mineral products for the year, including the precious metals. The term "undistributed" includes total values of such substances as are grouped to avoid disclosing private business, as in the case of single operations in a county. In the large and complete tables published by the State Mining Bureau, from which these figures are taken, the amount of value of each substance in said county is set forth.

It is, therefore, necessary, in some cases, to place the figures in the "undistributed" column.

1—Shasta	$3,510,728.00	29—Fresno	$201,057.00
2—Nevada	2,072,604.00	30—Yuba	166,865.00
3—Amador	1,849,846.00	31—Madera	147,907.00
4—Tuolumne	1,757,735.00	32—Contra Costa	141,440.00
5—Los Angeles	1,732,357.00	33—Sonoma	132,709.00
6—San Bernardino	1,644,152.00	34—Sacramento	131,438.00
7—Placer	1,535,525.00	35—San Francisco	129,595.00
8—Kern	1,129,573.00	36—Lake	102,096.00
9—Calaveras	1,024,507.00	37—San Joaquin	91,289.00
10—Trinity	1,010,769.00	38—San Luis Obispo	84,267.00
11—Siskiyou	769,125.00	39—San Mateo	77,000.00
12—San Diego	694,418.00	40—Marin	67,800.00
13—Ventura	654,063.00	41—Orange	65,600.00
14—Napa	555,966.00	42—Lassen	37,760.00
15—Butte	529,225.00	43—Stanislaus	22,298.00
16—Mono	519,421.00	44—Solano	20,635.00
17—El Dorado	512,300.00	45—Monterey	19,115.00
18—Santa Barbara	472,784.00	46—Tulare	15,900.00
19—Alameda	443,759.00	47—Tehama	9,400.00
20—Inyo	434,766.00	48—Del Norte	9,054.00
21—Sierra	399,582.00	49—Kings	8,450.00
22—Plumas	369,609.00	50—Mendocino	3,330.00
23—Humboldt	357,288.00	51—Colusa	1,093.00
24—Mariposa	337,411.00	52—Yolo	384.00
25—Santa Clara	334,848.00	Undistributed	219,990.00
26—Santa Cruz	270,636.00		
27—Riverside	247,022.00	Total	$27,289,079.00
28—San Benito	212,585.00		

It will be seen from this statement that fifty-two of the fifty-seven counties of the State of California produced last year one or more mineral substances of economic value. The most notable increase is occurring in copper, petroleum, and asphalt. The reason of the superior standing of Shasta County, as shown in the preceding table, is that the mine and smelting plant of the Mountain Copper Company is in that county, and that product, added to the gold, gave the county the highest standing. The next three counties on the list produce little aside from gold. The standing of Los Angeles is due to its oil fields, and that of San Bernardino largely to the borax deposits. Placer, Kern, Calaveras, and Trinity produce mainly gold.

WHERE THE MINERALS COME FROM.—All the asbestos produced in California in 1898 was from Riverside County, and the antimony came from Kern. Asphalt was produced in Kern, Santa Barbara, and Ventura counties. Bituminous rock came from Mendocino, San Benito, San Luis Obispo, and Santa Cruz. The borax yield was from San Bernardino and Inyo. Brick clays were utilized in Alameda, Butte,

VIEW OF CALIFORNIA MINING EXHIBIT AT MID-WINTER FAIR, SAN FRANCISCO, 1893-94.

Contra Costa, Fresno, Humboldt, Kern, Kings, Los Angeles, Madera, Marin, Mendocino, Monterey, Orange, Riverside, Sacramento, San Bernardino, San Diego, San Joaquin, San Luis Obispo, San Mateo, Santa Barbara, Santa Clara, Sonoma, Shasta, Tehama, Tulare, Tuolumne, and Ventura counties. Pottery clay came from Amador, Los Angeles, Placer, Riverside, and Santa Clara. All the hydraulic cement came from San Bernardino County. Coal was produced in Alameda, Amador, Contra Costa, Orange, Riverside, and San Benito. The copper was mined in Amador, Calaveras, Inyo, Nevada, and Shasta counties. The pyrites were all from Nevada County. Granite was quarried in the counties of Madera, Nevada, Placer, Riverside, Sacramento, San Bernardino, San Diego, Solano, Sonoma, and Tulare. Gypsum came from Fresno, Los Angeles, and San Benito; and lead from Inyo and Mono counties. Gold was mined in 1898 in the counties of Amador, Butte, Calaveras, Del Norte, El Dorado, Fresno, Humboldt, Inyo, Kern, Lassen, Los Angeles, Madera, Mariposa, Mono, Nevada, Placer, Plumas, Riverside, Sacramento, San Bernardino, San Diego, San Luis Obispo, Santa Barbara, Shasta, Sierra, Siskiyou, Stanislaus, Trinity. Tulare, Tuolumne, and Yuba. Lime and limestone were quarried in El Dorado, Kern, Mono, Riverside, San Bernardino, Santa Cruz, Solano, and Shasta counties. The principal macadam quarries are in Alameda, Los Angeles, Marin, Sacramento, San Benito, San Francisco, Santa Clara, Solano, and Sonoma counties. The magnesite all came from Napa County, and manganese all from Alameda County. Marble was quarried in Amador, Inyo, and San Bernardino counties. Mineral paint was mined in Butte, Nevada, and Stanislaus counties. The mineral springs which utilize the waters commercially are in Butte, Colusa, Contra Costa, Fresno, Lake, Monterey, Napa, San Benito, San Diego, San Luis Obispo, Santa Barbara, Santa Clara, Shasta, Siskiyou, Sonoma, and Tehama counties. Natural gas was utilized in Sacramento, San Joaquin, and Santa Barbara counties. Paving blocks, or basalt, were quarried in San Bernardino, Solano, and Sonoma counties. The platinum came mainly from Trinity and Siskiyou counties. Petroleum was produced in Fresno, Kern, Los Angeles, Orange, Santa Barbara, Santa Clara, and Ventura counties. Quicksilver came from Lake, Napa, San Benito, San Luis Obispo, Santa Clara, Sonoma, and Trinity counties. Rubble was quarried in the counties of Alameda, Humboldt, Madera, Marin, Monterey, Sacramento, San Diego, San Luis Obispo, and San Mateo. The salt came from Alameda, Colusa, Riverside, and San Diego counties. Sandstone was from Alameda, Los Angeles, and Yolo counties. The sepertine was from Los Angeles; slate from El Dorado; soda from Inyo; and sulphur from Los Angeles counties. Silver was produced in the counties of Amador, Butte, Calaveras, El Dorado, Inyo, Kern, Lassen, Madera, Mariposa, Mono, Nevada, Placer, Riverside, San Bernardino, San Diego, Shasta, Sierra, Siskiyou, Trinity, and Tuolumne. The principal silver-producing counties were Shasta, Inyo, Mono, and San Bernardino.

As far as "banner" counties are concerned, in the different mineral products, the following is the record for 1898, with the values of material from the county named: Alameda leads in the production of salt ($155,812), manganese ($2,102), and coal ($176,250); Butte leads in mineral paints ($3,000); El Dorado, in slate ($2,800); Humboldt, in rubble ($297,276); Inyo, in soda ($154,000), lead ($21,170), and marble ($12,000); Kern, in antimony ($1,200); Los Angeles, in petroleum ($1,462,871), brick clays ($188,386), gypsum ($18,500), serpentine ($3,000), and sulphur ($50); Madera leads in granite ($49,673); Napa, in mineral waters ($63,919), quicksilver ($472,972), magnesite ($19,075); Nevada leads in gold ($2,017,628), and pyrites

STARTING TO SINK A PROSPECT SHAFT, AMADOR COUNTY, CALIFORNIA.

($30,000); Placer has the most pottery clay ($12,000); Riverside leads in asbestos ($200); San Bernardino has the most borax ($1,120,-000), cement ($150,000), and limestone ($6,600); San Francisco quarries the most macadam ($129,595); San Joaquin utilizes the most natural gas ($57,289); Santa Barbara has the most extensive asphalt output ($351,400); Santa Cruz produces the most lime ($151,000), and bituminous rock ($113,893); the bulk of the copper comes from Shasta County ($2,465,830), and this county also leads in silver product ($171,768); the most paving blocks are from Sonoma County ($13,310).

THE VARIED MINERAL INTERESTS.—It has been impossible, in the space assigned, to give more than a mere glance at the mineral industries of this State aside from the gold interests, but some of the substances named in the preceding lists are of great economic importance. Moreover, they occur in many places other than those named. Only those counties which made a product of the named substances last year are given. There are, however, many deposits which are not developed, or are not at present in a productive state.

Of antimony, only the States of California and Nevada are at present producing any. In this State, deposits occur in Kern, Riverside, San Benito, Monterey, Tulare, Inyo, Mono, and Santa Barbara counties; but the only deposits now being worked are in Kern. Argentiferous galena is found in Inyo, Mono, Orange, San Bernardino, and Siskiyou counties. Asbestos is found in Riverside, Del Norte, Calaveras, Los Angeles, San Diego, San Bernardino, Yolo, Shasta, Tulare, Mariposa, and Inyo counties, but only in the first-named is it now being produced.

ASPHALT AND ALLIED BITUMENS.—California is the principal producer of asphaltum and allied bitumens in the United States. Deposits of asphalt of all varieties are widely scattered over the United States, but are only worked to any extent in California, Utah, and Kentucky. There are deposits of this substance being worked in California in the counties of Kern, Santa Barbara, and Ventura, and they are known also in Mendocino, Monterey, and San Luis Obispo. The asphalt from the famous deposit on Trinidad Island, so largely used all over the world, only runs from thirty-four to thirty-five per cent in bitumen, the rest being water and impurities; while the La Patera deposit, in Santa Barbara, this State, carries sixty-six per cent of bitumen, and that at Asphalto, Kern County, runs from eighty to ninety per cent. The California yield is largely on the increase, and much is now shipped abroad for paving purposes. The California output of asphalt has now reached about 25,000 tons yearly, of a value of nearly half a million dollars. There are still many very extensive undeveloped deposits, some of them showing millions of tons exposed ready for mining. Bituminous rock deposits always occur in regions where the asphalt occurs, but the reverse is not the case. The bituminous rock now mined comes from the counties of San Benito, San Luis Obispo, and Santa Cruz, but it is also found in Santa Clara, Monterey, Santa Barbara, Ventura and Mendocino counties, and probably in other localities.

BORAX.— Borax is produced in only two States in the Union, California and Nevada, and the bulk of the supply now comes from this State. The fields are very extensive. It is taken from the borax marshes, and also from a "vein," mined by shafts, etc., in San Bernardino County. That and Inyo County are the only counties now producing this substance. The figures of product are given in a preceding table.

OLD HORSE-POWER ARRASTRA IN KERN COUNTY.

CHROME.— For a number of years, the only product of chromic iron in this country came from California. Its occurrence is noted in a number of counties, but mainly in Alameda, Calaveras, Del Norte, Fresno, Glenn, Mendocino, Napa, San Luis Obispo, Santa Clara, Shasta, Sierra, Sonoma, and Tehama. The industry in mining this substance is not very flourishing, owing to the facility with which it can be imported from Asia Minor. Owing to the "pockety" nature of the deposits, it is difficult to determine their extent except by actual work, and the deposits or pockets are usually soon exhausted. The substance is generally mined on a royalty to the owner of the land where it occurs. The necessity of building roads to new deposits where, possibly, only a few hundred tons may be extracted, prevents many known ones from being worked. Cost of transportation to the only markets,

re and Philadelphia, is the main reason which prevents Cali-
supplying the entire demand of the United States. No chrome
y quantity was mined in this State last year, though sometimes
produce five or ten thousand tons a year.

COAL.—Coal is now being mined in Alameda, Amador, Contra
Costa, Orange, Riverside, and San Benito counties, Alameda and
Contra Costa being the largest producers. The coals of the State
are generally inferior in character to those imported. There are many
places in the State where coal is known to exist, but where it is not
utilized except locally. The output has increased of late owing to
the operations of the large Tesla Mine, or Corral Hollow field, near
Livermore, Alameda County. In addition to the counties named, it
has been found in Del Norte, Fresno, Humboldt, Los Angeles, Merced,
Mendocino, Monterey, San Benito, Shasta, Siskiyou, Sonoma, Sutter,
and Trinity counties.

COPPER.— Copper ore is found in many places in California, notably in the counties of Amador, Calaveras, Del Norte, El Dorado,
Fresno, Humboldt, Madera, Marin, Mendocino, Nevada, Plumas,
Riverside, San Bernardino, Shasta, Sierra, Siskiyou, Trinity, and
Tulare. At present it is being mined in Amador, Calaveras, Inyo,
Nevada, Shasta, and Fresno counties. Within the past two years the
copper output of the State has very largely increased owing to the
operations of the Mountain Copper Company in Shasta County,
described in another chapter in this book. The success of that company has stimulated prospecting and development in the direction of
copper ores, and hundreds of known properties long idle have now
become of value. Some very large transactions in copper have of
late occurred in the State, notably in the copper region in Shasta and
Trinity counties. Work is also being done on properties in Fresno,
Madera, Riverside, and San Bernardino counties, and some smelters
are being erected. The copper belt of Shasta County is the most
prominent in the State and the most productive. The deposits now
being worked so profitably were well known for forty years, but it
remained for foreign capitalists to exploit them, the California
mining people paying little attention to other things than gold mines.
Very extensive deposits are known in Plumas County, but distance
from rail transportation has kept them in the background, as has
been the case in other localities. At present, the value of the copper
product of California exceeds that of all other mineral substances

except gold, and it is very rapidly increasing annually. There are numbers of valuable deposits which only await the advent of capital to become productive and profitable.

STRUCTURAL MATERIALS.—Although there are several places in this State where cement may be made, notably in Santa Cruz and Solano counties, the only product is now from San Bernardino County. The output from that locality is increasing yearly.

CURRENT WHEELS, BRADY MINE, SCOTT RIVER, SISKIYOU COUNTY.

In structural materials, mined or quarried, California is gradually advancing. For many years little or no use was made of our natural resources in this respect, but since the era of modern buildings and improved streets and roads, the demand has encouraged owners of deposits to develop them and market the products. Naturally, those having the best facilities of transportation are the first to be opened, but there are many undeveloped properties from which an increased supply can be obtained at any time demand and prices may warrant.

Among the rock formations suitable for building purposes at present made use of in this State, the granites easily take the first place, being found throughout the length of the State, but as the question of transportation figures largely, only those localities have become prominent where there is railroad or water communication with the principal cities of the State. It is now being quarried in the counties of Madera, Nevada, Placer, Riverside. Sacramento, San Bernardino, San Diego, Solano, Sonoma, and Tulare. Some of the quarries are very extensive, and the output of this substance is largely on the increase.

As to lime, there are kilns to be found in many of the counties of the State, and the total output is very large. The principal quarries are in El Dorado, Kern, Mono, Riverside, San Bernardino, Santa Cruz, Solano, and Shasta counties. A good deal of limestone is used in the manufacture of sugar beets, and some is used for road making.

MACADAM AND MARBLE.—The business of quarrying rock for macadam and concrete is quite an important one in many counties of the State, notably in Alameda, Los Angeles, Marin, Sacramento, San Benito, San Francisco, Santa Clara, San Mateo, Solano. and Sonoma counties.

Many varieties of marble are found in the State, and several quarries are now being operated, notably in Amador, Inyo, Tuolumne, and San Bernardino counties. It is found, but not utilized, also in Calaveras, Los Angeles, Madera, Mariposa, Siskiyou, Trinity, and Tulare counties.

SANDSTONE AND CLAYS.—The manufacture of paving blocks from "basalt" is no longer the important industry it formerly was, owing to the increased use of asphalt and bituminous rock for paving.

Sandstone quarries are found in many localities, but sandstone is quarried now only in Alameda, Los Angeles, and Yolo counties.

California clays are utilized in the manufacture of pottery, terra cotta, sewer-pipe, fire-clay, architectural work, tiles, and building brick. These industries give employment to a large number of men. There are numerous deposits of fine clays and kaolin not yet opened. The pottery clay is mainly from Amador, Los Angeles, Placer, Riverside, and Santa Clara counties; and brick clay is found almost everywhere.

SLATE, LEAD AND GYPSUM.—Only a few slate quarries are now being worked, and the output is not large. Steatite is mined only in a few places, notably at Catalina Island, Los Angeles County. Travertine and onyx are quarried in Mono, San Luis Obispo, and Solano counties. Serpentine is quarried in Los Angeles and Amador counties. The lead produced comes from Inyo and Mono counties. All pyrites produced come from Nevada County. Iron is found in quantities in many localities, but owing to the lack of cheap fuel is not utilized. There are several deposits of gypsum, but they are being worked only in Fresno, Los Angeles, and San Benito counties. Infusorial earth is also abundant in several places, but very little is mined.

MAGNESITE AND MANGANESE.—Magnesite occurs in large deposits in several places in California, though it is not abundant elsewhere in the United States. The entire domestic production is from this State, though the only deposits worked are in Napa County; both raw and calcined ore are produced. The mineral is utilized for a bleaching agent in making paper from wood pulp, for furnace linings, and in making gas for soda fountains, etc. Deposits occur in Napa, Santa Clara, and Tulare counties. All the manganese last year came from Alameda County, but it occurs also in Del Norte, Marin, San Joaquin, Santa Clara, and Sonoma counties.

MINERAL SPRINGS AND NATURAL GAS.—California is specially rich in mineral springs, and large quantities of the water are bottled and commercially utilized. The mineral springs now being utilized in this way are in the counties of Butte, Colusa, Contra Costa, Fresno, Lake, Monterey, Napa, San Benito, San Diego, San Luis Obispo, Santa Barbara, Santa Clara, Shasta, Siskiyou, Sonoma, and Tehama counties.

Natural gas is found mainly in the oil regions, and is now being commercially utilized in Sacramento, San Joaquin, and Santa Barbara counties. The gas fields have as yet received very little attention.

PLATINUM, SODA AND SULPHUR.—Mineral paint, both iron and ochre, is being mined in Butte, Nevada, Mendocino, and Stanislaus counties.

Platinum is found in the "black sands" of the hydraulic mines in the northern part of this State and in the sands of the ocean beaches. The only product of the United States is from California. With it is found iridium and osmium, small quantities of each of which are produced each year.

Soda to the amount of about $50,000 worth yearly is taken from the waters of Owens Lake, in Inyo County.

There are several deposits of sulphur, but the only one used is that in Los Angeles County.

The salt industry of the State is of some importance, though most of the salt is made by evaporating the saline waters of the bays of San Francisco and San Diego. Some is made from the saline wells of the Colorado Desert, in San Diego County, and quarry salt comes from Colusa and Riverside counties.

QUICKSILVER.—The total product of quicksilver in the United States, from 1850 to 1898 inclusive, has been about 1,770,000 flasks of seventy-six and a half pounds each, and all of this vast quantity, with the exception of a few hundred flasks from Oregon, has come from the mines of California. It may be said, therefore, that practically all the quicksilver mined in the United States is mined in the State of California. The metal is found in many counties in deposits of larger or smaller extent, but is at present being mined in the counties of Lake, Napa, San Benito, San Luis Obispo, Santa Clara, Sonoma, and Trinity. The annual output is now between 25,000 and 27,000 flasks.

DREDGE ON BAR OR BENCH, 150 FEET ABOVE THE LEVEL OF THE AMERICAN RIVER, NEAR FOLSOM.

The bulk of the entire product of the State has come from the famous New Almaden Mine, in Santa Clara County, still a large producer. Some other mines at this time almost equal the annual output of this mine.

PETROLEUM.—California is now about the fourth in rank among the states in production of petroleum, and the petroleum industry is rapidly assuming an increased importance among the mineral industries of the State. In fact, it is now third in importance, only being exceeded by gold and copper. Within the past year, the developments in Fresno and Kern counties, added to those in Los Angeles, Ventura, and Santa Barbara, have given the industry a wonderful impetus. Just at the present writing petroleum is attracting more attention than gold mining among people of the State, and those in a position to judge think that, in time, petroleum will overshadow in importance even the gold mining industry. The fields are vast in extent, are widely distributed, and are only partly developed. Other articles in this volume deal with this industry in detail, so that little need be said in this chapter. It is noteworthy, however, that this State possesses the only submarine oil fields in the world, since, in Santa Barbara County, they are boring wells, and have productive ones, outside the ocean surf line some distance from the beach.

MINOR MINERAL SUBSTANCES. — Among other mineral substances found in this State, the following may be mentioned: Talc is found in El Dorado, Los Angeles, Napa and Riverside counties, and is being mined at present in Shasta and Ventura counties, the material being shipped to Oregon for the paper mills. Fuller's earth is being mined and utilized in Kern and San Bernardino counties, being used in the packing houses for clarifying fats, oils and grease. Baryta is found in Santa Barbara County, and emery in Sierra County. Natural carbonic acid gas is utilized from the workings of the New Almaden quicksilver mine in Santa Clara County. Tin is found in San Diego, San Bernardino, and Tuolumne counties. It was at one time mined in San Bernardino, but the venture did not pay and was abandoned. Mica occurs in many places, notably in El Dorado, Plumas, Lassen, San Bernardino, and Kern counties, but at present none is being mined. Chrysophrase, a semi-precious stone of great beauty, is being taken from a deposit in Tulare County and brings about $10 per pound in the rough. Small diamonds have been found in Butte, El Dorado, Amador, and Nevada counties, the best coming from Butte County. Zinc occurs in several places, notably in Shasta County, but has not been utilized. Native alum is found in Lake, Los Angeles, Sonoma, Alpine, and Napa counties. Native metallic arsenic occurs in Monterey County and a few other places.

Malachite occurs in several places, but is of little or no value. Quartz crystals of marketable value are found in Butte and Calaveras counties, and a small quantity is sold annually for ornamental purposes. So-called "rock-soap" is found in Ventura and San Benito counties. Some nitrate of soda has been found in San Bernardino county. Tellurium exists in several places, but principally in Shasta and Calaveras counties. Corundum has been found in Plumas County, quite an extensive deposit existing. Bismuth and baryta are both found, but not utilized. There is plenty of graphite, but it is not of the best quality. Small garnets are also found in several localities. There is also some nickel, but it has never been mined commercially. A deposit of bauxite has been discovered near Spenceville, Nevada

County. Good gems of red tourmaline have been found. There is an abundance of glass sand. There are numerous deposits of diatomaceous earth. The rubolite from San Diego County, a very beautiful mineral, is also now being utilized commercially. Volcanic ash and trachytic tufa are quarried in several places and used for building purposes.

CONCLUSION.—In all, there are over seventy mineral substances mined, quarried, or otherwise obtained in the State of California, and some forty-five of them are annually produced in greater or less quantities, the others being mined intermittently, or as demand requires. Many of these substances are left inert simply because other places produce them more cheaply; or because they are distant from rail or water transportation facilities; or, as is usually the case, because capital has not yet been attracted toward their development.

In considering these substances, other than gold, the writer thinks it proper to apologize for the brevity with which he has treated them, since he has by no means given them the space their importance warrants; but typographical exigencies required that the concluding paragraphs of this chapter should be made as brief as possible, so it has been necessary to condense very greatly the remarks originally intended to be made. In fact, it has been difficult to give any comprehensive and intelligent review of the mineral industries of a State where they are of such a varied character, within the limited space at command, and such descriptions as have been given are by no means up to the desired standard. This may be considered, therefore, merely a sketch in outline of the mineral industry of the State of California.

The Mother Lode of California.*

BY

ROSS E. BROWNE.

HE so-called "auriferous slate belt" of California, in which the Mother Lode occurs, begins south of Mariposa county and extends north-westerly along the western flank of the Sierra Nevada to the north line of Plumas county, and doubtless farther, but is thence covered with vast sheets of barren lava. As described, it is about 250 miles long and twenty to seventy miles wide, and forms about one-third of the precious mineral-bearing area of the State—so far the most productive part.

This belt is composed of a more or less conformable series of highly tilted slates and metamorphic schists, with intervening layers and occasional intrusions of igneous rocks. The general strike of the strata is north-westerly—parallel with the axis of the Sierra—and the dip very steep to the north-east. The cleavage of the slates corresponds practically with the tilted bedding planes.

Within this area are vast accumulations of auriferous gravels, and a countless number of gold-bearing quartz veins, commonly following the strike and dip of the strata, occasionally crossing them. Only a small per cent of the great number of these veins contain gold enough to pay the expense of mining and milling, though they all contribute value to the gravels. However, there are in the belt several zones and irregular areas, within the limits of which the veins contain more than the average amount of gold and have produced many valuable mines. The most important of these better mineralized areas are the Grass Valley and Nevada City district, the so-called

* Republished by permission from the Jubilee edition of the *Mining and Scientific Press*, January 29, 1898.

Mother Lode belt, and the east belt of Tuolumne and neighboring counties.

THE LODE.—The Mother Lode is centrally located in the "auriferous slate belt," and is remarkable for its great length and strength, extending over one hundred miles in a north-westerly direction through Mariposa, Tuolumne, Calaveras, Amador and El Dorado counties. It is a continuous series of parallel quartz veins, following more or less persistently a narrow belt of soft black slate. This belt is bounded laterally by harder slates, greenstones (diabase) and nearly allied

CROSS-SECTIONS OF MOTHER LODE

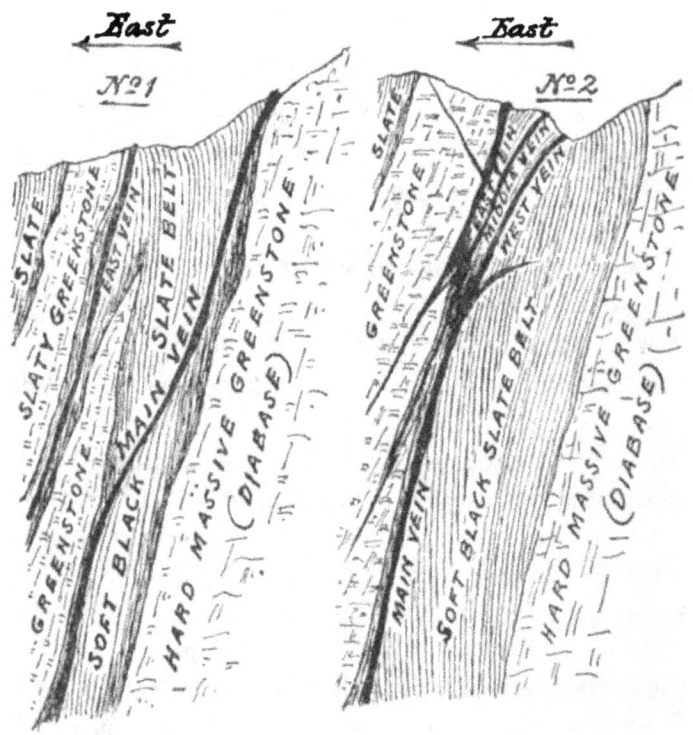

schists (amphibolite), serpentines and occasionally by masses of intrusive granitic rocks. The strike of the strata is between north and west, while the dip varies from 60° to the north-east to 80° to the south-west, but is most commonly 70° or 80° to the north-east. Near to the surface the tilted strata are usually bent downward with the slope of the hill. In some of the sections they stand vertical or even dip to the west.

The veins follow the slates very closely in strike, but cut the strata at acute angles on the dip, being 10° or 15° flatter. The average dip is between 50° and 70° to the north-east. The veins are generally in the black slate either wholly or at one of its contacts with greenstone or serpentine, but occasionally they pass, for short distances, entirely into the amphibolite schist or slaty portions of the greenstone.

CROSS SECTIONS.—Several typical cross sections of the lode are herewith presented.

CONTINUITY.—The actual continuity of a single vein throughout the entire one hundred miles length is not apparent, and it is not believed that such will ever be definitely established. With more development it is probable that continuous veins will be traced through Tuolumne and Amador counties, where the lode is well con-

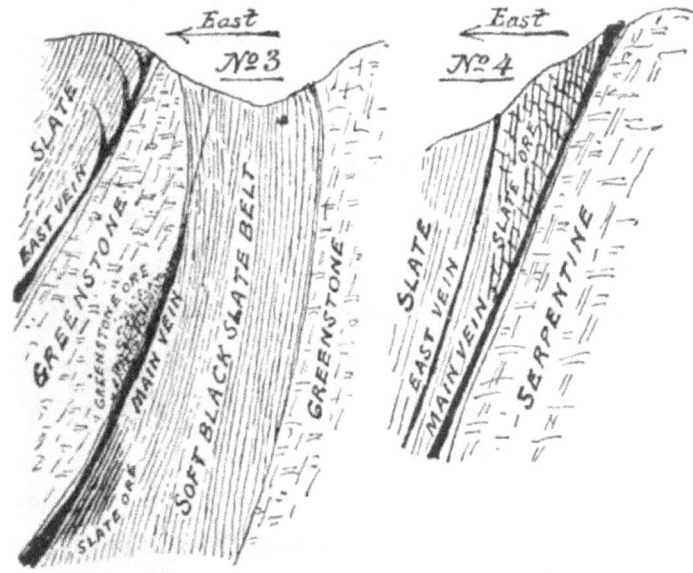

fined within narrow limits, but there are difficulties in the way of connecting these through the central part of Calaveras county, where the branches are widely scattered.

Probably there occurred, at some time while the mountain masses of the Sierra were uplifting and readjusting, one great break and fault following the length of the soft slate belt as a line of least resistance. This break is doubtless represented in some sections by one or two large fissures only and in others by a great number of small fissures following the strata, overlapping and interconnecting somewhat obscurely.

The filling of such fissures would form one great vein with a system of interconnected branches. But there is no authority for assuming that the various veins of the belt represent but one period of fissuring, faulting and filling. There may have been several distinct periods, and upon this depend some important questions of identity and the propriety of applying the term "Mother Lode" to the main vein as the feeder of all others of the section.

THE QUARTZ BODIES.— The main vein always carries a strong gouge either on the foot or hanging wall. The quartz bodies forming the ore shoots occur in lenticular masses from a few inches to fifty feet or more in thickness, from one hundred to one thousand feet long in strike and often of greater extent in the direction of the dip. They are irregularly distributed along the line of the great fissure. Between these bodies the line of fissure is marked by gouge matter, smooth walls, etc., and is easily followed. The walls are more or less ribbed and polished — evidences of motion.

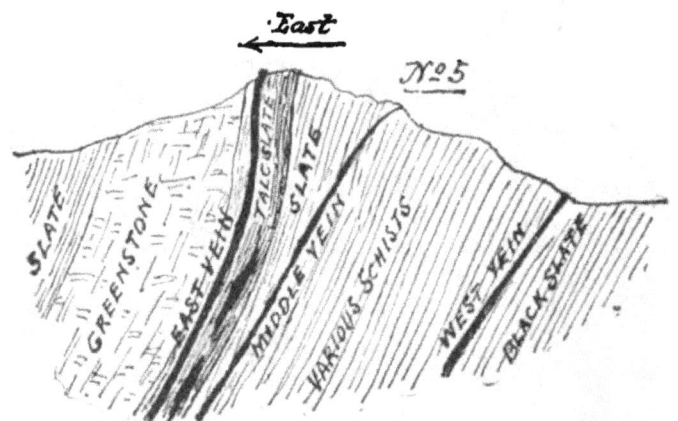

A green mineral, known as mariposite, colors great masses of the veins, and characterizes a large portion of the lode, notably in Mariposa, Tuolumne and the south end of Calaveras counties, to some extent in El Dorado, but not to any important extent in Amador.

Concerning the trend of the ore shoots in depth, various local rules are put forward, but all that may be said in a general way is that they go down on the dip and are longer in that direction than with the strike, but are otherwise quite irregular in outline.

THE ORES.— The ore bodies are mainly quartz, more or less mixed with stringers of slate or bunches of greenstone. They are at times partly colored with mariposite, and usually include some seams of calc-spar.

The ores contain free gold and gold-bearing sulphurets — mostly iron pyrites, with more or less copper and arsenical pyrites, and occasionally galena and zinc-blende. In large quantities the ores carry from $1 to $25 per ton in free gold, and from one to four per cent of sulphurets valued at $40 to $125 per ton. Ores containing less than $7 per ton are considered low grade, $7 to $12 medium grade, above $12 high grade. The average pay ores contain one or two per cent of sulphurets worth $75 to $100 per ton. Rich sulphurets generally mean good values in free gold — poor sulphurets the contrary.

There are occasional occurrences of rich telluride ores, notably in the northern part of Tuolumne and southern part of Calaveras counties, but they seldom form any important part of the ore output of the mine.

The gold is mostly fine and pretty well distributed along the length of the ore shoot. There is apt to be quite as wide a variation in values across the thickness of the vein as along the length of the shoot.

Coarse gold is generally exceptional, but has been found in considerable quantities in some of the sections — for example, in the southern part of Calaveras county, where more than $1,000,000 worth of coarse gold is said to have been obtained by the early miners.

So-called ribbon rock is characteristic of a great part of the lode. It is quartz ore with a banded structure, due to many black slaty seams parallel to the walls of the vein. The structure is more pronounced near to one of the walls, usually the foot. In well mineralized sections of the lode the ribbon rock contains the better values — the slaty seams being rich. It occurs as well, however, in poorly mineralized sections, where it contains little value, of course.

Quartz often extends from the vein irregularly into more or less shattered portions of the wall rocks — in large part by substitution, it is believed. This quartz occasionally carries sufficient valuable mineral with it to form considerable bodies of profitable ore, usually low-grade, however.

In some sections of the lode there are extensive bodies of "mineralized" talcose slates, directly associated with one of the main fissures. These sometimes contain several dollars per ton in fine gold and have been mined to some extent near to the surface.

The hanging-wall slates, near to the surface, are in places fissured and traversed by a great number of small seams of quartz, carrying coarse gold. Many such bodies have been mined by open-cut work, but peter out or become unprofitable in depth.

THE PRESENT SURFACE WORKS OF THE KENNEDY MINE AMADOR COUNTY.

At the northerly end of El Dorado county, where the Mother Lode belt divides and apparently scatters, there are a number of so-called seam diggings — occurrences somewhat similar to the above, but not directly connected with a large fissure, at least not apparently. These are belts of soft, finely laminated slate, somewhat talcose, seamed with innumerable veinlets and stringers of quartz, and containing considerable coarse gold. Most of the seams cut the laminæ at very acute angles, while others are nearly at right angles. These belts are 100 or 200 feet wide and have been mined to a considerable extent by open-cut work, especially at the Georgia slide and Spanish Dry Diggings.

The average thickness of the underground ore bodies, so far mined in the lode proper, has been probably five or six feet, varying from one to fifty or more. The average yield has been about $10 per ton, varying from $3 to $25. But it must be borne in mind that the greater part of the ore has been extracted from the richer sections of the lode. With the new enterprises now under way, and the larger milling capacities, the average yield will doubtless be reduced to $5 or $6 in time.

THE GOLD PRODUCT. — The total product of the lode is difficult to estimate with any satisfactory degree of approximation.

Several of the mines have yielded between $5,000,000 and $12,000,-000 each — the Keystone, Old Eureka, Utica, Plymouth Consolidated and Kennedy — several others $2,000,000 or $3,000,000 and a great number between $100,000 and $1,000,000 each.

There is more work done on the lode in Amador than in other counties, and the yield there has been over $35,000,000 and probably less than $50,000,000. The product of the entire lode, exclusive of the placers partly derived from it, has doubtless been more than $70,000,000 and less than $100,000,000.

For a number of years past the lode has been yielding about a fifth of the State's annual gold product, and its entire yield has been between a quarter and a third of the quartz mining product of the State.

The more productive mines have been very profitable.

TREATMENT OF ORES. — The treatment of the ores is, of course, by simple stamp milling, amalgamation of the free gold, concentration and subsequent chlorination of the sulphurets. The tendency of late has been to use coarser screens and thus increase the capacity of the mill and avoid unnecessary sliming of the sulphurets. Formerly two tons to the stamp per day was deemed a good average; now it is three or four.

The process applied to these ores is cheap and efficient, costing from 50 cents to $1 per ton, and yielding from 90 to 95 per cent of the gold content.

COST OF PRODUCTION.—The cost of production by the prevailing methods of working is shown by the records of a number of important mines on the lode, which have been continuously operated for the past ten or fifteen years. From these it appears that the average expense, after establishing the plant and sinking the shafts far enough to begin the development and extraction of ore, has been about $5 per ton of ore milled. This includes, besides the cost of immediate mining and milling, the dead work required to keep the ground opened ahead, the minor additions to the plant, the repairs, etc.—it is practically the difference between product and dividend.

The dead work once completed, it has at times been found profitable to break, hoist and mill ores of as low grade as $3 per ton.

Several of the better equipped mines are now operating at a materially less cost—less, it is claimed, than $3 per ton. This may be so, but figures based upon accounts covering short periods should not be too readily adopted as averages.

REDUCING THE COST.—The importance of reducing the operating expenses per ton of ore milled is even greater than might at first appear. The quartz bodies of the lode yielding between $2 and $5 per ton are much more numerous and greater in extent than those yielding more than that amount. With the operating expense at $5 per ton there was only a small part of the length of the lode promising an attractive return for the expense of opening, but with figures reduced to $3, the area of vein promising profit is much greater—doubtless several times greater.

In considering how the cost per ton may be reduced, the following approximate subdivision of running expenses is of interest: First, cost of operating the mining plant, including the power supply, labor of topmen, repairs to machinery, etc.—20 or 25 per cent of the total operating expense; second, cost of labor and supplies consumed in making the underground excavations, tramming, etc.—25 or 30 per cent; third, cost of timbering in supporting and maintaining the underground excavations—25 or 30 per cent; fourth, cost of treating the ores, milling and chlorination—15 or 20 per cent; fifth, cost of management, 5 per cent.

It will be seen that the bulk of the expense is connected with the mining, *i. e.*, the work of excavating, protecting the openings and delivering the ores to the surface — the cost of treating the ores is relatively small.

It must be said, furthermore, that the prevailing methods of mining are far less satisfactory than the prevailing methods of treating the ores, and that the improvement of the former offers by far the greater opportunity for important saving.

THE SURFACE PLANT AT THE EAST OR NEW SHAFT, KENNEDY MINE.

The methods of opening the mines for deep work, the hoisting capacity of the shafts, and the manner of protecting the openings (the timbering), particularly, call for improvement.

METHODS OF OPENING.— The mines of the lode have generally been opened, and, with few exceptions, are still being operated through inclined shafts closely following the vein. The reasons for this are evident. The ground near to the vein is soft and easily excavated, and good progress in sinking can be made at small first expense. The approximate following of the vein on its dip enables, from the start, frequent inspection and test of the ores as the work progresses.

The typical incline has two hoisting compartments and a ladder-way. The first cost in sinking and timbering is from $30 to $50 per running foot, and the progress from 10 to 25 feet per week.

The hoists are small, double-geared, and provided for running both with steam and water power. The water power is generally cheaper, but steam relay is necessary in case of accident to the water supply.

Pine wood is used as fuel, costing $4 to $7 per cord. Water costs 15 or 20 cents per miner's inch per twenty-four hours (15,000 gallons). It is claimed that pine wood at $5 per cord is about as cheap to run the works with as water power under 175 feet fall at a cost of 20 cents per inch.

The rock is hoisted in large self-dumping skips, carrying loads of two or three tons.

The amount of water encountered in the lode is small, amounting to a flow of 1 to 10 miner's inches (10 to 100 gallons per minute) in each mine. This is mostly surface water, not extending to very great depth. As a rule, if all the water above the 700 or 800-foot levels is caught up and delivered to the surface, the balance of the ground in depth remains practically dry. This water is generally handled in the same way as the rock, in preference to pumping.

Owing to the soft swelling ground near to the vein, the timbering requires close attention and frequent repairs. The dip of the lode is somewhat irregular, varying occasionally from 45° to 70° in a single section, and in trying to follow it, even approximately, there occur bends or marked changes in the inclination of the shaft. The ultimate result is great expense of repairs and material reduction of the safe speed of hoisting.

These disadvantages are not felt in the beginning, but invariably become matters of serious concern when considerable depth is attained. After reaching depths of 1500 or 2000 feet, the cost of keeping a single incline in barely passable repair is occasionally as great as $4000 or $5000 per year, and the safe speed of hoisting is lowered to 300 or 400 feet per minute, reducing the capacity to such an extent that there are required two or three such inclines to supply a 60-stamp mill. In fact, the limit in depth, where low grade ores may no longer be profitably worked, is reached long before there appears any serious drawback due to natural causes.

VERTICAL SHAFTS.*—Messrs. Thomas & McClure of the Gwin mine have made a step in advance. They put down a vertical shaft to tap the vein under the old works at a depth of 1500 feet, and are now supplying a 40-stamp mill with 150 tons of ore per day at less than the usual cost. When their mill is enlarged. they will doubtless reduce the running expense per ton of ore very materially.

The Merced and Oneida companies are also opening their mines with vertical shafts, and it is understood that the Kennedy and Wildman companies are contemplating the sinking of deep vertical shafts to operate through. The Oneida shaft is down nearly 1500 feet and is designed to tap the vein at 1750 feet. It is equipped with a hoist easily capable of supplying an 80-stamp mill with ore from a depth of 2500 feet or more. The works were designed and erected by Mr. H. C. Behr and are the most capacious on the lode.

The advantages of the vertical shaft for deep working are manifest. It is removed from the lode, in firmer rock, is easily kept in repair, and is adapted to rapid hoisting. A single shaft may be made to do the work of several crooked inclines and thus make a great saving in the operating expenses.

It is true that much greater lengths of crosscut are required to reach the vein, and this disadvantage is prohibitive in developing flat veins such as those of Grass Valley and Nevada City with dips of 25° to 45°; but in the case of the Mother Lode, with an average dip of 60° or 70°, it becomes less significant.

The cost of the Gwin shaft is understood to have been not more than that of the average incline—$35 or $40 per foot. The Oneida shaft is larger, in harder rock, and had much more water to contend with; it cost about $65 per foot.

The Oneida shaft was laid out with the length of the collar parallel to the croppings of the vein and the strike of the slates. It was expected that the strata would pitch to the east, but they were

---

* Since the above was written, the Wildman and Kennedy Mining Companies have started vertical shafts far to the east of the vein apex. The Kennedy is a three-compartment shaft designed to intersect the lode at a depth of 3,500 feet. It was laid out with the length of the collar at right angles to the vein and the strike of the slates, and has already been sunk 400 feet deep, making an average progress of twenty feet per week, being in very hard gray slate. The ore vein of quartz at the 2,200-foot level is thirty or forty feet thick, and has inspired the confidence necessary to justify the expensive vertical shaft designed to cheapen the cost of operation on a large scale.

The Gwin shaft has reached a depth nearly 1,800 feet and has recently made an additional valuable development.

The Oneida shaft has been carried down to the 1,900-foot level, cutting at that depth a large vein of ore. The immediate erection of a mill is now contemplated, being justified by the assurance of a large quantity of pay ore.

found vertical through the depth of the shaft, and the ground was exceedingly difficult to break in consequence. Machine drills could not be used to advantage, it being impracticable to put down machine holes otherwise than parallel with the seams and strata, giving the powder little chance to break and throw the rock. The progress made by hand drilling with three shifts of six men each was forty to eighty feet per month; average sixty. The amount of water handled is from two to ten miner's inches, according to the season. Had the collar of the shaft been laid out to cut across the strike of the slates, better progress would certainly have resulted, and the experience is worth remembering.

GALLOWS FRAME AT THE ONEIDA MINE, IN PROCESS OF ENCLOSURE.

It may be said that the incline is the preferable means of prospecting, developing and working the mine to moderate depth, while the vertical shaft is much the better for operations to great depth.

Beginning with a virgin mine on the lode, the start is very properly made with an incline following the vein. After operating to depths of 800 or 1000 feet it is found that the expense of repairs is rapidly growing, while the capacity of the incline is rapidly diminishing. Then is the time to stop and inaugurate a better system of work, if the showing justifies it. But an oversanguine spirit, impatience and improvidence are apt to prevail. If in good ore, it is generally preferred to follow the old system as long as it pays expenses, with the hope of finding something still better. An unfortunate turn destroys confidence and credit, and the mine, which might have paid well under better management, is closed indefinitely. This is doubtless the history of a number of the mines. Some of them were better managed — others have paid notwithstanding bad management—still others never had an attractive showing.

DEVELOPMENTS.—In order to illustrate the proportion of the lode already worked, if we should plot it all upon a longitudinal section, it would impress us as insignificant compared with the area still unexplored in depth. The greater part of such area is, of course, likely to be poor, but there are many sections in it attracting test, and some with values practically established.

Twenty or thirty shafts have been sunk on the lode to depths exceeding 1000 feet—two of them exceeding 2000 feet. The developments made have inspired the utmost confidence in the permanency of the main fissure, and its mineralization—in fact, it is commonly asserted that the lode improves with depth, but this claim is doubtless oversanguine.

In Amador county the Kennedy mine, after developing low and medium grade ores down to a depth of 400 or 500 feet, passed through a comparatively barren zone, and at a depth of 900 feet encountered a high-grade ore body extending down 1000 feet farther and yielding several millions of dollars. Quite recently a new body of high-grade ore was opened at a depth of 2200 feet.

The neighboring company, the Argonaut, encouraged by these developments, but with a very poor showing on the surface, put down a deep shaft, and at a depth of 1000 feet made an apparently equally valuable discovery.

The Utica mine, in Calaveras county, proved richer in depth than at the surface; and the same is said of a few others.

But there are several important mines of the lode developed to depths of 1200 or 1500 feet with much poorer results below the 800 or 900-foot levels than above.

All these tests have naturally been made in sections with a good showing on the immediate surface or near at hand.

CHANCE IN DEPTH.—The facts doubtless warrant the assumption of just as good a chance of discovery at great depth as near to the surface. The chance is certainly favorable in most of the sections which have been good on the surface—depending, of course, to some extent upon local conditions.

There appear excellent reasons for confidence in the main fissure

—the lode proper—to greater depth than the miner is likely ever to reach.

WIDTH OF BELT.—As already stated, the Mother Lode belt is alternately widespread, containing many parallel veins, and narrowly confined, with only one or two important veins. The width generally recognized as the area, within which the main fissure may be sought, varies from one hundred feet to a mile or two.

The greater number of the more valuable discoveries have been made in the narrower sections thus defined, but several notable exceptions prevent the adoption of a rule. All that may be said properly is that great strength of fissure is one of the important

COMPLETED HOISTING WORKS AT THE ONEIDA MINE.

factors in making a great and lasting mine, and is more apt to be found in the better confined than in the widespread section—but the conditions which favor the strongest mineralization are little understood and doubtless complicated.

A FIELD FOR DEEP MINING.—The rich spots on the surface of the lode, the poor man's diggings, have long since been exhausted, but, unfortunately, many of the claims awaiting proper test are still held by men unable to open them and unreasonable in their expectations. Such men retard development.

There remain to be tested a number of sections which were good on the surface but became poor at moderate depths, and many more which show only low grade ore on the surface.

CONCENTRATORS IN THE GWIN MINE DEVELOPMENT COMPANY'S MILL, CALAVERAS COUNTY, CAL.

On the whole, there is an excellent field for deep mining in the better mineralized sections of this Mother Lode, but it is one that should be entered only with large capital and a rational mining spirit.

LITERATURE.— The geological features of the lode are best presented by Mr. Fairbanks in the State Mineralogist's Tenth Annual Report, 1890, accompanied by a map showing approximately the distribution of the various enclosing rocks, and the line of most important claims; also by Messrs. Lindgren and Turner in the published atlas sheets of the United States Geological Survey. The latter are considered the most authoritative regarding the surface formation, but the scale is too small to illustrate the location of individual mines.

There is a certain amount of information to be had concerning the mines of the lode in the United States Reports on Mineral Resources, in the various reports of the State Mineralogist, and in the files of the MINING AND SCIENTIFIC PRESS. Probably the best information to be had concerning the early workings is in the volume of Mineral Resources of 1868.

It must be said that in view of the importance of the subject there is a notable dearth of accurate and reliable information to be had in published form. A correct map of the underground workings, with a detailed history of the values of the ores, and the more important experiences in mining, would be of great value in the future. Many expensive mistakes will doubtless be made owing to lack of information which might be furnished by a suitable system of record.

It is believed that a good purpose would be served if the State should compel the filing of such information where it would be open to public inspection. It is greatly to the interest of the mining industry to discourage deception and narrow-minded secrecy.

Electric Power:

Its Generation and Utilization for Mining Work on the Pacific Coast.

BY

A. M. HUNT and WYNN MEREDITH.

N writing the following paper, the authors have borne in mind that it is to be placed before mining engineers, many of whom have slight acquaintance with the modern developments of electrical engineering. No attempt is made to discuss the very technical matters connected with alternating current transmission and apparatus, but so far as possible it is intended to give an idea of the extent to which electric power has already been applied in mining operations; the success, both economic and mechanical, which has attended such applications; and to answer some of the questions which naturally occur to the mining engineer who wishes to determine whether, under the circumstances in which he is placed, he can develop a more or less distant source of power, transmit, and successfully utilize it for his work.

A very common and familiar question asked by those unversed in electrical matters is: How far can electric power be transmitted, how large a wire will be required, and what will be the loss?

This question is for all practical purposes parallel with the following: How far can water be pumped, what size of pipe will be required, and what will be the loss?

The limitation on distance, in the case of pumping water, is the pressure which may be employed, bearing in mind the necessity of having pumping apparatus and pipe line capable of standing the pressure supplied. The size of pipe will be a factor of the amount of water to be delivered, the pressure employed, and the distance. The amount of lost energy will be dependent on distance, size of pipe, and quantity delivered. Economic considerations of cost of apparatus and pipe line determine the commercial feasibility of any such proposition.

In the case of transmission of electric power, the parallel is as follows:

The limitation on distance is the pressure or voltage which may be employed, bearing in mind the necessity of having generating apparatus, and pole line capable of standing the pressure or voltage supplied. The size of wire required will be a factor of the amount of energy to be delivered, the pressure or voltage employed, and the distance of transmission. The amount of energy lost will be dependent on distance, size of wire, and the amount of energy delivered.

Again, economic considerations of cost of generating apparatus, and pole line, determine the commercial feasibility of any such proposition.

The mining districts of the Pacific Coast are in most instances located in the mountain ranges or in the foothills, where numerous streams, having rapid fall, offer abundant opportunity for the development of power.

The State Mineralogist of California, in his report, published in 1893, makes the following comment:

"Already the work of transmitting the power generated by this fall of water to distant points, through the agency of electricity, has been entered upon in this State, the experiment made having proved eminently successful. That great benefits will grow out of this new use of electric current may be expected, there being many mines upon which the introduction of water is not practicable, but to which the power produced by a water wheel might be carried through the medium of an electric conductor."

During the past six years, the expectations thus expressed have been realized to quite a large extent, and from the present outlook, the developments of the next few years should greatly exceed those of the past.

The relatively low cost at which electric power can be delivered in many localities, as compared with the cost of steam power, where the rapidly diminishing wood supply is the only source of fuel, reduces the cost of operation, and enables mines to be worked at a profit which would otherwise be worked at a loss. New mines have been opened in places where the cost of steam power would have been prohibitive, and where water power direct could not be made available. Some of the old ditch companies have found it to their interest to put in electric plants, and sell electric energy to their customers instead of delivered water.

ELECTRIC LOCOMOTIVE. HIDDEN TREASURE DRIFT MINE, PLACER COUNTY.

Success has accompanied the efforts to apply electric power to practically all branches of mining work where power is needed. It offers higher flexibility for bringing the power to the point of required use, than any other means. There is no limitation imposed on the location of the machinery, as is so often the case where water power is used direct, and mill buildings, etc., may be located solely with reference to convenient and economic handling of ore. Mine owners and mining engineers have been inclined to be skeptical as to the adaptability of electric power for the various classes of work required around mines, and the introduction of the power involved a large amount of pioneer and missionary work on the part of the electrical manufacturing companies of the country. Special forms of apparatus had to be devised to meet the requirements of the miners, and, while mistakes have been made, it can be safely asserted that, where mines have been operated by electric power, and the installation has been made with proper engineering precautions, the results have been smooth and successful operation, and increased economy of production.

The ability to successfully transmit electric energy to a distance has been so thoroughly demonstrated by actual practice as to admit of no question. A very few years ago, whenever such a claim was made, the classic experiment of transmitting one hundred horse-power from Frankfort to Lauffen, in Germany, was usually cited. At the present time, transmission plants have become quite numerous, there being a large number on the Pacific Coast, which leads in distance and pressure of transmission.

A plant in Southern California is transmitting power into Los Angeles over a line eighty-one miles long, at a pressure of 33,000 volts.

Work is now in progress on a plant for transmitting current into San Francisco, over a line one hundred and fifty-six miles long, at a pressure of 40,000 volts, and there are numerous installations transmitting energy over lines varying from five to thirty miles, at pressures varying from 5,000 to 15,000 volts.

Extreme distances of transmission are commercially practicable only when the cost of power from other sources is high, and where a considerable amount is required.

The advances that have been made in transmission work have been due to the development of apparatus permitting the use of high pressures, thus bringing the cost of pole lines down to a point, by reason of the possibility of using smaller wires, where interest charges do not so largely affect the cost of delivered power.

Until within a comparatively few years, no successful motors for operation with alternating currents had been worked out. It was not possible to construct successful direct current machines to deliver energy in quantity, at pressures much, if any, exceeding 1000 volts. When, therefore, a successful alternating current motor was produced, the problem of utilizing electric energy at a distance became a simple matter, as the alternating current admits of being transformed, either up or down in voltage, by the use of simple and efficient devices. As a result, all transmissions of electrical energy over distances exceeding a very few miles are, as a matter of practical necessity, made by the use of alternating current.

It is impossible to lay down any general rules to govern the selection of the component parts of a transmission system, as each case constitutes an independent problem, and must be considered with reference to local conditions, and the requirements of the work to be performed.

While not a part of the electrical apparatus, the hydraulic end, where water power is used, is such an important element that a few words concerning it seem justified.

Most of the water powers on this coast are developed with what are abnormally high heads, which has led to the almost exclusive adoption of impulse wheels, and the heavy pressures and high velocities have often been sources of trouble. At the outset in this work of utilizing water powers, it was assumed that the serious difficulties encountered would be with the electrical end of the plants. The writers, however, feel justified in saying, that in every case that has come under their notice, the most serious disturbances and interruption of service have come from weakness or poor design of the hydraulic apparatus and installation. The old idea that anything in the way of a water wheel was good enough to run an electric plant has been modified, and too much care and attention cannot be given to this important feature.

In this connection, as in all other features of such a plant, it should be borne in mind that it is important to provide against interruption, either by duplication or subdivision of units, so that breakdown of one unit does not entirely shut down the plant.

Practically, all long-distance transmissions, where power forms even part of the load, are made with polyphase alternating current apparatus. A polyphase alternating current, as a source of power, may be considered as analogous to a multi-crank engine which has no dead centre.

Self-starting single-phase motors have been built in small sizes, but it may be stated that single-phase motors require to be brought up to a synchronous speed with the generator supplying them before the load is thrown on, and if they fall out of step with the current supply, they will stop. It requires some extraneous source of power to bring them up to speed, and they must run at constant speed, or at a speed exactly proportional to the number of alternations of the current supplying them.

These features are also common to what are known as synchronous polyphase motors, but what are known as induction polyphase motors are so designed as to admit of starting with load.

One of the first features of a plant to be determined is the pressure or voltage at which current is to be generated and transmitted. The decision must be based on the distance, the amount of energy to be delivered, and the permissible loss in transmission.

A diagram is appended to this article, on which are given directions for use, which gives by inspection the size of wire required to transmit any number of horse-power, any distance, at any loss, and at any voltage, within ranges of ordinary sizes of plants and ordinary practice.

This diagram has been prepared by Mr. R. S. Masson, of San Francisco, to whom the authors are indebted for its use.

Generators can be, and are, wound for pressures as high as 10,000 volts, and if the amount of power to be transmitted and the distance are such that the size of wire is not excessive, the generated pressure may be placed directly on the line without the intervention of step-up transformers. There may, however, be reasons of sufficient importance to warrant generation at lower voltage. Questions of voltage having been determined, which also determine amount of line conductor required, decision must be made as to method of driving generators. In our coast plants, the head under which the water is utilized is almost always great enough to permit the water wheels to run at speeds which admit of connecting generators direct, without involving abnormal cost of generating apparatus.

As to whether generation shall be two or three-phase, engineers differ, some claiming superiority for one, and others for the other. The truth seems to be that each system has its advantages and its disadvantages, and the question as to which is preferable depends on considerations at the receiving end of the line, and on certain technical points which cannot be discussed in a paper of this nature.

As to transmission circuit, it seems unquestionable that it should be what is known as the three-wire, three-phase system. A special method of transformer connection readily permits changing current from two to three-phase, and vice versa.

In the generating station, details of switch-boards and other apparatus have to be determined and worked out, and on such design depends, in a considerable measure, the satisfactory operation of the plant.

During the past year, aluminum, as a material for line conductors, has become a factor in the market. Copper has advanced so materially in price that aluminum becomes a successful competitor on the

TRIPLEX, MOTOR DRIVEN STATION PUMP.

basis of cost. While it has been used in comparatively few cases, and has nowhere been tested for a very great length of time, there seems to be no good reason why it should give rise to any trouble.

In the earlier days of transmission work, it was considered that the insulators used on the lines must be made of porcelain, to withstand the pressure used. The porcelain insulators used on some of the first lines on this coast were a rather frequent source of trouble, and there has been a gradual change toward the use of glass insulators, until at the present time almost all installations are made with glass, at a considerable saving in cost and with better results.

Pole lines may be run over very rough country, and no grading is required. It is desirable to fell all trees that may endanger the line by falling, and to clear off all brush for fifty feet on either side of the line.

One of the earliest transmission plants for mining work was installed by the Standard Consolidated Mine of Bodie. The fuel cost at this mine was very heavy, and it was very desirable to obtain a cheaper source of power. The superintendent of the mine, Mr. T. H. Leggett, now in South Africa, made up his mind that electric power could be successfully utilized, and in 1892 contracted for a generating plant of 160 horse-power and a motor of 120 horse-power. The source of power was Green Creek, distant from the mine some 12½ miles.

At the time the contract was made, it required a great deal of confidence to take such a step. The engineers of the large electric companies were still discussing whether direct or alternating current should be used for such a transmission. Mr. Leggett decided to use the alternating current, and the motor was a synchronous single-phase one, requiring a starting motor. There were certain minor difficulties encountered at the outset, but the plant was eminently successful. Mr. Leggett, in a description of the plant published by him, states that the plant cost not over $38,000.00 complete, and that $2,100.00 was saved in one month over the old steam plant using wood.

Part of the plant was later destroyed by fire, and has been replaced. As a pioneer plant, it did much to give the miners confidence in this new form of power.

Other installations gradually followed, until now you can hardly find a mining district on the coast that has not an example.

Electric power has been applied to the operation of stamps, rock crushers, air compressors, concentrators, haulage locomotives, hoists, pumps, ventilating fans, dredging, and exterior as well as interior lighting.

It is wise, if possible, to separate the intermittent and steady work. The starting and stopping of intermittently operated motors is apt to affect the speed of other motors running on the same circuit.

Electric motors normally run at a comparatively large number of revolutions, and it is usually necessary to reduce this speed by gearing down.

For the operation of stamps and concentrators, motors are eminently adapted, and it is claimed by some who use them, that there is an actual increase of output obtained, owing to the so nearly perfectly uniform speed of driving. The motor takes up a comparatively small amount of space, and can be belted for the driving work. Its application is so simple as to require no extended comment.

Rock crushers are preferably operated by independent motors, and through belting rather than solid gears, so as to introduce an element of elasticity. Air compressors can be readily driven by motors, either through gearing or belts. While solid gears have been used, the most successfully motor-driven air compressors use rope transmission.

With the exception of the series direct current type, all electric motors tend, when operated under the same electrical conditions, to run at a fixed, nearly constant speed, regardless of load, if it does not exceed their capacity. The amount of energy consumed is proportional to the load, although the speed does not vary. This has led to the design of an air compressor which delivers a variable quantity of air, according to the demand, while running at a constant speed. A compressor of this type is being operated at the North Star mine in Grass Valley, Cal., driven by a two-phase synchronous motor, through gearing. It is a two-stage compressor, and the inlet valves of the initial cylinder, which are of the Corliss type, with trip, may remain open during the whole, or a portion of the delivery stroke, the position of the trip being governed by the rise in pressure of the discharge above that normally required. It operates very satisfactorily.

In another instance of a two-stage compressor, the discharge pressure, on reaching a predetermined amount, opens a valve connecting the receiver to the atmosphere, so that only the volume of the small cylinder of free air is discharged against the final pressure.

Both of these are examples of the tendency to adapt the machinery driven to the conditions governing that which drives.

Electrically driven pumps have been installed in a number of instances. The triplex-geared type seems to be best adapted for electrical operation, and several illustrations are given of this type. One of these shows a connection by means of a bell crank and connecting rod for operating a sinking pump from a station one.

The main objection to the operation of pumps by motors is well stated in the following paragraph from a Bulletin of the California State Mining Bureau, by Mr. Hans C. Behr, as follows:

"Electricity is not well adapted as an economical means of transmitting power to reciprocating pumps which have a variable duty, unless they can be run intermittently, permitting the water to accumulate in large reservoirs or tanks during the time of stoppage. The method of applying electricity to pumping machinery will have to be much improved before it can bear out the claims of efficiency made by electrical companies."

To overcome the objections noted by Mr. Behr, Mr. George E. Dow, of San Francisco, has developed an extremely ingenious pump, which, while operating at constant speed and length of stroke, takes

TRIPLEX, MOTOR DRIVEN STATION PUMP, WITH SINKING PUMP
DRIVEN FROM A BOB.

water in proportion to the amount supplied to the sump, and the work done, exclusive of pump friction, is proportional to the water delivered. Connected with the displacement space of the pump is an inlet pipe with check valve, which admits air on the suction stroke. The amount of opening of this pipe is controlled by a cock actuated by the difference in level of water in the sump. The pump, on its suction stroke, takes in both air and water, the relative proportion of each depending on the amount of opening in the air inlet pipe. On the delivery stroke the air is first discharged through a check valve, of peculiar construction, against atmospheric pressure. As soon as the water reaches this valve it closes, and the water discharges into the column.

The indicator cards, taken from the water cylinders, show very clearly that the work done is proportional to the water delivered. During the first portion of the stroke, the only work done is to expel the air against atmospheric pressure, which amounts to very little. A motor-driven pump of this type should operate very satisfactorily and efficiently.

The electric motor is poorly adapted for driving anything in the line of slow moving Cornish pumps.

Where an old mine, that has a steam plant, is to be supplied with electric power, a very satisfactory though not highly efficient plan is to instal a motor driving a compressor, using the air in the steam cylinders. The boilers can be used as storage for air, and if the air is heated before going to the cylinders of the engines, a fair degree of efficiency can be obtained. Where such a plan is adopted, steam operation can be taken up on comparatively short notice in case of an emergency.

Electric haulage locomotives have been used with notable success in drift gravel mining, and could be advantageously employed in many instances.

Electrically operated hoists have been less successful than any other class of electrically driven machinery. This is due to the character of the load and the requirements as regards speed. To avoid severe strains on the various parts, the load must be started gradually. It must be possible to hoist rapidly or slowly. It must be possible to stop the hoist with precision at levels and stations. The most successful hoists have been those operated by direct current motors, similar in type to those used for railway work, and with similar methods of control. These have done excellent service in cases where moderate power is needed.

An alternating current hoist has been installed lately at the War Eagle mine, in British Columbia. The motor is capable of delivering 300 horse-power at 300 revolutions, and is geared to a drum hoist. The maximum load is eight tons, including the load, cage and rope, the speed being 720 feet per minute for this load.

A full description of this plant was published in an article in the "Journal of Electricity," in June, 1899, and contains many items of interest. The author of this article, Mr. Geo. P. Low, makes the following statement: "One who is interested could spend hours watching the operation of this hoist. It is easily handled by one man who finds himself with much less to do than has the motorman on an electric railway * * * * * * * At times, when men are on the cage, the hoist is "kicked" along by the momentary application of power to the motor, which enables it to be run at a much slower speed even than that possible with the controller on the first notch. At other times, in hoisting ore, a dead load of five tons of which is almost always carried, the motor will be brought up to speed in a very few seconds, and this without any abnormal inrush of current."

An alternating current hoist has been contracted for at the Griffiths mine, in California, of 200 horse-power, which has a method of control differing from that at the War Eagle mine. The claims made for it by the installing company warrant the belief that it will be successful.

Dredging boats, for operating river-bed claims and gravel banks, have been very successfully used in California for several years. A number of these use electric power, in localities where power from other sources would be prohibitive in cost.

The lighting of underground workings and mill buildings has become very common. Many mines, using steam-power, have put in engine-driven units for this work.

An attempt has been made to introduce an electrically driven percussion drill, but it has not proven successful.

There are many minor operations around a mine where electric power can be advantageously applied. Electric motors of small power can be easily transported to any point where power is needed. They require no expensive and heavy foundations, and the cost of running wires to them is small. They can be used to operate saws for framing mine timbers. In one case, where a hoisting plant is located about 1500 feet from the mill building, the ore is transported to the latter in cars by rope haulage, the motive power being a stationary motor at the mill. The ease with which the power may be brought to the point of use constitutes an important point in its favor.

A partial list of a number of the plants installed on this coast is given below, with class of work done. Only a few cases are noted, but it gives some idea of the advance made in the last five years.

Mt. Gaines Mine, Hornitos, Cal. A 180-horse-power motor, driving an air compressor. A 30-horse-power motor, operating mill. A 10-horse-power motor, driving rock breaker. All alternating current three-phase motors made by the General Electric Co.; Generating plant, Stanley Co.

Spring Creek Electric Power Co., Redding, Cal. Generating plant of 200 horse-power. Three-phase alternating current. General Electric Co., supplying power to mines in vicinity.

Lightner Mine. A 150-horse-power motor, driving mill. A

ELECTRIC HOIST. NATIONAL CONSOLIDATED MINE, REDDING, CAL.

100-horse-power motor, driving compressor. A 10-horse-power motor, driving crusher. All two-phase, alternating current. Westinghouse Electric and Mfg. Co.

Griffith Con. G. M. Co., Diamond Springs, Cal. Generating station, 330-horse-power alternating current. A 40-horse-power motor for stamps.   A 20-horse-power motor for rock crusher.   A 30-horse-power motor for pump.   A 200-horse-power motor for geared hoist to lift 5,500 pounds at 600 feet per minute.   This hoist is not yet installed.   Westinghouse Electric and Mfg. Co.

Nevada County Electric Power Co.   Generating station contains 2000-horse-power two-phase alternating current, Stanley Electric Co.'s machinery.   It lights the towns of Grass Valley and Nevada City, and distributes power to a number of the mines in those districts, among which are the following :

North Star Mine.   A 440-horse-power Stanley motor, geared to a Rix variable volume air compressor.

Allison Ranch Mine.   A 350-horse-power Stanley motor, with belt to countershaft and rope drive to air compressor.

Homeward Bound Mine.   A 180-horse-power alternating current Stanley motor, driving a General Electric direct current generator.   Direct current operates street-car motors geared to old steam hoist.   A direct current motor is belted to an old Cornish pump.

Pennsylvania Mine.   An 80-horse-power alternating current motor, driving stamps, concentrators and rock crusher.

W. Y. O. D. Mine.   Same equipment as Pennsylvania.

There are a number of other mines supplied with power by the Nevada County Electric Power Co.

San Francisco and San Joaquin Coal Co.   Three 25-horse-power electric locomotives.   Four 20-horse-power and one 30-horse-power motors, driving fans.   30 horse-power of small motors, miscellaneous work.   General Electric Co.

Utica G. M. Co., Angels Camp, Cal.   This company is now installing a generating plant, of about 675 horse-power, to supply current for motors of about 300 horse-power at Stickles Mill, and the same amount at Utica.   Westinghouse Electric and Mfg. Co.'s alternating current machinery.

Electric locomotives, General Electric type, are in use at the Hidden Treasure mine, the Mountain Copper Co.'s mine, the Bunker Hill and Sullivan mines, in Idaho, and the Roslyn coal mines in Washington.

Standard Consolidated Mine, Bodie, Cal.   Operates a 240-horse-power generating plant and motor.   Stanley Electric Co., alternating current.

West Kootenay Light and Power Co., British Columbia. This company has a generating plant with a·capacity of 4000 horse-power, which is being rapidly taken up for mining work. It furnishes current for the War Eagle hoist, before noted, a 400-horse-power motor for driving compressor at the same place, and a number of other motors at mines in the vicinity. The machinery was supplied by the Canadian General Electric Co.

Arizona Copper Co., Clifton, Arizona. This company is operating an interesting plant. The electric generators are driven by Crossley gas engines, using Dowson gas. The capacity of plant is 400 horse-power, which is soon to be increased, and distributes current to a number of scattered motors. A high degree of economy is claimed. The electric machinery was furnished by the Westinghouse Co.

The Yuba Electric Power Co. is installing a plant on the North Yuba river that will eventually have a capacity of 15,000 horse-power, according to present plans. It will distribute power for miscellaneous uses, a considerable amount being taken for mining work.

From such data as the writers have been able to get together, the aggregate amount of the power used exclusively for mining work on the Pacific Coast is probably about fifteen thousand horse-power. The data, however, is not complete, and the amount is increasing very rapidly.

Notes on Dredging for Gold.

BY

R. H. POSTLETHWAITE.

OR many years this class of mining was looked upon with great suspicion by miners, especially in the Western States. The few dredges which had been tried were generally built on similar lines to those used in harbor work, and were quite unsuited to the very different conditions met with when operating for gold; hence, numerous costly mistakes were made before the present type of dredge was evolved, giving a serious set-back to the industry.

The fields of gold dredging can be fairly divided into two classes, which correspond to a certain extent to the two classes of quartz mining, viz.: small high-grade mines, and large low-grade mines. A third class, namely, a large high-grade mine, need not be discussed, as there are few such dredging properties, and what there are can easily take care of themselves.

A very good illustration of a large low-grade proposition is the country below the gorge of the Feather River, in Butte County, California, comprising some 5000 acres. The false bottom, generally composed of lava ashes, has been overflowed by an average of thirty-five feet of auriferous gravel, every yard of which carries more or less gold. The average value over the whole area will run about 15 cents per cubic yard; the bed-rock, being of a soft character, can be easily cleaned and, practically, the whole of the values in the ground obtained. The gold is uniformly flaky and light, but with the present improved tables, hereinafter described, there is no difficulty in saving it. The gold runs about 922 fine and is worth $19.00 per ounce. The boulders in the gravel are few, and very large ones never met with, the only obstructions being trees and stumps. Where these stumps and trees are thick, it pays to have Chinamen go ahead of the dredge for the purpose of cleaning the ground; for, although the dredge can handle them, it causes some delay and lessens the daily capacity of the machine.

Two years ago, the Risdon Iron Works built a dredge to operate on this ground; since then the same firm has built two more, and has under construction another two, for the same locality.

One very important question, over which there has been considerable discussion, has been practically determined by the "Kia Ora" dredge. This dredge is operating in an orchard, some 3000 feet from the Feather River. A small excavation was made, about 11 feet deep, by 80 feet square, the dredge being built in the dry hole; a pump was erected at the river and, by means of a flume, supplies about one-quarter million gallons of water a day to the dredge. This quantity, although the natural water level of the country is in many places eighteen feet below the surface of the ground, is sufficient to overcome the seepage, and suffices to keep the water sufficiently clean to operate in. The cost of pumping is now $5.50 per day, but when electric power is brought on the ground, this cost will be reduced to less than $2.00 per day. The dredge has been operating successfully ever since the start and, after two months' run, has now a face opened out of over 300 feet, with ample tailings room behind it, thus demonstrating that a dredge can be designed to keep itself free in inland ground.

As regards the advisability of operating on the river or of starting inland, I believe that, where possible, all the advantages are in favor of the latter. The cost of pumping is very small, and the advantage of being able to make your own water-level is often very great. In addition to this, the dredge is entirely independent of floods, which, if not actually dangerous, are always a detriment to working.

A short description of the Risdon dredge may be of interest, after which I will point out what I believe are its advantages over other types.

The dredge is of the continuous bucket type, the buckets holding either 3¼, 5, or 7 cubic feet, according to the size of the dredge. The average speed of dumping is thirteen buckets per minute, thus giving a theoretical gross capacity of 93, 144, and 200 cubic yards per hour respectively. The actual capacity varies very much according to the nature of the ground, and the skill and attention shown by the winch man.

One of the $3\frac{1}{4}$ cubic foot dredges has, by actual measurement, during a two months' run, lifted 80 per cent of the gross capacity. The bank measurement is, of course, very much less than this, as the ground swells considerably when broken out. This same dredge has, however, lifted 8000 cubic yards, bank measurement, in 122 hours' running time. The above figures are useful in showing what can be, and what actually has been, done.

The buckets are built up of steel plate, riveted onto steel links; the links are bushed with manganese steel, which wears well and, when worn out, is cheap to replace. The pins are also made of manganese steel. The buckets are made up into a continuous belt by means of heavy connecting links, which render the belt strong, heavy, and capable of standing very hard usage. The buckets are carried on a heavy girder ladder, which is pivoted at the top end, and suspended by sheaves and wire rope at the lower end. By means of this ladder the buckets, which run over a five-sided cast-steel tumbler, supported in the ladder, are lowered into the material. The buckets dump onto a delivery plate, shod with heavy steel bars, the material passing by gravity into a revolving screen or grizzly. This screen is perforated, the size of the holes depending on the class of gold found, and is set on an incline so that the material gradually passes through it and at the same time, owing to the circular motion, climbs the side and falls back on itself. A heavy stream of water from a perforated water pipe, running the whole length of the screen, plays on the material whilst passing through, thus thoroughly disintegrating and washing it. The gold, black sand and all fines pass through the perforations into a distributing box, whilst the coarse material travels by gravity into an elevator, or stacker, which dumps it at the required height above the surface of the water. This elevator is also of the continuous bucket type, the buckets being close connected and running on rollers carried by two steel "I" beams. The gold and fines pass through a number of adjustable doors on to the various sections of gold-saving tables. These tables are made of cast iron, thus obviating the cracking and shrinking which wood is subject to, and are covered with cocoa matting and expanded metal, the finest gold saver yet discovered. Any one section can be cut out and washed up without interfering with the operation of the dredge. Nearly all the gold is caught within a very few inches of the head of the table, which is 8 feet long with an area of 200 square feet. The sand and waste fines pass from the tables into a sluice fitted with angle-iron

DREDGE AT POKER BAR, TRINITY COUNTY.

riffles, and are thus conveyed over-board astern of the dredge. This sluice is more of a conveyer than a gold saver, as the tables practically catch all the gold.

During a period when $2,500.00 was taken off the tables, only $16.00 was taken out of the sluices. The most careful panning will seldom yield a color in the tailings.

The motive power to operate the dredges is in most cases steam. A compound condensing, vertical engine, indicating about 40 horsepower, running the whole dredge, except the electric light plant and winch, which have separate engines. The power is transmitted to the buckets by ropes and gearing, a friction clutch being placed on the countershaft so that, instead of breaking anything, in the event of the buckets running up against solid, hard bed-rock or other obstruction, the clutch merely slips and continues to slip until the winch man eases the strain, when the bucket belt starts again.

Steam is supplied by a horizontal return tubular boiler, burning $2\frac{3}{4}$ cords of wood per day. The condenser is placed in the suction of the main pump, thus doing away with the necessity of a circulating pump; a vacuum of twenty-five inches is obtained.

The operating crew consists of two men per shift, one engineer and one winch man. The winch man has complete control of the dredge, his winch having six independent barrels: one being for the ladder line, one for the head line, and the other four being for lines connected to the four corners of the boat and fastened to dead men on shore. By means of these lines the dredge can be made to take up any position required.

I will now discuss shortly the merits and demerits of other types of dredge, and will endeavor to point out the advantages which the Risdon continuous-bucket type has over all others.

The shovel dredge, for instance, is a most excellent machine in its place, but is not best adapted for gold dredging. It is necessarily operated with spuds and cannot therefore be rapidly moved. This is a serious drawback, especially where very heavy boulders or submerged trees are likely to be met with, as in such case it is of the greatest benefit to be able rapidly to shift the point of attack. Again, when stacking tailings to a great height, it is of great importance to be able to move the stern of the dredge without shifting the bow, or vice versa; this can readily be done when operated by lines, but only with difficulty when spuds are used. The resiliency in the head line also relieves the digging buckets of much unnecessary strain, and enables bed-rock to be more easily cleaned. Again, when cleaning bed-rock, particularly if hard and uneven, the comparatively small bucket of the continuous belt machine can work to much greater advantage than the wide-mouthed shovel of the dredge under consideration. Then, again, a shovel dredge is only suited to dig in shallow ground, whereas a ladder dredge can and does operate successfully at forty-five feet and deeper, if necessary, below water-level, and twenty feet above. The question of a continuous feed also comes very prominently to the fore when a close gold saving has to be made. I do not believe a really satisfactory automatic hopper has yet been devised which will feed uniformly to the tables a yard and a half of heavy gravel, dumped into it intermittently, as is necessary in the case of the shovel dredge, and without such fairly uniform feed a close saving cannot be made. The difficulty of keeping the door of the shovel tight is a very serious objection, but has now, I believe, been to a certain extent overcome.

The crew necessary to operate a shovel dredge is from two to three times that required for the Risdon continuous-bucket dredge. The wear and tear on the shovel may be somewhat less than that on the buckets, but is certainly not sufficient to effect the extra wages. The continuous-bucket dredge and the shovel are the principal types of dredge used for gold dredging at the present time. The hydraulic-suction dredge has, however, in times past, had its supporters, and immense sums have been expended on it generally with very unsatisfactory results, although there are a few cases where it can be made to work to advantage. For example, in digging sand and conveying it long distances it probably has no equal; but for lifting heavy gravel and boulders, and for the picking up of gold, it has of necessity generally proved a failure, except in very rich ground.

There have been numerous attempts with pneumatic caissons of various kinds, but they have never proved of any value when it comes to handling comparatively low-grade ground in large quantities.

Some makers of dredges are content to make all their gold saving in one sluice, without grading the material. This may answer where only heavy gold is encountered, but every miner knows that the finest gold cannot be saved in a sluice which has to carry heavy material

A flat screen is also in some places used in lieu of the revolving cylinder, but I fail to see how such a screen can disintegrate and wash the material as well as the revolving cylinder; moreover, the wear and tear of a shaking table, carrying some two tons of gravel, must be very considerable, whereas the wear and tear on a revolving cylinder is not excessive, the motion being circular and uniform instead of reciprocating.

It is absurd for any dredge manufacturer to claim a dredge can be built which will not wear out. The wear and tear, if the dredge is handling much heavy gravel, is necessarily considerable, and I believe it is better, and in every way more economical, to make the wearing parts of the most suitable material for the purpose, and small and easily renewable.

To operate low-grade ground, it is absolutely essential to have a dredge which will handle large quantities of material at a low cost per yard. Some of the Risdon dredges are now handling ground at less than 5 cents per yard, including all running costs and $100.00 per week to depreciation fund, which has proved more than sufficient to cover all wear and tear.

A high-grade proposition with possibly a short season requires, I think, a somewhat different treatment, and economy, per cubic yard, must be to a certain extent sacrificed.

There are dredges, a large percentage of which are giving good returns, now operating in most of the Western States, and I am satisfied that during the next few years a very great increase in the number will be made, as it has been clearly demonstrated that a large area of ground, formerly considered valueless, can now, with improved type of dredge, be operated at a good profit.

California, possibly, may not be the richest dredging field in the world, but, in consequence of the climatic conditions which enable operations to be conducted throughout the entire year, and in consequence of the comparative uniformity of the distribution of values through the auriferous gravels, will, I believe, prove the banner dredging State of the Union, and will produce more permanent, uniform and, in the aggregate, larger dividends in proportion to the capital invested than any of the so-called very rich fields, where operations can only be conducted during a comparatively short period of the year. The Alaskan placer claims, for example, may possibly prove very much richer than the Californian, but, as an offset to this, operations in that locality can only be conducted for four months in the year; moreover, the cost of erecting a dredge in that country is at least treble what it is in this State, and the risk which the investor runs of losing his entire investment and of failing to discover permanent pay-gravel, is many times greater there than here. As a result

RISDON GOLD DREDGE ON FEATHER RIVER, BUTTE COUNTY.

of such conditions, a man with a limited capital cannot afford, nor will he be inclined, to take the risk of dredging in Alaska, more especially when he can find permanent and safe dredging properties right at home, in California.

There are thousands of acres of auriferous gravels in California, in running rivers, in old river channels, and in ancient deposits many hundreds of feet above the present channels of the rivers, averaging all the way from 10 to 25 cents per cubic yard, susceptible of being worked by the modern dredge, and which will yield handsome returns and big interest on the investment. The majority of these deposits, however, until the introduction of the modern dredge, were too low grade to be worked at a profit and were therefore allowed to lie idle. Now, however, all such deposits, from being valueless have suddenly become good, salable properties, and are eagerly sought after by dredging men. A new era has now dawned for California, and in course of a few years, I believe, all these lands will be taken up and developed, and when this shall happen, California will have taken the premier position and will have become the leading gold-dredging country of the world, and her wealth will, in consequence, be very materially increased.

In conclusion, it may be well to point out that one of the greatest advantages which dredging has over most other classes of mining is its comparative safeness as an investment; for, if the initial prospecting be carried out systematically and honestly, the value of the ground may be very closely determined. Thus, seeing that the cost of handling is now well known, and seeing that the value of the ground can be very closely determined, gold dredging may be considered a business investment rather than a speculation; a position which quartz mining can never attain to, as the most careful prospecting cannot possibly determine the permanency of the pay shoot.

Deep Mining at the Utica,

Angels, Calaveras County, California.

J. H. COLLIER, JR.,

CLASS OF '99, COLLEGE OF MINES, UNIVERSITY OF CALIFORNIA.

THE Mother Lode, or mineralized belt, at Angels, in Calaveras County, California, is three miles wide. At least a region of that width has been, and is being, prospected which has shown considerable mineralization. The lode proper divides into three lodes, or groups of veins, at this point. On all of these branches, mines have been operated with more or less success.

On the most easterly of these three groups, and therefore known locally as the East Lode, is situated the Utica-Stickle Mine, in the town of Angels. This lode was in an early day worked as open cut. The cut is still open for some distance, showing the strike of the vein very clearly. The cut would also give the impression that the mine was a simple fissure deposit. Underground development, however, has shown it to be a much more complicated deposit, at least to mine.

It might more properly be called a gold-bearing, mineralized zone rather than a simple vein. The zone consists of a large mass of crushed diabase which, under pressure, has developed a slaty cleavage. The crushed mass has been more or less altered into a schist containing considerable mariposite and white mica. In this mass, the ore occurs as large bodies of massive quartz of a brownish-gray color; as masses of quartz veins from one to three inches wide, interspersed with micaceous schists in a distinctly banded structure; as masses of more or less altered diabase, containing infiltrated silica often in reticulated veinlets; and as impregnations in massive diabase; all containing free gold and auriferous pyrites.

In working upward on the ore bodies, they sometimes change from massive quartz to a schistose character, and thin out toward the hanging wall. Crosscuts run into the footwall, then show the ore-making next the footwall, and soon widening out to the original width on its upward course, thus showing that the ledge was cut off by a large horse of diabase, split from the hanging wall and fallen to the footwall. In this way, the ore body is made up of several parallel lenses or masses, at different points dipping at a high angle to the eastward with a greater southerly pitch, ranging in all from ten to over one hundred feet in width.

With this outline of the general nature of the deposit, the difficulties of mining, more particularly stoping, such a deposit will be better understood. Not only must the stopes be properly supported and economically worked, but the ground must be thoroughly prospected as the work proceeds.

The prospecting must be thorough to ensure the most economical use of the long levels through which the ore is removed. These levels are often driven in extremely hard diabase, and are not only costly in driving and fitting with tracks and compressed air pipes, but are often very expensive and difficult to keep open, on account of the settling and crushing of the rock.

The prospecting, on account of the varying character of the ore, offered many difficulties, but the supporting of the stopes offered much more serious difficulties by the extreme width and crushed character of the ore body. After many difficulties and a serious cave, simple timbering was abandoned in the stopes. At the present time, timbering and stowing are used together. The timber is put in place as fast as there is room, and no opening measuring more than sixteen feet vertically is made without stowing. By this means, Mr. Wm. Miller, the foreman, has succeeded in carrying the stoping and prospecting along together, the waste from the prospect crosscuts in the wall rock being used for stowing.

SHAFT SINKING.—Before giving a detailed explanation of the excavation of the ore, it will, perhaps, be more logical to speak of the shafts. In all there have been four shafts sunk on this property. The North Utica, seven hundred feet deep, sunk a short distance back in the footwall, is still a good shaft. The South Utica, about nine hundred feet deep, was sunk on the ledge, and, after being kept open to hoist ore through only, at a large expense, had to be abandoned, and now gives some difficulty in stoping around it. The Stickle Shaft, one thousand feet deep, was sunk on the ledge and has been almost entirely re-timbered twice, at considerable expense, at one time necessitating the shut-down of the Stickle Mine and Mill for about a month.

Profiting by the experience with the other shafts, the new Cross Shaft, which it is intended to sink to a depth of fifteen hundred feet, on the Stickle Mine, has been sunk at a safe distance in the footwall from the stopes. As this is the newest shaft, and represents the best practice, I will give a detailed account of it.

After the removal of the surface soil, churn drills were used to put in holes two inches in diameter at the collar, tapering to one inch and a quarter at a depth of six or seven feet. The holes were charged with five or six sticks of No. 2 Judson dynamite, an inch and an eighth in diameter, and eight inches long. The powder was exploded by means of safety fuse and percussion caps.

After reaching solid rock, Ingersoll Eclipse machine drills were used. Four drills were run together, two in each end, hung on bars extending horizontally across the short way of the shaft, about three feet from the bottom. When the drilling was completed, one machine was removed from each bar and fastened by means of a rope to the bottom of the bucket and hoisted to the surface. The remaining machines were then turned parallel to their respective bars and clamped. Each bar, with its machine, is then hoisted to the surface also. In setting up the machines, the reverse operation is performed. In this way, the time and labor required to handle the machines is reduced to a minimum. The holes were drilled systematically in eight rows, of three holes each, across the short way of the shaft, with a V cut in the middle, consisting of six holes. The best results were obtained by drilling the holes from six to seven feet deep. The cut, with the next row on each side, was shot first and, if it broke well, the others were charged and fired as soon as sufficient rock had been removed to reach them. Each hole was charged with from six to nine sticks of the No. 2 Judson dynamite. At first, safety fuse and percussion caps were used to explode the charges, but later the charges were fired by electricity.

Near the surface, where the ground is loose, the shaft is built up of twelve by twelve-inch timber, the ends of the wall plates and end pieces being halved together. After reaching solid rock, the shaft is timbered in sets five feet apart, which are five feet wide and fourteen feet long in the clear. By means of the center braces, the shaft is divided into three compartments, each four feet four inches by five feet.

The system of timbering used is what is called the horn set (Plate I). The wall plates and end pieces are framed from ten-inch by

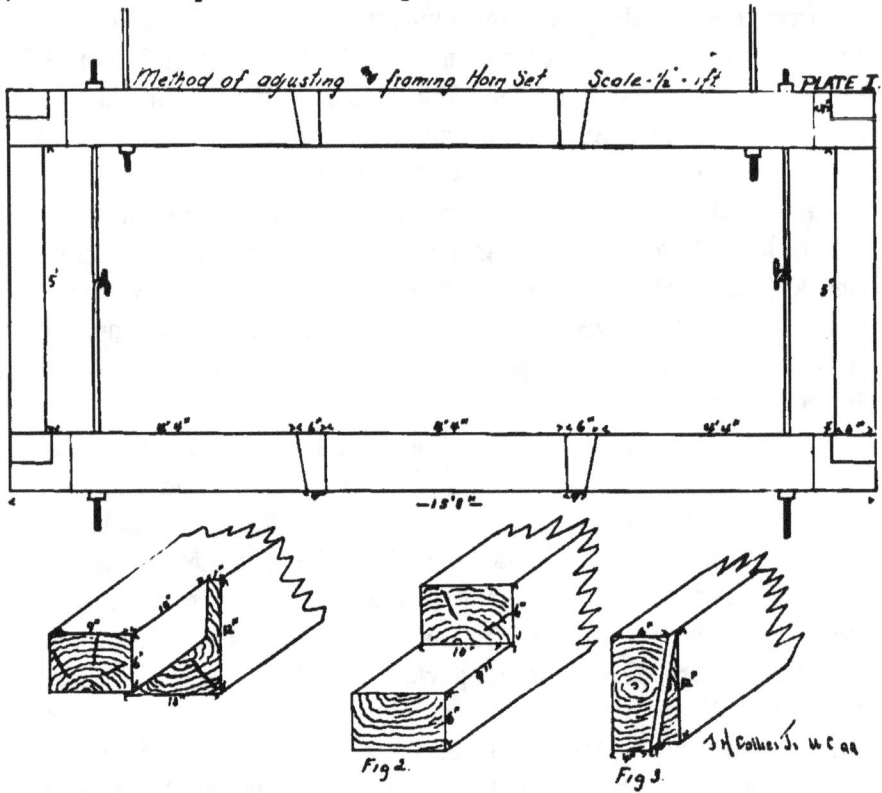

twelve-inch Oregon pine. The end pieces are framed at either end with a ten-inch horn, nine inches wide and six inches thick, leaving a shoulder one inch wide, which engages the inner side of either wall plate (Plate I, Fig. 1). The wall plates are framed with nine-inch horns at each end, ten inches wide and six inches thick (Fig. 2). The center braces are of six by twelve-inch timber, and are dovetailed into the wall plates at either end (Fig. 3). The wall plates have a total length of fifteen feet six inches, and the end pieces six feet eight inches. Exactly a foot from each end of the wall plates is a saw cut, square across the twelve-inch face, for convenience in plumbing when placing the timbers in position. Near each end of the wall plates are two auger holes, an inch and an eighth in diameter and six inches apart, through which are passed hanging bolts; one passing upward to the set above, the other downward to support the set being put in place. These hangers are made of inch iron, about four feet long. On one end is a hook, on the other a thread is cut for about a foot from the end.

The new wall plate is let down the shaft lengthwise, by means of a chain on the bottom of the skip, which is fastened just above the middle of the timber. The timber thus hangs vertically in the shaft, and, when it has passed the timbers in place, is easily turned into a horizontal position, as the weight is nearly balanced on the chain. A hanger is then inserted in each end with the hook upward, which engages the hook hanging downward from the corresponding end of the plate above. The other wall plate is similarly placed, and the two are then drawn to their approximate positions by means of the nuts below the plate on the hanger. The end plates are now dropped into place, and the corner posts, which are eight inches square and five feet long, are successively stood in place and held by drawing the wall plate and end piece up against it by means of the hanging bolts. The center braces are then put in place and the set lined by means of a plumb line over the saw cuts in the set above and the new set. The set is then blocked and wedged securely into place by blocks placed behind the wall plates opposite center braces and end pieces, and behind end pieces opposite wall plates. By this precaution only compressive strains are brought on the different members of the set. Plate II shows the horn set in place.

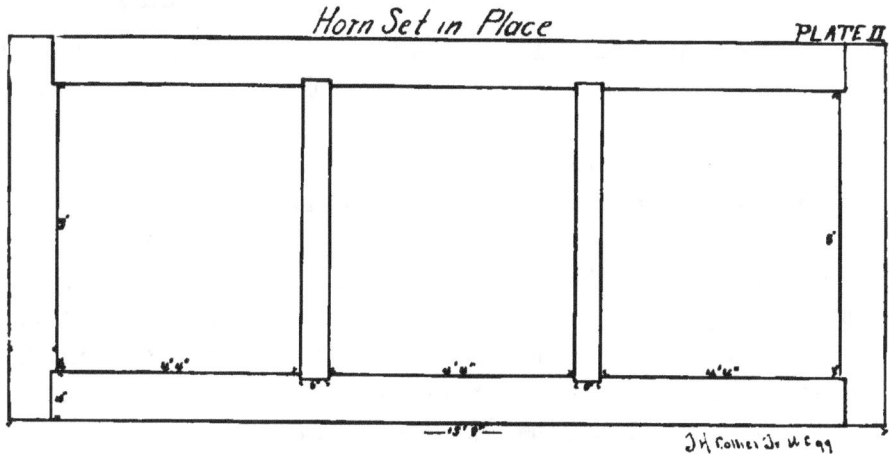

Horn Set in Place — PLATE II

The hangers are left in place until two or three sets are put in below, and then removed to be used again. Skip guides of six by six Oregon pine are fastened to the end pieces and center braces by means of lag screws, with the heads sunk into the guide, so as to prevent the shoe on the skip catching on them. The shaft is lined with two-inch plank, in twelve-foot lengths, behind the wall plates and end pieces.

A cheaper form of timbering used in the inclined winze in the Stickle Mine is shown in Plate III. The framing is simpler and no

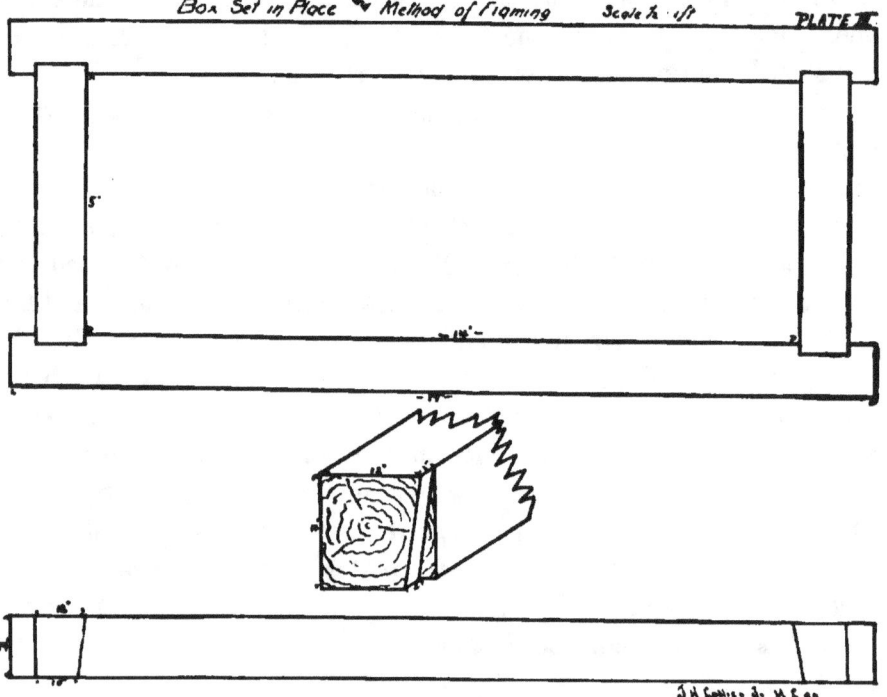

center braces are used. It answered very well for the purpose, but would not be satisfactory in a working shaft, especially in heavy ground.

In sinking the Cross Shaft, the following men were employed: Six machine men and six helpers, divided equally on three eight-hour shifts; two engineers on twelve-hour shifts: two machine men with helpers and three timber men, who worked only when the machines were running and were allowed full time each month. It was found that the timber men could work to the best advantage while the miners were drilling. All hands received three dollars per day. At the end of seven months, the shaft was sunk and timbered to a depth of eight hundred and twenty-five feet.

The special method used in sinking the Stickle Shaft has many strong points. The only serious defect was the problem of ventilation, aggravated by the large amount of "powder gas" caused by the blasting. This defect detracted very seriously from what otherwise would have been an extremely successful method in regard to both time and economy. The shaft was down some five hundred feet when the Utica Company purchased the mine. They immediately started to sink this shaft and, as the Utica was deeper than this shaft, a level was run over and a raise started. The sinking was carried on practically as already described, except that ground was encountered which gave difficulty in drilling down holes, while the up holes in the raise gave no trouble at all.

The raising was cheaper on account of the economical handling of the rock, which is brought about as follows: As soon as the raise had progressed far enough, a chute door was put in at the bottom. A line of stulls was put in near one side of the raise across the short way, on which was fastened, on the side toward the larger compartment of the raise, a lining of six-inch lagging. The larger compartment was arranged for a chute, the smaller for a man way, the ladders being nailed to the lagging.

After drilling a round of holes after the same system as used for a shaft of the same size, the drills and tools were stowed in the manway. The top of the man way was covered with heavy timber, leaving a small opening for the miners to pass through. The end of the compressed air hose was adjusted so as to throw the air into the head of the raise and be safe from the blast. The fuse was fired, and the last man pulled a piece of timber over the remaining opening as he descended. The broken rock then fell into the large compartment, called by the miners the "bull-pen." After two or three hours, the air would clear enough so that the miners could return to their work. The miners meantime had removed twenty or thirty cars of rock, so that sufficient working space remained at the head of the raise.

This method made it possible to remove the rock at the smallest possible cost, as no shoveling was necessary. After connecting with the shaft above, timbering was commenced, the rock being withdrawn below as the timbering advanced above.

DEVELOPMENT WORK.— In exploiting the mine, crosscuts were driven into the ledge every hundred feet. After encountering ledge matter, drifts were run, taking out the rock from wall to wall. In this manner the full extent of the ore body was determined, as well as its average value.

These drifts and crosscuts are never timbered, unless absolutely necessary, until it is determined whether or not they are to be used in the excavation of the ore. Permanent drifts are timbered up substantially with round timber, from sixteen to eighteen inches in diameter. The ordinary tunnel sets are used with seven-foot legs, six feet apart at the bottom, leaning in at the top to support a five-foot cap.

The excavating in the levels is done entirely with machines. From nine to twenty holes are drilled in a round, depending on the nature of the ground. The V cut is commonly used, generally with twelve holes and with the cut vertical. In this system, three horizontal rows of four holes are drilled. The middle two holes of each row are inclined toward each other and come near together or, in very hard ground, unite. These two holes of each row are thus made to explode together. The outer holes of each row are inclined slightly into the wall, in order to prevent the narrowing of the drift.

The upper row always, and sometimes the middle rows, is given an upward pitch. The holes will then keep clear in most ground if drilled dry, but when water, under pressure, can be obtained, it is an advantage to keep the holes clear by this means, using a hose and a small nozzle which can be inserted beside the drill in the hole. The bottom row always, and sometimes the middle row, is given a downward pitch sufficient to make the holes hold water well. The water mixes with the drillings and is thrown out as mud by the drill, from time to time, thus keeping the hole clear. The water may be thrown into these holes with a cup or by means of a hose, as before. The water under pressure is best, as it keeps the hole washed almost perfectly free from drillings, thus greatly reducing the time required to drill the holes.

From this last statement must be excepted the case of extremely soft rocks, such as talc and micaceous schists. In these rocks the drill cuts so fast that the dust does not have time to become mixed with the water, and so packs very tight around the shank of the drill. This stops the progress of the drill, and no further headway can be made until the hole is cleaned and water again admitted to the bottom of it. As an example of an extreme case of this difficulty might be mentioned a piece of ground in which a top hole of proper pitch could be drilled dry to a depth of seven feet in a half hour, while a hole of same depth with a downward pitch, drilled wet, required fully two hours to drill. In a case of this kind, only the bottom holes are drilled wet.

Some ground is found to break better with the cut horizontal. This is done by drilling four horizontal rows of three holes each, with the holes of the second row from the top pitching downward, to meet the corresponding holes of the third row, which pitch upward.

In ground which is difficult to break, it is found necessary to increase the number of cut holes from six to eight. In still more difficult ground, the back holes also have to be increased, so that as many as twenty-four holes are found necessary. It is also found necessary, sometimes, to increase the width of the drift, in order to give the cut holes a greater cut. Whatever system is used, the holes are given as nearly as possible the same load. The most central are arranged to be shot first, followed by the other cut holes in pairs. The middle holes of the adjoining rows, on either side of the cut, come next, followed by the remaining holes in each row.

If the ground is extremely schistose, or if there is a slip or gouge seam following along one wall, nine holes may be sufficient for the first breaking to the slip or the side of the drift. If there is no gouge seam or slip, the drift may be kept straight by breaking alternately to one side and the other.

The easiest and most difficult ground to break have only a slight difference in structure. They are both schistose, and have an extremely knarled structure. But the difficult ground has been filled with infiltrated silica, which has so hardened it that it has to be literally burned out with powder.

Occasionally soft and running ground is encountered, which gives some trouble. The difficulties offered by this sort of ground are overcome by an ingenious system of poling, which will be described in connection with the stoping.

STOPING. — In the excavating operations, an enormous amount of timber is used. This timber is all round and comes in sixteen-foot lengths, from twelve to twenty-six inches in diameter. Poles are from six to twelve inches in diameter, and come in twenty-foot lengths, which are cut in two for use in the mine. The posts for the sets in the stopes are generally made from the timber of the largest diameter. The logs are cut in two, in the middle, for stope timbers. Posts are framed to fourteen inches on two opposite corresponding sides of each end, to a distance of three inches from each end. The caps are framed on one side at each end, leaving seven feet between the joggles and about six-inch horns.

A stope is started by breasting out the ore to the full width of the deposit. Crosscuts are run into both walls, to be sure that all the ore has been removed. The opening is then timbered with eight-foot stope sets. If the ground is solid, no timbering is done until the whole mass of the rock covering the area of the stope has been removed. The posts are then set in the solid rock, with six-inch spreaders and twelve-inch round brace-sprags between the posts at the bottom. A floor is then laid over the spreaders.

If the rock is loose or soft, one set is put in at a time as fast as room is made for them. In the soft ground heavy sills are laid, as shown in Plate IV, to give a solid foundation for the posts. Sills are

PLATE I

not, as is generally supposed, an advantage in working up under an old stope. Good floors laid across the spreaders, even though they have been in place so long as to be badly decayed, are found to be more serviceable than sills. The sills are seldom in place when reached, and have to be caught up securely, or they are liable, by their own movement, to start a serious run in the waste above them. After the sill floor is opened and timbered (as shown in Plate IV, which represents a stope thirty-five feet wide), a raise following the footwall is run up to the level above.

This raise is necessary for the proper ventilation of the stope, as well as the economical introduction of timber and waste into the stope; the timber and waste being thrown down the raise into the stope. The raise is located in the most convenient part of the stope and, if possible, where there is a seam of gouge on the footwall, which greatly lessens the cost of the work by lessening the difficulty of breaking the ground. If the rock is hard and solid, machines are used and the raise is timbered with full-sized stope sets, if timbering is necessary; so that, as the stope is carried up, the timber of the stope is joined on to that of the raise. If the ground is loose, the work is all done by hand and the raise is built up solid with round timber, halved together at the ends, making the raise four feet square in the clear.

The second floor above the level is now started from the raise, the ore falling to the sill floors where it is shoveled into cars. After this floor is excavated, a set for the full length of the stope is lagged on tops and sides with half-round slabs, made by cutting twelve-inch logs, eight feet long, in two lengthwise. This set is then kept open for the gangway, along the line of which the chutes and man-ways, leading up into the stope, are started. All the remaining space to the top of the sill floor set is now filled with waste. This waste is obtained from three sources within the mine: first, from the vein rock, by hand sorting in the mine; second, from crosscuts run in the wall rock from the different floors of the stope, for the double purpose of prospecting the ground and supplying the waste; and, third, from the dead work in different parts of the mine, which is brought in through the raise from the level above.

From this time the chutes and man-ways are carried up by means of cribbing (shown in Plate V), to within one set of the back of the

PLATE V.

Cribbing for Chutes & Manways   Scale ¾·1ft

stope. When the ground is heavy, similar cribbing is used to help support the ground. In wide and heavy stopes, a row of such cribbing is generally put in extending the full length of the middle of the stope, and is stowed with waste as rapidly as possible.

The different floors are started successively from the raise. As soon as a floor has been advanced far enough from the raise to prevent the mixing of the quartz and waste, filling is commenced by throwing waste down the raise. When the floor is completed, the lower floor, remote from the raise, is stowed with waste from the crosscuts. In order to supply sufficient material for this purpose, the crosscuts, which start out with small dimensions, are widened out into large chambers. In this way no opening in the stope is kept open to a greater vertical height than sixteen feet.

When a level is being worked, a large mass of rock is left in place in the ledge, opposite the shaft, until the last thing before the level is abandoned. This precaution protects the shaft, which is already weakened by the cutting of the station.

When the vein matter is badly crushed and broken up, an ingenious system of lagging, called poling, is used. Poles are run out over a cap or sprag — depending on the direction of the work, whether with or across the vein—which supports the ground until the timber is in place (Plate VI). In some cases the poles are a necessity to

PLATE VI.

Poling

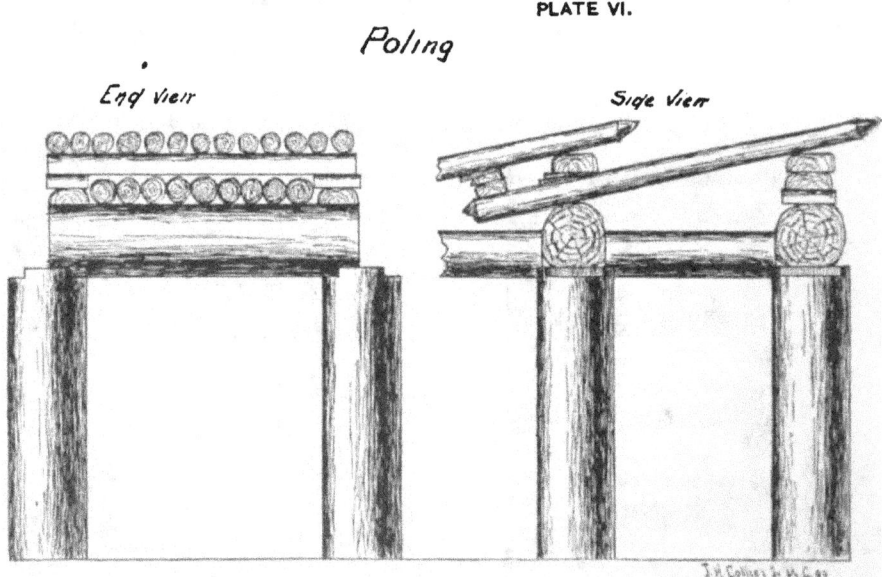

prevent the ground from running, in others they are put in more for a protection to the miners working beneath them. In the latter case the trouble is caused by detached pieces which continually slack away, due to the presence of water and slight settling.

In running ground, the poling has to be performed with great care. The ends of the poles are sharpened and worked in over the cap all together by working the rock loose around the points with a bar and driving the poles ahead with a hammer. As the poles are advanced over the cap, the broken rock is raked back from the points of the poles and left standing on the natural slope to prevent its running. Also side poles are put in, if necessary, beginning next the roof and working them in as the rock is removed, and the face breast boarded to prevent the loose rock running around the points of the poles. Proceeding in this way, room is made for one set at a time.

The details of this work are shown in two sets of drawings (Plates VI and VII). When the set of timbers is put in beneath the

PLATE VII.

Poling

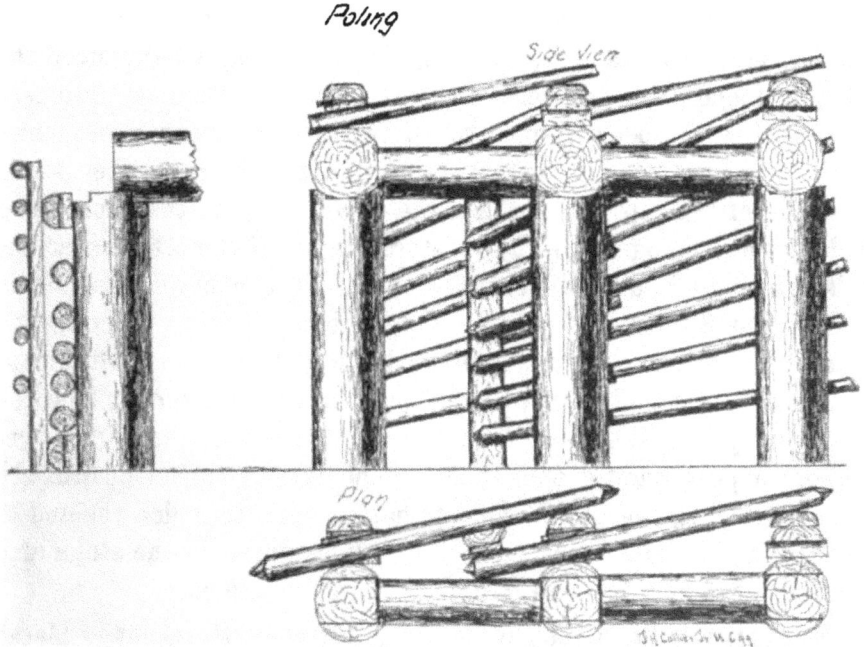

poles, a bridge-piece is put in supporting them, so that they cannot come down on the cap before the next poles are started. After the poles for the next set are run out over the cap, wedges are inserted between the poles and the bridge-piece, and driven until the weight is taken off the short blocks supporting the bridge-piece. The blocks are then replaced by poles. Where the ground will permit, pieces of two-inch plank are inserted crosswise behind the poles, this adding greatly to the efficiency of the poles.

The system seems to have all the merits of the square set system used on the Comstock, at the same time being cheaper and more flexible. It is cheaper because the cheaper round timber is used and the framing is much more simple. Much of the framing is done in the mine while waiting for the partially framed timber to come from the surface, which always causes more or less delay when several crews are to be supplied through the same shaft.

It will be remembered that the caps were only framed on one side before being sent into the mine. There are several reasons for this. As the logs are not always exactly sixteen feet long nor cut square on the end, it is found necessary to leave a space between the ends of the caps of about two inches to allow for this irregularity. This is brought about by framing the ends of the posts to fourteen inches, and the caps so that they measure seven feet between joggles which, if the log was exactly sixteen feet, would leave six-inch horns. If the side-pressure was great enough, this space might be reduced the full two inches, thus reducing the hitch formed by the two adjacent caps (if they had been framed) to twelve instead of fourteen inches. Again, the ends of the posts not being exactly square, they would raise one of the caps higher than the other. Again, the caps, being made from the logs, have the natural taper of the tree and so the ends are not of equal dimensions, the difference in diameter being often as much as six inches. As the posts in the stope do not settle equally, use is made of this irregularity to keep the line of the caps horizontal. The end of the cap of the largest diameter is placed on the lowest post. When the next set above is put in, the joggle over the higher post is cut deeper than the one over the lower post of the under set. It is, therefore, evident that the size, to which the end of the cap is framed, depends on the circumstances in the stope and, therefore, can only be done to advantage in the stope.

The general dip of the ore body is nearly vertical, but in places where it bulges the wall rock sometimes has a pitch of forty-five degrees. The timbering is adapted to this case by first leaving a horn on the cap extending over the post nearest the footwall. When this horn must be so long as to greatly increase the weight of the cap, a short butt cap is used instead. This butt cap is framed the same as an ordinary cap on one end, on the other it is beveled to fit the wall rock. The length of the butt cap is increased as the wall recedes, until the distance between the post and the wall rock becomes great enough for a full set. A hitch is then cut in the foot-wall for another post. While the caps on the footwall side have been lengthened, those on the hanging wall side have been correspondingly shortened. Full sets are also dropped on the hanging wall side as they are taken up on the footwall side. By these means, the system may be applied to a ledge which has a comparatively small angle of dip as well as considerable width.

The method, as a whole, has as much rigidity as the extreme width and the constant settling of the ledge will permit, combined with a sufficiently low cost, to allow of its being applied in the excavation of large bodies of comparatively low-grade ores.

HOISTING AND TRAMMING. — Each shaft is supplied with a double-acting water-power hoist. Two Dodd water-wheels, nine feet in diameter, are placed on the pinion shaft in reversed position, so that power may be applied to run the hoist in either direction. The water is applied through three nozzles, so that the power can be readily proportioned to the load. The nozzles are served by a ten-inch wrought-iron pipe, which delivers the water under a head of four hundred and twenty feet, giving a standing pressure of one hundred and eighty pounds, which falls to one hundred and thirty when the hoist is in operation.

Wire ropes, one inch in diameter, are used on five-foot drums. The ropes, which last from three to eight months, are inspected daily, and at the least sign of failure are condemned and removed. The rope passes from the drum over a seven-foot sheave wheel at the top of a gallows frame, eighty feet high.

Two skips are used, each covered by a sheet-iron bonnet and supplied with an automatic safety clutch. When the skip descends faster than a certain rate of speed, which can be regulated, a parachute under the bonnet raises and releases the clutches which grip the guides, and so arrest the descent of the skip.

The rock skip is constructed of sheet-iron, with a capacity of two tons. It is loaded from the chutes in the shaft at the different levels. At the surface it is dumped automatically by means of a curved switch, which turns the sheet-iron box, containing the ore, out of the line of the guides over a chute leading to the rock breaker. The frame remains on the guides and, as it ascends, it turns the box bottom up on the axis at the bottom of the frame. The skip frame is then lowered, which turns the box back again into position, where it is locked automatically.

The water skip is also made of sheet-iron, three feet square and eight feet long, holding about five hundred gallons of water. The tank is filled through a butterfly valve in the bottom of the tank. When it reaches the surface, one wing of the valve is opened automatically by a system of levers, which are actuated by a bumper on the gallows frame, and the water is discharged into a tank at the side of the shaft.

On the levels, ton cars are used to transport the ore from the stopes to the shaft. The cars are constructed of sheet-iron and run on light T iron rails laid on three-inch ties. The tracks are laid along the line of the chutes leading up into the stopes. The cars are loaded from these chutes and run out to the shaft where they are dumped into a chute, which has a capacity of fifty or sixty tons. This chute has a door opening into the shaft through which the skip is loaded.

PUMPS. — The drainage of the mine is further provided for by a pumping plant consisting of two seven-inch bucket and one eight-inch plunger pumps. This plant is run by a six-foot Dodd wheel, geared to a long reciprocating rod in the shaft, which supplies the power to the pumps at the several stations in the shaft. The capacity of this plant is about seventy-five gallons per minute, from a depth of nine hundred feet. Most of the water is handled, however, by the water skip.

To supply a sixty-stamp mill with about three hundred tons per day, requires two shifts of forty-five men each, consisting of miners, helpers and shovelers, working on ten-hour shifts. The two shifts alternate on night and day shift, changing every two weeks. Each shift consists approximately of twelve miners, each with a helper, about fifteen shovelers, and six men running the cars on the level. This division of the working force is not constant, as it depends on the conditions in the mine, which are constantly changing. When the broken rock is plenty and easy to get at, some of the miners may be put to shoveling pay rock, and some of the shovelers put to work on the stowing. On the other hand, if rock is short, some of the shovelers may be added to the force of miners. The miners are paid three dollars per day; the helpers, shovelers and car men are paid two dollars and a half per day. In addition to these two shifts, a crew of ten timber men is also necessary, receiving three dollars per day.

I would have been glad to add data as to the cost of the excavation of the ore, but as I was not in possession of any official data, I thought it would not be best to give an estimate without the Companies' permission. For the opportunity to collect the information which this paper contains, I am indebted to the kindness of Mr. Wm. Miller, under whom I have worked for about two years, as well as during the vacation periods of my Junior and Senior years at the University of California. I am also indebted to Mr. C. D. Lane and Mr. Alvinza Hayward, through whose courtesy not only I, but several others in the College of Mines have been able to gain considerable valuable experience in the mines at Angels.

The Genesis of Petroleum and Asphaltum in California.

BY

A. S. COOPER, State Mineralogist.

T can be tentatively stated that fossil bitumens are principally derived from terrestrial and marine vegetation, deposited in sedimentary strata, then changed to carbonaceous matter, and afterwards distilled by the heat of metamorphism.

In other words, the bitumens are Nature's coal oils and tars distilled in Nature's still, generally with infinite slowness when compared with a modern tar still or retort.

Although some of the hydrocarbons can be produced synthetically in the laboratory, still it is to

A. S. COOPER.

be believed that nearly all, if not all, of the accumulations of fossil hydrocarbons owe their existence to the vital principle, that they are derived directly or indirectly from organized beings — animal or vegetable, probably the latter, for reasons given hereafter.

In the decomposition of vegetable tissue, when air is wholly or partially excluded from it, as, for example, when buried in the ground, the constituent elements of the vegetable tissue re-arrange themselves mutually into new products, either with or without the co-operation of the elements of water, the oxygen gradually uniting with the carbon to form carbonic acid, which separates and leaves as a residue substances rich in carbon and hydrogen. It is in this way that bituminous coal, peat and brown coal (lignite) have been formed from vegetable matter.

With carbonaceous material in the same deposit, the same series of strata or in the same stratum, there are differences of composition. The varieties of carbonaceous materials may have been produced from different kinds of plant form, from which coal has been derived, and the peculiar conditions of the districts where the plants flourished, before their downfall and inhumation or submersion. The changes that have taken place in the original plants, during their passage from woody fiber into coal, are ascribed to evolution of a part of their hydrogen and oxygen, as there are less of these elements in coal than in wood. This will be observed by viewing the following table:

ORGANIC.	FORMATION.	C.	H.	O.
Wood	Recent	52.65	5.25	42.10
Peat	Recent	60.44	5.96	33.60
Lignite	Cretaceous and Tertiary	66.96	5.27	27.76
Brown Coal	Cretaceous and Tertiary	74.20	5.89	19.90
Coal	Secondary	76.18	5.64	18.07
Coal	Older	90.50	5.05	4.40
Anthracite	Crystalline	92.85	3.96	3.19
Graphite	Crystalline and Archæan	100.00		

There is no strict line of demarcation between the above-named organic matter; the one below gradually merges into the one above.

The older the formation, the greater the amount of carbon contained in the coal, the amount of hydrogen and oxygen having been diminished. This fact may be ascribed chiefly or in part to the degree of heat and pressure to which the lower and older coal strata have been, and still are, subjected.

Graphitic and anthracitic varieties of coal are metamorphic coal produced by heat, the volatile matter being vaporized, which, probably, afterwards condensed in fissures and porous rock above, as bitumen, petroleum, etc.

Graphite occurs in the Archæan, but none of the other coals are found in these rocks; neither is bitumen or petroleum oil found in the Archæan formation, but they are found in all the other formations lying above the Archæan. From this it would seem that the bitumens originated in or above the Laurentian rocks, and not below them. The absence of the bitumens and petroleum in the Archæan formation is evidence against the theory that petroleums are produced by the action of water upon red-hot carbonized iron in the interior of the earth. It is at least conclusive evidence that this process is not at present in operation.

Graphite is, in all probability, the ultimate stage in the series of changes which vegetable matter undergoes, passing through the conditions of heat, lignite and mineral coal to end in graphite. Several modifications of graphite may be produced artificially. When cast-iron is melted with an excess of charcoal, it dissolves a portion of the carbon. This carbon, when the iron is allowed to cool, slowly crystallizes out in the form of large and beautiful leaflets of graphite.

Anthracitic varieties of coal are associated with folded and metamorphic rocks. Anthracitic condition of coal may sometimes be traced to local effect of igneous rocks. As rocks grow less and less metamorphic, the more bituminous is the coal contained in them.

Graphite is disseminated in strings, veins and beds through hundreds of feet of the lower Laurentian strata, and its amount is calculated to be equal in quantity to the coal seams of an equal area of the carboniferous rocks.

In Central Scotland, where the coal fields have been so abundantly pierced by igneous masses, petroleum and asphaltum are of frequent occurrence, sometimes in chinks and veins of sandstone and other sedimentary strata, sometimes in the cavities of the igneous rocks themselves. In West Lothian, intrusive sheets, traversing a group of strata containing seams of coal and oil shale, have a distinct bituminous odor when freshly broken, and little globules of petroleum may be detected in their cavities. In the same district, the joints and fissures of a massive sandstone are filled with solid, brown asphalt, which the quarry men manufacture into candles.

Graphite has probably the same vegetable origin as mineral coal. It is now generally conceded to be of organic origin, the result of metamorphism of some of the products of destructive distillation of vegetable tissues. In general arrangement and microscopic structure, the layers of graphite correspond frequently with coal and some bituminous deposits.

The majority of coal mines in California are in the cretaceous rocks and the lower tertiary. The rocks underlying the tertiary formation and the lower parts of the tertiary formation are in many instances metamorphic, the cretaceous and tertiary rocks having been changed to metamorphic rock by hydrothermal action, and, if they contained carbonaceous material, petroleum was distilled, which ascended in a vaporous condition and was condensed in the unaltered rocks. After condensation, it was carried upwards by gas and hydrostatic pressure, and, in some instances, by rock pressure.

The presence of nitrogen in the California petroleum oils has been adduced as a proof that they are of animal origin; this proof is not conclusive, as most all of the coals and carbonaceous shales yield nitrogenous products when subjected to destructive distillation. As a proof of the animal origin of petroleum, it has been stated that pools of petroleum have been found which were filled with live maggots.

Many hundreds of pools of petroleum in California have been examined without discovering any maggots or other forms of life in them, although there have been swarms of flies and other insect life in their neighborhood. In the vicinity of pools of petroleum, flies quickly discover and lay their eggs on spoiled or moist food stuffs, meat, decaying meat, meat broths, dead animals, in manure heaps, etc. Taking into consideration the persistency of flies to deposit their eggs upon anything capable of sustaining the life of their young, it is to be believed that all pools of petroleum would be filled with maggots if flies were present and petroleum was a suitable food for their larvæ.

Maggots feed upon animal and vegetable tissues and liquids, and not upon their fats and oils.

When animal hydrocarbons are deprived of all animal tissues and liquids, they are not molested by flies. Petroleum is also found to be an excellent maggot killer.

There are only four ways of preserving animal remains from decay and putrefaction: first, in a case from which the air has been completely removed and excluded; second, by antiseptic agents; third, subjected to a heat exceeding 100° C.; fourth, subjected to a cold under 0° C.

To secure and preserve any quantity of animal remains by any of these four ways, cataclysms of nature must occur. Geology does not show that any large quantity of animals have been inhumed in the earth by cataclysms; even provided animals were buried in this manner, there is sufficient oxygen present to start decay, and, after this has commenced, putrefaction ensues.

It is also claimed that the decay of organic matter is not due simply to oxidation, but to the action of organisms, ferments or enzymes, which attack the organic substances and decompose them into their original elements or into simpler compounds.

The ordinary rate at which sedimentary rocks are deposited is too slow to effect the inhumation of animal substances, so that they will be excluded from the air and preserved.

But few animal remains are found encased in ice, and these are not buried beneath the earth; there is no other way in which they can be preserved by cold.

There is no way in nature that animal remains can be maintained at a temperature of 100° C. from the time of their death until changed to bitumen.

Very seldom are animal remains preserved by antiseptics in nature, and, if so preserved, their aggregate amount is small. All the ways that can be employed to preserve animal substances are destructive to animal life. The presence of life and the conditions and way necessary for the preservation of animal substances cannot exist in the same place at the same time.

In order that petroleum oil may be derived from animal remains, it is first necessary that such remains be placed and preserved in a condition for such change.

In the present age, the inevitable end of animal after-life is to furnish nutriment to the scavengers of the earth and sea, or their decomposition by putrefaction.

There is no reason why, in former ages, this was not the ultimate fate of animals.

The products of the decomposition of nitrogenized animal substances are as follows: The oxygen of the substance unites with the carbon to form carbonic acid, while the hydrogen divides itself between the nitrogen, the sulphur and the phosphorus, and forms ammonia with sulphuretted and phosphoretted hydrogen.

The surroundings would indicate that the fossils in the petroliferous formations of California lived and died, and were embedded in the same manner as the mollusca of the present day; in fact, a large number of the fossils in these formations are the prototypes of existing mollusca and must have lived and died in the same manner. There is nothing showing that after death the mollusca existing at the present day are preserved for the future manufacture of petroleum. This is also true of the infusoria.

The fossil animals do not show any stage of transition between animal matter and petroleum oil.

Large quantities of fossil shells exist in the shales and sandstones of California; these are usually filled with silt, or the shells have been changed to silica and the interior filled with silica, or molds and casts exist, the carbonate of lime having been leached away. They only contain bitumen when the interspaces and cracks in the adjoining rocks are filled with bitumen. There is no carbon in these fossils, except in some instances the carbonate of lime constituting their shells. There is no carbonaceous matter from which bitumens could be derived by any known process. There are no more fossils in the rocks which are impregnated with bitumen than in those which are destitute of bitumen.

Bitumens are found in the unaltered sedimentary rocks of California sandstones, shales, limestones, etc. There is no bitumen in the metamorphic rocks. The only substances originally contained in these rocks, that could and can be converted into bitumen, are organic vegetable remains in the form of coal, lignite and carbonaceous shales. These carbonaceous substances are not changed to bituminous unless distilled by heat.

Most all of the known beds of coal, lignite and carbonaceous shale belong to the cretaceous or tertiary formations.

If bitumens exist in them at all, the surrounding and accompanying phenomena show that these accumulations of bitumen are secondary.

A primary deposit of petroleum is in the rocks in which it was formed.

A secondary deposit is where it has migrated from the rocks in which it was formed and accumulated in other rocks.

If petroleum is produced by destructive distillation, it cannot remain in the rocks in which it formed.

The petroleum of California is not confined to any particular geological horizon in the Coast Range, but may exist in any of the sedimentary rocks lying above the altered rocks; therefore, paleontology is of but little value in determining its location.

The chief guide to the discovery of bituminous accumulations is the character of the rocks constituting the formation and their structure and position.

There is no evidence tending to show that these accumulations or deposits are primary; and, if secondary, they may occur in any porous or seamed sedimentary strata of any age lying above the altered rocks.

The age of a formation may assist in the discovery of primary mineral deposits, but migratory fluids and gases will circulate through any porous or seamed strata and accumulate in such places as are rendered suitable through structure and position, irrespective of the age of the rocks.

The accumulations of bitumens in the domes and summits of the anticlines, and the existence of tar and gas springs prove that they were, and are at the present time, migratory, and that the principal direction of their migrations was, and is, upward.

In California, the upper cretaceous and the eocene, miocene, pliocene and quaternary formations, when not metamorphosed or otherwise changed, consist of soft shales, sandstones, conglomerctes and lime-stone. The large majority of these rocks are soft shales and sand-stones. The rocks of one age resemble those of another

In California, the miocene seems to contain the largest amount of petroleum; this is owing to the fact that, in ascending from its place of origin below, it had not reached the pliocene in the same quantity as it did in the miocene; the pliocene, being the farthest away from the place of its origin and the eocene, does not have the same amount of exposures as the other formations, and, when visible, the formation is so broken, and tilted to such high angles, that the oil has escaped.

The fossil shells are vastly more numerous in the tertiary rocks of California than the remains of all other kinds of animals put together. Mollusca, similar to those of California, abound to an extraordinary degree in the tertiary in other places besides California, but usually they contain no petroleum.

In none of these ages are animal remains found in a state fit for the future manufacture of petroleum. Neither can there be found any substance that would suggest a state of transition between the fauna of these ages and bitumen.

But suppose, for argument's sake, that it is admitted that petroleum oil is derived from animal remains, there is no reason why the fauna of one of these ages should not be the origin of petroleum as well as the fauna of another of these ages. Oil is not found in the quaternary which can be claimed as being indigenous.

In California there are large areas of fine sand containing many fossil shells which contain no bitumen; if they had at any time contained bitumen, there would still be some bitumen remaining in them, as there are many shells lying in such a position that the bitumen could not escape by draining from, or being floated out of them. If the petroleum originated in these shells, each shell would contain a modicum of oil, and would not be completely filled with petroleum or entirely destitute of the same, these being the conditions in which they are usually found.

If these fossil shells are full, these deposits must be partly or wholly a secondary one, as no mollusk could produce a quantity of oil equal to the size of its body; and if they were destitute of oil, it would tend to prove that they are not the source of oil.

It certainly cannot be claimed that animal remains in anticlines produce oil, while similar remains in the synclines are unproductive.

The presence of nitrogen in the bitumen does not prove that petroleum containing nitrogen is of animal origin, as nearly all plant remains contain nitrogen in greater or less quantity, and yield nitrogenous compounds by distillation.

How can the different kinds of bitumen be accounted for if they are derived from animal remains? Is the oil that originates from one species of mollusca different from that which is derived from another?

In Russia, the petroleum consists principally of the naphtenes series; in Pennsylvania, of the paraffine series and a paraffine base; in California, of the olefine series with an asphalt base.

Adjoining fossil shells, when bituminized, contain similar bitumens; for instance, if one shell is found enclosing a bitumen containing six per cent of sulphur, the adjoining shells all contain bitumen containing six per cent of sulphur. This large amount of sulphur was not originally present in the body of the mollusca, consequently it must have been derived from some other source, and if derived from some extraneous source, the percentage of sulphur in the bitumen in each shell would not have been so constant—some would contain more sulphur than the other. This would lead to the belief that the bitumen was sulphurized before entering the shells and, therefore, a secondary deposit.

Then there are varieties of asphaltites, uinteite, albertite, grahamite and elaterite.

It does not seem possible that the bodies of the mollusca, which so closely resemble each other in composition, should produce so many dissimilar bitumens.

When petroleum is found in fossil shells, it is also found in the porous or seamed strata in which the shells are embedded.

The character of the bitumen in the shells is the same as that which is in the porous strata, excepting that it is sometimes in a more liquid condition, owing to the fact that a greater opportunity is offered for the evaporation of the bitumen in porous strata than in closed shells.

Owing to the buoyancy of the bitumens in fresh or mineral water, their migrations are usually upwards until checked by impervious strata, or until they have reached the highest point to which water will float them, or until they reach the surface of the earth.

Very seldom do they descend, and then descent only occurs when the strata to which they have ascended are uplifted above permanent water by orogenic movements. They may migrate for a thousand feet, and, as stated above, generally upwards.

In the Ojai Valley is an accumulation of asphalt, lying like a huge black tear upon a hillside, which has slowly issued from the small springs at its upper end. Scattered throughout the Coast Range are similar deposits.

DISTILLATION.— The maximum quantity of liquid hydrocarbons is obtained from the solids by a process of distillation under high pressure and low temperature, combined with rapid condensation.

Temperature and pressure exercise a considerable influence on the nature of the products of distillation. The method of cooling also exercises great influence in the re-arranging of the molecules and upon the nature of product of distillation.

Slow cooling or quick cooling makes no difference on some substances, but the difference between slow and rapid cooling has a marked effect on others.

If we reduce the heat of water from a high to a low temperature, it will not affect the constitution of the water; whether we lower the temperature slowly or quickly, the result will be the same. Water that is slowly cooled from the boiling point down to the freezing point will have the same properties as water that is rapidly cooled down from 212° to 32° F.; but that does not hold good with all substances. For example, if we take three bars of steel of equal dimensions and make them all red-hot; then, if we slowly cool one bar down to say 50° in the air, and if we cool the second bar by slipping it slowly into cold water, and if the third bar be suddenly cooled by plunging it into cold water, the effect of cooling these three bars of steel will produce a different effect in each bar. The bar that is slowly cooled in the air will remain comparatively soft and be of fibrous texture, malleable and ductile, capable of being bent double without breaking. The second bar will be harder and more elastic, and can only be bent a small degree without breaking. The third bar will be very hard and brittle and cannot be bent, and if struck with a hammer will fly to pieces, the fracture showing crystalline structure. Now, these three bars were simply deprived of the same number of degrees of heat. They were reduced from an equally high degree of heat to an equally low temperature, but we find that the difference in time occupied in cooling makes a vast difference in the molecular structure and properties of the metal.

These facts are well known to all workers in metal, but it is not so well known that in cooling mixed gases, especially the hydrocarbons, from a high to a low temperature, the effect on the constitution of the gases varies with the time occupied in cooling, and that difference between quiescence and agitation during cooling has different effects on the molecular constitution of mixed gases, especially with mixed hydrocarbons.

Place two retorts, of the kind used for making illuminating gas from coal, in one oven, so that they will get equally heated up to a bright, red heat, and let us charge each retort with one hundredweight of good coal, which will make 500 feet of gas. Now, let us cool the gas from retort number one slowly by passing the gas slowly through a number of pipes placed vertically in the open air, in the manner usually done in gas works; let us cool the gas from number two retort rapidly by passing the gas through a multitubular condenser surrounded with a freezing mixture, and we will find the result to be that 500 feet of permanent gas and five pounds of tar will be delivered by number one retort, and we shall get about 300 feet of permanent gas and thirteen pounds of tar from number two retort. There we find that rapid condensation reduces two-fifths of the gas to a liquid state; and if we were to distill under high pressure and low temperature, we should, with rapid condensation accompanied with agitation, reduce nine-tenths of the gas to a liquid state. To produce permanent gas from coal, we should distill, under low pressure and high temperature, and cool the vapors rapidly under agitation.

When gas is violently agitated during the cooling process, a greater quantity is condensed into the liquid state than when kept in a quiescent state during the cooling process.

When gas, much cooled, is passed through a coke tower down which heavy oil is trickling, this oil will absorb the light hydrocarbons of the gas.

The boiling points of the hydrocarbons of petroleum are altered very considerably by foreign, or even by the traces of foreign substances being present. The presence of different substances during distillation has an influence on the distillate. It has been found that, when a mixture of chlorine with hydric chloric is passed through an ordinary charged gas retort, it acts as a hydrogenating or dissociating agent, producing a tar very rich in benzol. On the other hand, zinc chloride in the presence of hydric chloride greatly increases the yield of heavy hydrocarbides from coal, and can convert some of the lighter constituents of the tar when distilled therewith into heavy ones.

The nature of the product also depends on the material of the retort. A rough surface will facilitate chemical changes.

By repeated distillations solid paraffine can be gradually changed into liquid kinds of paraffines and olefines.

Mineral tar, by repeated distillation in the presence of superheated steam, will be converted into gas if the united vapors of the steam and tar are decomposed by heat after each distillation.

Sulphur, when present in a still with paraffine, retards its ebullition. The introduction of sulphur into a paraffine boiling at 140°–156° C. may retard its boiling point as far as 180°-200° C., according as it may exist in a greater or less quantity. In consequence of the presence of the sulphur in the still retarding the ebullition, the vapors of the paraffine are heated to a point far above their boiling points; therefore, they are decomposed to hydrogen and carbon. The liberated hydrogen combines with the sulphur vapors, forming sulphuretted hydrogen.

Shales containing $5/15$ per cent of sulphur yield scarcely any paraffine on distillation.

YIELD OF GAS, OIL, ETC., FROM SHALES AND COALS AT HIGH AND LOW HEAT.

		GOOD SHALES		BOGHEAD COAL		GAS COAL	
		High	Low	High	Low	High	Low
Volatile	Gas	13.65	2.54	37.32	4.83	20.49	6.49
	Ammonia water	3.65	6.47	2.43	3.23	3.09	7.24
	Tar or oil........	11.04	17.65	20.65	50.29	17.08	26.45
	Sulphur..........	.99	......	.18	....	.29	....
	Water at 212° F.	2.82	......	.80	....	4.15	....
		32.15	26.66	61.38	58.35	45.10	40.18
Coke	Fixed carbons...	4.16	10.81	9.01	12.40	45.00	49.93
	Sulphur	1.05	......	.06	....	.34	....
	Ash	62.64	62.53	29.55	29.25	9.56	9.89
		100.00	100.00	100.00	100.00	100.00	100.00

What is known under the general name of petroleum includes a series of hydrocarbon oils varying widely in physical properties. Some are limpid fluids with many intermediate grades, others are found viscid and tar-like.

Hydrocarbons generally exist in three different conditions: first, the gaseous condition, wherein the equivalents of hydrogen are equal to, or greater than, the number of equivalents of carbon; second, in the liquid state, where the equivalents of carbon exceed the equivalents of hydrogen; third, in the solid state, where the carbon exceeds the hydrogen in still greater ratio than in the liquid state.

Their color by transmitted light ranges from a light yellow through orange and red to a reddish brown, so dense as to be translucent only in thin films, while by reflected light it passes from a light dusky color to a dark green and to a black. They differ as markedly in odor and also in other properties, some having a very disagreeable smell, while others are considered even pleasant.

There is a wide range in their gravity. The greater the quantity of carbon in proportion to the hydrogen any one of them contains, the greater is its specific gravity, the higher its boiling point and density of vapor. In the same oil field, the same series of strata, and the same stratum there are differences of composition.

The following are the commercial names of the products of distillation of crude petroleum: cymogene, rigolene, gasoline, naphtha, benzine, kerosene, maltha and paraffine. There is no well marked division line between any of the above named products, but they gradually merge one into the other.

Their division is simply one of caprice. These hydrocarbons are extremely complex and different in composition. The proportion of carbon and hydrogen is extremely variable. There seems to be no end to the different combinations of hydrogen and carbon.

The great diversity in the physical and chemical conditions of the bitumens can be attributed: first, to the organic remains from which it was distilled—different kinds of terrestrial vegetation and marine vegetation—by natural process these organic remains may have been changed into peat, lignite or coal before distillation; second, the degree of temperature to which organic remains are subjected during distillation; third, the pressure to which it is subjected during distillation; fourth, the time consumed in effecting distillation; fifth, the presence of different substances during distillation—sulphur, lime, water, oxygen, nitrogen, etc., which render their properties very different; sixth, the condensation of the bitumen after distillation, whether rapid or slow, agitated or quiescent; seventh, the material of the still; eighth, to repeated distillations; ninth, evaporation; tenth, sulphuration, oxygenation, etc. Electricity may also play an important part.

Maltha, asphalt, jew pitch, mineral pitch and brea are hydrocarbons which contain either sulphur, oxygen or nitrogen. They may contain one or more or all three of these elements in varying proportions. They can be produced synthetically by sulphurizing, oxidizing or nitrogenizing petroleum oils.

Oxygen, sulphur or nitrogen, when chemically united with a hydrocarbon such as some of the petroleum oils, produces a resin-like substance, to-wit: asphaltum.

One of the solid asphaltums, when taken from the ground, is brown, owing to its porous condition, caused by the evaporation of the petrolene. It melts at 245° Fahr. When it is melted, it becomes black or blackish green. Another asphaltum, when taken from the ground, is black in color, shaded with brown, or red and dark green when sulphur is present. When purified, it assumes an indigo blue reflection. It is opaque, scentless, tasteless and fragile, breaking with a conchoidal fracture, which has a glassy brilliancy. By rubbing, it acquires a resinous electricity. Its specific gravity varies from 1.100 to 1.247. At ordinary temperature it is readily reduced to a powder. In a condition of extreme subdivision it takes a brownish tinge. It melts at about 105° to 108° C.

Immediately above its melting point asphaltum is volatile, and, if the temperature is carefully raised, it disengages in abundance white vapors which belong to oils, which become thicker in proportion as the operation is prolonged little by little, but slowly the volume of vapors diminishes, gas ceases to form, and a deposit of carbon, slightly bituminous, is reached, which is solid and has the appearance and often the hardness of jet. When subjected to quick distillation in passing on toward the dark red, it sets free at the same time a mixture of brown oils, sometimes sulphuretted hydrogen and sometimes ammonia, while the retort retains about one-third of its weight of loosely compacted carbon. It is entirely insoluble in water; it gives up to absolute alcohol a small quantity of a yellow substance which exhales the odor, and has the appearance of, resin. Ether extracts from it a brownish black substance called petrolene. The portion left by the alcohol and the ether is asphaltene.

The yellow substances, petrolene and asphaltene, do not exist in asphaltum in defined proportions. Sometimes the petrolene will represent two-thirds or more of the asphaltum. On the other hand, asphaltene composes almost exclusively other asphaltums. The properties of the asphaltums vary according to different proportions of these three principles.

YELLOW RESIN. — Absolute alcohol dissolves yellow resin without dissolving petrolene or asphaltene; it is also readily soluble in the solvents of petrolene and asphaltene. It has the appearance of a resin. When the solution in alcohol of the yellow resin-like principle is treated with liquid ammonia, it produces an abundant white precipitate, while small globules of petrolene spring up from the bosom of the liquid and come floating in greenish yellow lentils to the surface.

PETROLENE. — This is brownish black in color and has a soft and glutinous consistency. It is insoluble in absolute alcohol. Ether, benzine, benzene, acetone and the fat oils dissolve it and the yellow resin, but leave the asphaltene intact.

Petrolene is also soluble in the solvents of asphaltene. The specific gravity of petrolene at 21° C. is 0.891. It burns with a very sooty flame. It has but very little taste.

A highly concentrated solution of caustic soda or caustic potash, when hot, dissolves petrolene; if some diluted sulphuric acid is poured into the liquid, a brown gelatinous substance is precipitated.

A current of chlorine precipitates the petrolene from its solution in benzine or in turpentine in brown and viscous flakes. These precipitates contain chlorine, and do not give anything further to alcohol or ether.

Hydrochloric acid precipitates petrolene from its solution in benzine in thick flakes; sulphuric acid, in a solid and viscous deposit, which is transformed in time into a brownish red.

In making the experiment with sulphuric acid, the acid must be carefully and slowly added.

If asphaltum be kept at a temperature of about 250° C. by means of an oil bath, until it no longer loses by weight, the petrolene is evaporated from the asphaltene.

ASPHALTENE. — Heavy petroleum oil, carbolic acid, turpentine, chloroform and bisulphide of carbons dissolve asphaltene without residue. It burns like resin, leaving coke. It is black, brilliant, and breaks with a conchoidal fracture. In the fire it only becomes soft near 300° C., and decomposes before completely melted. When torrified upon a platinum plate, it diffuses an odor of burned fat, afterwards of a sharp taste, which reveals an acid. It is solid, hard and fragile. When pulverized, it presents a mass of purple color, oftener of a brownish red. It develops, by friction, resinous electricity. In some of the asphaltums, analysis has disclosed a large proportion of oxygen; in others a large proportion of sulphur. A current of chlorine precipitates the asphaltene from its solution in petroleum oils and turpentine.

Asphaltene is not sensibly affected by caustic potash or caustic soda in a concentrated solution in water.

THE FORMATION OF ASPHALTUM BY THE RESINIFICATION OF PETROLEUM OILS. — When petroleum oils are left for a long time in the presence of oxygen gas, or the atmosphere which contains oxygen, and in the light, they absorb oxygen; some carbonic acid is set free, water is formed, their odor becomes weakened, and they likewise become viscous while they assume a darker and darker color. When petroleum oil is heated in a current of oxygen, it undergoes a quick change and turns into petrolene.

If a current of sulphuretted hydrogen is conducted into boiling petroleum, a very mobile sulphurized liquid is distilled, having an unbearable odor of onions. If this treatment is repeated with the new compound, a second portion of the sulphuretted hydrogen comes to reinforce the former, and the odor of the liquid becomes that of garlic. When this sulphurized oil is evaporated, a resin is formed.

If petroleum oil is changed by the compression of sulphuretted hydrogen, and then the sulphuretted hydrogen be decomposed to sulphur and water, the oil will be sulphurized and resinified.

If petroleum oil is distilled in the presence of sulphur, the oils will be decomposed and sulphur compounds formed in the shape of a resin. Petroleum oil treated with nitrous gas absorbs it with a slight development of heat; the petroleum becomes thicker and is partly converted into resin.

All the petroleum, upon which azotic acid is caused to act, furnishes yellow resins.

If petroleum is boiled in a concentrated solution of nitrate of potash or of soda, the nitrate converts the bitumen into resin, and the liquid becomes a brownish red.

POLYMERISM OF ASPHALTUM.—By exposure to daylight, asphaltum polymerizes, that is, it acquires a higher molecular weight, retaining the same atomic proportions. The stronger the daylight the more rapid polymerization takes place. When polymerized, its molecule consists of two or more simple molecules united to form a complex molecule. It can be changed from a state of polymerization to its original or simple state by heat.

Polymerization is much more rapid and conspicuous with asphaltene than petrolene; the part of asphaltum soluble in alcohol does not polymerize.

Polymerization is more rapid and greater in asphaltum containing sulphur than in asphaltum containing oxygen. When polymerized, the physical properties of the asphaltum are changed; it has a greater specific gravity, it is harder and more brittle; but the most marked change is that of becoming insoluble, or, to speak more exactly, of being dissolved with greater difficulty in its solvents than when not polymerized.

Asphaltum, on account of this photochemical action, is used in photography.

If a moderately concentrated solution of asphaltum, in spirits of turpentine or chloroform, be placed in a transparent bottle and securely corked, and then exposed for some time to the light of the sun, resinous substances separate and gradually appear, which dissolve with greater difficulty in these solvents. If heated, they dissolve; the greater portion of these separated substances adheres firmly to the sides of the bottle; a smaller portion remains in suspension in the solution.

Colored resinous substances will form in the California petroleums of commerce, if exposed to light, in the manner described above.

Solutions of asphaltum, which are to be employed in photography, must be kept in the dark.

Asphaltum is employed in photography in the following manner: When the solution of asphaltum with turpentine or chloroform is spread over a plate, and left in a dark room until it becomes nearly dry, which will require a few days, and the plate exposed in a camera, or placed under an object in contact with it, the time necessary to make the print varies very much, and can only be ascertained by experiment. When printed, the development is effected by quickly flooding with spirits of turpentine, which will at once dissolve the asphaltum, which has been protected from light, and partly dissolve that portion which has been exposed to the light. As soon as the subject is seen to be fully developed, a gentle stream of water from a tap is allowed to flow over it to wash off the turpentine. If all operations have been conducted rightly, a very delicate and perfect picture in asphaltum is the result.

Anticlines, synclines, monoclines, centroclines and quaquaversals, and also faults, exert a great influence in the accumulations of gas, petroleum oil and water. Especially is this true in California, where the dips and undulations along the strike of the anticline are exposed and well defined.

Although a description of the different inclinations and curvatures of strata would seem elemental, a thorough knowledge of the effect of these inclinations is necessary for an understanding of the laws governing the accumulations of bitumen in California. In the Eastern States the slopes of the domes frequently do not exceed twenty feet to a mile, whereas in California the strata stand at a very steep angle with the horizon, frequently being overturned.

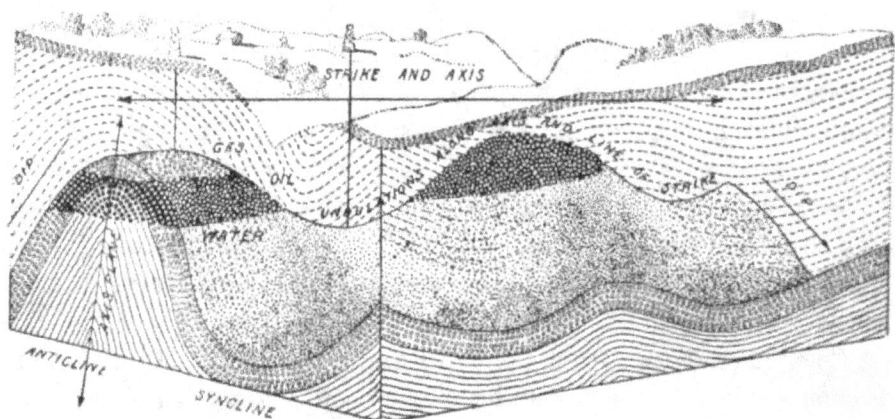

Fig. 1.—PLANO-SECTION SHOWING INCLINATIONS.

When a group of strata is bent into a curve like a saddle, with its convexity turned towards the earth, it is called an anticlinal curve. Such a condition of strata is shown in Fig. 1 above the word "anticline." A synclinal curve is exactly the opposite of an anticlinal curve. When the strata is folded or curved, so as to form a trough, the concave side of which is turned from the earth, it is called a synclinal curve. This is shown in Fig. 1 above the word "syncline." In both anticlines and synclines, the line in each bed, along which the change in the direction of the dip takes place, is called the anticlinal or synclinal axis of that bed, and the planes containing all the axis of an anticlinal ridge, or a synclinal trough,

are called axis planes.   The axis plane usually approaches verticality.
Anticlines and synclines frequently nose out, or coalesce.

When an anticline undulates along the line of its axis, dome-like
elevations occur, from the summits of which the beds dip away in
every direction.   In this case the strata are said to have a quaquaversal dip.   An anticline is an elongated dome.

A quaquaversal, or dome, is a nest, usually of a great number, of
different strata composing numerous overlying, gigantic, inverted
funnels, the strata of the formation forming the sides of these rock
funnels, all of which tend to guide and convey the ascending gas and
oil to the apex of the dome.

When a syncline undulates along the line of its axis, basin-shaped
depressions occur, towards the bottom of which the beds dip from all
sides.   This is called a centroclinal dip.   A syncline is an elongated
basin.

A fault, or dislocation, is a fissure or crack in the crust of the
earth, accompanied by the elevation of the mass upon one side of the
fault, while the other side remains stationary, or sinks down.   Anticlines and synclines are often truncated by faults, and may be so
faulted as to form the segments of a sphere or cone.   If an oil-
bearing bed, ascending to the north, be interrupted by an east-and-
west fault, the further ascent of the oil northwards will be arrested,
and then an abundant supply of oil may be obtained by boring on the
south side of the fault, while, for a considerable distance to the
north, water will occupy the formation, to the exclusion of the oil.
This is more apt to be the case where the throw of the fault is sufficient, so that the edges of the porous strata are covered by impervious
strata.   Selvage frequently occupies the line of faults, generally
caused by the movement of the two sides of the fault on each other,
which have ground up the materials of the rock, forming a sheet of
matter impervious to the flow of oil or water, or the faults may be
filled up with mineral matter of various kinds, which are also impervious to oil or water.

When a formation contains permanent water, the accumulation of
petroleum oil will be found near the upper part of the dome, as is
shown in the plane section, Fig. 1.   The oil floats on the surface of
the water, and if natural gas is present, it will be found above the
oil.   These three substances arrange themselves according to their
specific gravity, the lightest on top.

Fig. 2 is a view of an anticline. The camera was pointed in the direction of its strike. The black line represents the plane of bed-

FIG. 2. — MESA DEPOSIT, SISQUOC, SANTA BARBARA COUNTY.

ding, which was once horizontal, but is now curved in the form of an arch.

The unaltered rocks of California cover an area of forty thousand square miles.

The bitumens are found, in greater or less quantities, in all of the unaltered rocks of California of the cretaceous and tertiary periods, and sometimes in the quaternary rocks in the form of natural gas, petroleum oil, and solid and liquid asphaltum. The difficulty is the discovery of accumulations at particular places large enough to justify developments.

The unaltered rocks consist principally of alternating beds of shale, sandstone, and conglomerate, varying in thickness and resting upon metamorphic rock. The sandstones and conglomerates act as reservoirs for the accumulations of bitumen, and the shales as encasements for these reservoirs. Some of these sandstone beds are more than three hundred feet in thickness, as will be described hereafter.

Anticlines exercise great influence upon the accumulation of natural petroleum oil and other bitumens.

The main anticlines of California oil regions bear northwestly and southeastly. The summits of these anticlines have been denuded, exposing metamorphic rock. Numerous smaller anticlines branch in all directions from these main anticlines, and generally nose out in the valleys.

Smaller and lower anticlines also run rudely parallel with the main anticlines.

When the uplifts, by orogenic movements, have been great, the apexes of the anticlines are frequently denuded, the bitumens either being washed away or, as more often happens, drained into the dips of the anticlines.

The petroleum-bearing strata are exposed to a greater geological depth in the outcrops of the strata of the main anticlines, that show a metamorphic core, than in the lower anticlines that are but slightly denuded, consequently, there are visible many seepages of oil and flows of gas from the outcrops of oil strata on the sides of the main anticlines, while in the lower anticlines the same oil strata lie far below the surface of the earth. In the valleys in which bituminous strata are overlaid by quaternary rocks, the bituminous deposits may exist at such great depths that they cannot be reached by drilling.

In the northern part of the State a large amount of tertiary rocks have been washed away; these rocks grow thicker and thicker to the southward, until in the southern part of the State they are of great thickness. The unaltered rocks in the northern part of the State are geographically higher and more broken than in the southern part. Owing to these different conditions, there is more and better storage room for the bitumens in the southern part of the State than in the northern.

The unaltered rocks of the Coast Range are more broken and contorted, and have a much larger outcrop than those forming the foothills of the Sierra Nevada on the east side of the San Joaquin Valley. Therefore, there are more visible evidences of bitumen on the west side than on the east side of the valley.

There can be no question but that the cretaceous and tertiary rocks, which are the oil-bearing rocks, of California, underlie the quaternary deposits of the San Joaquin and Sacramento Valleys. Gas wells exist in these valleys; this shows that the lighter and more volatile parts of the petroleum oil have been preserved, consequently, the heavier parts of the oil exist.

RED SHALES AS CONNECTED WITH THE GENESIS OF BITUMEN IN CALIFORNIA. — Shales were, and are, deposited in still and salt water. The iron contained in these waters and organic remains, both animal and vegetable, and other materials constituting the shales, were deposited contemporaneously.

If the iron was the peroxide of iron, ferric oxide (Fe_2O_3), by contact with organic remains, it was deoxidized and reduced to a protoxide, ferrous oxide (FeO), by the absorption of one equivalent of its oxygen; when the peroxide was reduced to a protoxide, carbonic acid, produced by the decomposition of organic matter, then united with the protoxide, forming carbonate of iron ($FeCO_2$).

The carbonate of iron imparts a bluish or greenish color to the deposit.

The accumulation of iron, in the presence of an excess of organic matter, retains the form of ferrous carbonate. In all coal measures, of all periods, whether carboniferous, jurassic, cretaceous or tertiary, or in all cases where there is organic matter in excess in a state of change—in all strata, whether older or newer, in which there is organic matter in excess in a state of change (not graphite)—the iron is in the form of carbonate protoxide, or ferrous carbonate ($FeCO_3$).

Sulphide of iron, ferric sulphide (FeS_2), is subsequently formed and deposited instead of carbonate of iron. The sulphates of lime ($CaOSO_3$), and of magnesia ($MgOSO_2 + 7HO$), and other sulphates which exist in sea water, when subjected to the action of decaying organic matter, out of contact with air, are deoxidized and converted into solubles, from which sulphuretted hydrogen gas is set free by the carbonic acid gas produced by the decomposition of organic matter. Sulphuretted hydrogen converts the soluble compounds of iron into sulphide of iron. The color of pyrites is brass yellow.

The presence of protoxide of iron, and of iron pyrites, in these shale beds, arises from the considerable amount of organic substances exercising a reducing action. The water, flowing from the mountain heights, where there are no organic substances, exercises at first an oxidizing influence, in virtue of which the rocks over which they flow are decomposed.

The suspended substances carried down by these rivers, and the detritus swept along their beds, come, after a time, in contact with organic substances, by means of which the peroxidized iron compounds are again reduced. Consequently, the iron thus carried into the sea is, for the most part, in the state of protoxide, either combined with silica or with carbonic acid, the silicate being suspended, and the carbonate dissolved in the water.

When unaltered by oxidation, the carbonate of iron, with varying amounts of lime, clay or sand, is dark grayish-blue or green, or even white, in color.

When unaltered by oxidation, the sulphide of iron is brassy yellow in color.

From the preceding explanation, it is safe to say that, at the time of their deposition, the carbonaceous shales were not red; and as long as they are not submitted to oxidizing influences, they will not become red.

When carbonate of iron is exposed to the oxidation of the air, it forms limonite (hydrous ferric oxide), which is usually of a brownish yellow, or brownish red color. These iron ores are found in all stages of transformation. On the outcrop, they are limonite; under dense cover, carbonate. While going from the ontcrop inward, the limonite constantly decreases in proportion to the carbonate. In the alteration of the compact carbonate, the line of chemical change and color is usually very sharply defined, and the limonite covering can often be entirely removed from the enclosed core of carbonate by a blow with a hammer, the limonite covering preventing the carbonate core from being oxidized by the air. In shales charged with gray carbonates of iron, the following reaction takes place by the action of the air: the carbonic acid is released, and part of its oxygen oxidizes the iron.

Gray shales containing finely divided pyrites, or bisulphide of iron, are converted by heat into bright red, the sulphur being released, leaving the shales charged with red oxide.

The color of burnt ferruginous shale is entirely due to the amount of iron present. Gray shales containing less than one per cent, or one and one-half per cent of iron, change by heat to various shades of cream color, or buff; while those containing two per cent to ten per cent, or twelve per cent of iron, produce, by heat, pink and bright red bodies. The depth of the color depends merely on the amount of iron present, the buff shades gradating into the deeper shades of red.

A group of stratified rocks usually consists of various species, arranged in alternating beds, a series of beds of many hundreds, or even thousands, of feet in thickness, containing strata of shale, lime-stone or sandstone.

Some strata are seamed or porous, and easily penetrated by fluids, serving as conduits and reservoirs for fluids. Some strata are nearly impervious to fluids, while others are practically so, frequently serving as incasements for the conduits and reservoirs formed in, and by, porous and seamed strata.

All stratified beds have been originally deposited in a horizontal position, or approximately so. While these beds were in the hori-zontality of their deposition, and incased by impervious strata, there was little or no circulation of fluids within their porous or seamed strata. When they were tilted and inclined to the horizon, at angles varying from the horizontal to nearly absolute perpendicularity, and their porous and seamed strata exposed to the entrance of fluids, by denudation, fracture or otherwise, and an exit for the fluids was sup-plied, or produced, at a lower level than the place of its entrance, the circulation of fluids commenced, slowly at first, and gradually in-creasing as the inclination and exposure of the different strata be-came greater. The course of the circulating fluids was complex and anfractuous.

Water, supplied by the rainfall of the region, enters at the out-crop of the porous and seamed strata. If the porous and seamed strata are incased in impervious strata, the greater the depth to which the strata extend from the place of entrance of the water, the greater the pressure will become. In some instances, this pressure will be very great, forcing the water into comparatively impenetra-ble rocks.

The water, percolating and circulating through the porous and seamed strata, by its solvent action, accumulates mineral ingredients. These waters, saturated with minerals, coming in contact with other minerals existing in the shales, by chemical reaction produce heat. This heat contracts and fractures the shales, permitting a freer cir-culation of water.

This chemical heat distills petroleum from carbonaceous shales, and oxidizes the carbonate and sulphide of iron, producing the red colors of the shales, and water of different temperatures, charged with mineral ingredients, will frequently rise, by hydrostatic pressure, through fissures and faults, etc., to the surface of the earth, forming mineral springs. These springs are often accompanied by bitumen.

But very few fossils exist in these red pyrogenous shales, as they have been obliterated by the solvent action of hot water, or by the chemicals held in solution by the circulating waters; or, if the molds or casts of their external forms existed, they have disappeared from the same causes, or they have been crushed and distorted beyond recognition.

RED SHALES IN CALIFORNIA.—Red shales in California are the effects of chemical heat. Strata which have been, more or less, altered by the action of heat, emanating in the strata from chemical reaction, consist of burnt shale, porcelain jasper, earth clinkers, slag and white shale.

BURNT SHALE.—Its color is usually red, sometimes gray, yellow or brown, and graduating from cream color to brilliant red. It is clay, or shale burnt, but not so much changed as to form a porcelaneous mass.

PORCELAIN JASPER.—It is shale, or changed into a kind of porcelain by the action of heat. It is dark red, yellow or striped yellow and red.

EARTH CLINKER OR SLAG.—This is a shale, converted into a kind of clinker slag. It is black brownish or reddish, and it has occasionally a tempered steel tarnish.

Sometimes it shows iridescent colors. It is vesicular, usually amorphous, but occasionally possessing the prismatic form of artificial coke.

WHITE SHALE.—During these chemical fires, carbonic acid, sulphuretted hydrogen and aqueous vapors are formed; these exhalations, in passing through the shale, bleach and decompose it. The silicates are decomposed by the continuous action of aqueous vapors, at 212° Fahr., sulphuretted hydrogen, air and the alkalies, magnesia and lime, are nearly removed, and metallic oxides are carried away. The vapors convert the shale into a white clay, or nearly white, when a small quantity of iron still remains. By the removal of the alkalies, magnesia, lime, and metallic oxides, the quantity of alumina and silica increases. The absence of the bases, such as lime and iron, in these bleached shales, give growth to a different vegetation from that which grows where these bases exist. This difference in vegetation is a good index to deposits of petroleum and bitumen. The removal of these substances makes the shale incapable of sustaining vegetable life; the absence of, or scarcity of, vegetation is indicative of this action. Immense beds of these white and altered shales frequently occur in the vicinity of bituminous deposits, generally running in the direction of the anticlines.

Not infrequently the marine shales, through which hot silicated waters have percolated, and from which the bases, such as lime, magnesia, iron, etc., have been carried away by the solvent action of these waters, contain diatoms in large numbers, whereas the adjoining shales, which have not been leached, do not contain diatoms in any notable quantities.

Diatoms abound in the hot springs of California and Yellowstone Park. In the hot springs of the Yellowstone Park, deposits of this kind are now forming over many square miles, and are five or six feet thick.

Why should they not originate and abound in percolating hot silicated saline waters, and be deposited in the interspaces and joints of the shales through which the water percolates?

Isolated bodies of diatomaceous earth in California would indicate that they originated and were deposited from springs.

At the Buena Vista Oil Springs, in Kern County, quaternary deposits of infusorial earth exist, the stratification of which is horizontal; it has either been denuded from the leached and adjoining formation, and deposited in still water, or else it originated in quaternary waters, and then deposited; probably the latter is the case, as these strata do not show the presence of other material from the adjoining formation.

From the immense amount of mineral matter which has been carried away by the solvent action of water, thousands of tons of fossil shells, silica, magnesia, iron, etc., and the large area now occupied by the whitened shales, the flow of mineral water, at some former time, must have been very copious as compared to the flow at the present time.

The illustration (Fig. 3) shows an outcrop of red shales and porcelanite near Mount Solomon, Santa Barbara County.

PHENOMENA ATTENDING RED SHALES.—The red shales are discovered by their bright colors, by the heat of the earth in their vicinity, and sometimes by smoke. Sulphurous and other vapors frequently occur. These vapors, in their course upward, are condensed, and incrust the fissures of the rocks, and even the surface of the ground.

Mineral springs, hot and cold, issue from the ground in their vicinity. The earth is charged with salts and minerals occasioned by the percolation and evaporation of these mineral waters. Shales, through the joints of which these mineral waters have flowed, have become impregnated with salts, and the salts, subsequently to the flow, have become vitrified by heat. They are further known by the issuance of warm or cold natural hydrocarbon gas, by seepages of bitumen in their neighborhood, by fissures, joints, and porous rock

FIG. 3. — OUTCROP OF RED SHALES AND PORCELANITE.

filled with asphalt, and by the almost total absence of fossils in the burnt shale porcelanite and clinkers, which have been obliterated by hot water and heat. Before chemical heat commenced, these shales did not contain over two per cent of carbonaceous matter, not sufficient for them to be set on fire at the surface.

PHYSICAL CHARACTERISTICS. — When unburnt, these shales are easily split along their lines of lamination, but when burnt to a tile

red, or to a greater degree, their fissility is partly destroyed.   When
unburnt, their lines of lamination are plainly visible, but, when burnt,
their lines of lamination are obscured or obliterated.   When unburnt,
they have a clayey-like smell when breathed upon, but this physical
characteristic is partly, if not altogether, lost when they are burnt.
When unburnt and suspended so as to freely vibrate, they have a dull
sound when struck, but, when burnt, they become resonant.   In this
characteristic they resemble brick.   Chemical fires destroy, or partly
destroy, their lines of lamination, their fissility, and their argillaceous
smell when breathed upon, but increase their resonance.   These
characteristics do not occur when these shales are discolored by the
oxidation of the iron, naturally contained in them, through the agency
of water without heat.   Serpentine cups filled with a pigment made
from these bright shales are dug from the graves of the aborigines.

About thirteen miles east of Santa Barbara City, an excavation was
made on the bluff of the ocean for the road-bed of the Southern Pacific
Railroad, and the gray shale, charged with chemical substances and
carbonaceous matter, taken from the excavation, was thrown over
the bluff, forming a conical-shaped pile, composed of pieces of shale
containing from one to eight cubic inches.   Water could easily penetrate the broken shale, and air could easily circulate through the
mass.   When the winter rains fell upon this pile, chemical action
commenced, producing sufficient heat to vitrify and weld the pieces
of shale together.   A large part of this shale was burned to a red
porcelanite, and the remainder was colored a buff shade, graduating
into the deeper shades of red.   Above the railroad track, the face of
the shale bluff has been cut off to the angle of repose by the railroad
company.   This smooth surface is a good place to observe the action
of chemical heat and attending phenomena.

LA PATERA MINE.—La Patera Mine lies nine miles west of the
city of Santa Barbara.   Its relation with a lake and the ocean is
shown in Fig. 4.   The lake contains about sixty acres.   Along the
periphery of the lake, the stratification of the shale dips towards the
lake at an angle varying from 30° to 40°.   The composition and
the arrangement of the component parts of the soil are the same
upon the island as upon the mainland.   The shale must have existed
at a level shown by the dotted lines in Fig. 4, and subsided after the
deposition of the soil, otherwise the soil would not have been deposited upon the island in a manner similar to that of the mainland.

This subsidence was probably occasioned by the contraction of the underlying shale, produced by chemical heat. Some idea of the contraction of the shale by burning may be learned from the contraction of brick through burning. Before burning, and when in a dry condition, a brick is eight inches long; when burned—not vitrified—it is seven and seven-eighths inches long, and when vitrified it is seven and three-fourths inches long.

If the basin of the lake near the La Patera Mine had been formed by erosion of the land by sea or surface water, the shale would have

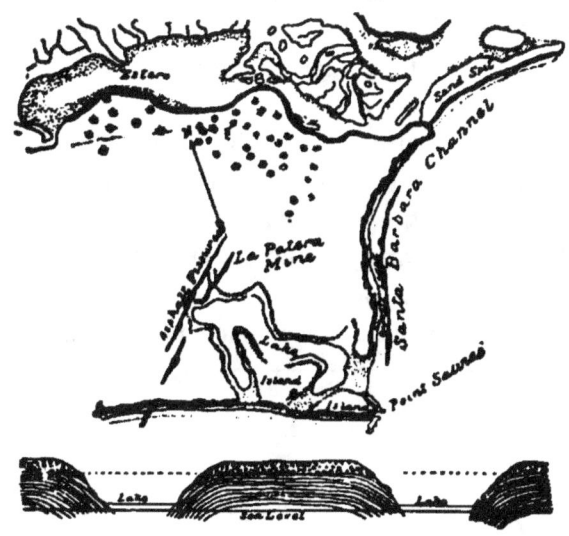

FIG. 4. — LA PATERA MINE, SANTA BARBARA, CALIFORNIA.

been squarely cut off and not contorted so as to dip towards the lake. In the excavation at the mine at the depth of 100 feet, a temperature of 105° Fahr. is generated in the shales by chemical heat. Circumjacent to the lake are fissures filled with hard asphalt, through which comminuted shale and mineral water are disseminated.

Off the shore, petroleum rises from submarine springs, covering a large surface of the ocean with a thin film of iridescent oil, the odor of which can be detected at a long distance. Ledges of hard asphalt exist in the ocean, below high tide, which run nearly parallel with the shore. Surface wells show the existence of water highly charged with mineral substances, in which petroleum is discovered. So far, no potable water has been found near this mine.

Six miles west of Santa Barbara, on the Calera Rancho, and on the ocean shore, an area of twenty acres has subsided some twenty-five feet; of this subsidence, four feet occurred in five years. This subsidence has occurred through the contraction of the shale. The surface of the subsidence is rifted and seamed, and from these rifts and seams sulphurous and other vapors ascend. The ground is hot. The bluff is composed of burnt shales, showing tints from a cream color to a brilliant red. Water containing salt seeps from the base of the bluff. Shales with carbonaceous material, shales saturated with bitumen, and smoky-looking shales surround the hot places. Near the hot places, heavy petroleum oils ooze through the shales. To the eastward and westward heavy and thick petroleum tars ascend through the cracks and seams and joints of the shale. Some of the seams of shales, containing a small proportion of bitumen, have hardened to such an extent that they resemble dark flint, and will cut glass.

Lyell gives the following: "Captain Mallett quotes Guinillar, as stating in his description of the Orinoco, that about seventy years ago a spot of land on the western coast of Trinidad, near halfway between the capital and an Indian village, sunk suddenly, and was immediately replaced by a small lake of pitch, to the great terror of the inhabitants." A similar subsidence at an earlier period may

FIG. 5. — LAKE FORMED BY SUBSIDENCE OF LAND.

probably have given rise to the great pitch lake of Trinidad, the cavity having become gradually filled with asphalt. There are a number of places in California near these red shales from which natural gas issues. Some are hot, showing that they are formed at a high temperature.

Fig. 5 is a view of a small lake formed by the subsidence of the land near Mount Solomon, in Santa Barbara County.

It may not be out of place to mention in this connection the occurrences of red shale in other parts of the world in which bituminous deposits are known to exist.

RED SHALES IN THE ISLAND OF TRINIDAD.—The formation of the island of Trinidad consists of clay, loose sands, shales, limestones, calcareous sandstones, indurated clays, porcelanites of brilliant red colors, with pitch deposits and lignite here and there.

The only substances containing sufficient carbon and hydrogen for the formation of asphalt, and likely to be inclosed in the strata, are vegetable remains. They are particularly abundant at La Brea, where most of the asphaltic beds have been originally carbonaceous and lignitic shales. Mineral springs abound throughout the island. In a series of loose sands, clay, and shale, lies Pitch Lake, seemingly

PITCH LAKE AT TRINIDAD. W. I.—From an old print.

occupying a depression in the strata. To the southward of the lake the shore is made up of bold cliffs, the strata of which consist of indurated clays. They also present thick veins of porcelain jasper. Strata of loosely coherent sandstone also abound, some of which are impregnated with bitumen. Rounded pebbles of pitch and porcelain jasper form a beach at the foot of the cliffs. A species of coke is occasionally observed along the shore, with a porous structure and the prismatic form of the artificial product, but, of course, much denser, on account of the large proportion of earth. Near the lake is a red, yellowish substance, semi-baked, evidencing a considerable degree of heat which attended its formation. Part of the impurities in the Trinidad asphalt consist of comminuted red clays or shales, with some sand.

It is evidently not adventitious at the surface, but must have been thoroughly incorporated, and brought up from the depths with the bitumen, judging from the constant amount, dissemination and character in all parts of the deposit. Water, containing all the mineral ingredients of strong thermal water, is found in the Trinidad asphalt. The presence of borates, iodides, and so many forms of sulphur compounds, and other characteristics, show that the water must be of the same origin as that of many thermal springs. This water, in all unaltered pitch, shows that the formation of the pitch and water must have been simultaneous, and cannot be considered adventitious.

It would be impossible for water, in any adventitious way, to become intimately mixed with the bitumen, so as to form, practically, an emulsion. Near the center of the lake is a body of pitch softer, blacker and newer than that of the remainder of the lake. Gas constantly issues from the cracks in the bitumen. Those phenomena show that asphalt is being distilled at the present time. The porcelanite and red shales must have been formed by heat, created in the strata themselves, as these shales are burned uniformly, in no place showing a greater degree of heat than in another. They are, probably, formed in the same manner as similar rocks in California. The depression, in which Pitch Lake lies, was, probably, made by the subsidence of the surface of the earth, caused by the heat contracting the underlying shale. There was no focus to this heat, no central point. If there was, in the material next to the central point, the evidences of heat would be great, gradually decreasing as you went from the focus; this is not shown. To illustrate, bricks of very different qualities are to be found in the same kiln, for as the fire is applied below in arches, the lower bricks in their immediate vicinity will be burnt to great hardness, or, perhaps, vitrified; those in the middle will be well burned, and those on the top will be too little burned. Even then the bricks the farthest from the fire would not have been burned to this extent, if it were not for the numerous flues left between the bricks in the construction of the kiln. The intense heat of a furnace is confined by a foot-wall of fire bricks. Three feet of lava will confine the heat of melted rock underneath. From the uniform burning of these shales, the heat must have originated in the shales themselves. A good, clean red heat is required for the burning of brick; it is fair to suppose that this temperature is required to produce red shales.

Porcelanites and vesicular clinkers are scattered throughout these red shales; they are not centralized. No fumeroles connect these porcelanites with a central fire. Moisture was concerned, as is evidenced by their even burning. Moisture was the vehicle of heat, or the burning would not have been so uniform in its effect if disseminated by conduction or radiation. The parts of the shale burned to porcelain resemble earthenware and stoneware; to burn earthenware and stoneware, a clean, white heat is required.

Arborescent forms of huge scale of these hydrothermal shales extend their ramifications throughout the earth in the vicinity of bituminous deposits. This form also goes to show, that their burning was accomplished by chemical action, with the presence of moisture, and not from radiation or conduction from a focalized fire. Nearly all readily solvent substances, and all volatile substances, have been removed from these red shales, or where solvent substances now exist in them, they are different from those that were in them at the time of their formation. The noticeable bright red colors of these shales could not have been produced by the oxidation of the iron in the shales by water alone; heat must have been present to produce them. This heat would be sufficient to distill any carbonaceous substance contained in them.

In the metamorphic rocks of the San Rafael range of mountains of Santa Barbara County, it can be plainly seen that these porcelanized shales were converted to serpentine. All gradations from shale to serpentine can be found; shales reddened by heat, porcelaneous masses still retaining the structure of shales, and porcelain partly converted to serpentine. There are no visible signs of metamorphic action now in operation where the serpentine is exposed to view, but in the tertiary shales and sandstones to the west, especially in the hills lying north and south of the Los Alamos Valley, this metamorphic action is going on at the present time.

These red shales lie above and precede greater metamorphism, such as is exhibited by serpentine and quartzite, and, probably, metamorphic granite. This can be seen where erosion has been great enough to expose the contact between metamorphic rock and red and unaltered shales. When burnt, the cracks and seams in the shale are large, showing the extent to which they have contracted. Near the surface, the red shales are vesicular, and look as if they had faulted. These red shales are very much distorted and contorted; many of the contortions have a radius of but a few feet. No carbonaceous matter exists in these red shales. No bituminous substances exist in these red shales unless they have entered subsequently to their burning. There is also an absence of sulphur. Carbonaceous and bituminous substances, and sulphur, are disseminated throughout the strata adjoining the red shales.

At the time of their deposition in sea water, and before any alteration had taken place, the red shales should contain the chloride of calcium, carbonate of calcium, carbonate of iron, carbonate of magnesia, sulphate of calcium, etc., vegetable matter in a fine state of subdivision, and, in some places, large deposits of vegetable matter; this organic matter, in time, becoming carbonaceous shales and seams of coal. When these shales are heated with a chemical heat, the following described vapors and gases are given off: sulphur dioxide, carbonic acid, sulphuretted hydrogen, carburetted hydrogen, distilled from the carbonaceous substances, nitrogen derived from the same source, etc. These vapors are forced into the circumjacent formation, which is highly charged with mineral matter originally present in the shale, and which had been deposited from hot water. Chemical actions arise when these vapors, or their condensations, come in contact with the minerals existing in the shales, causing heat. There remain in the shales, after the vapors and gases are eliminated by heat, oxide of calcium, oxide of iron, etc. Mineral waters and gases coming in contact with these substances produce heat. These chemical actions and reactions are complex and numerous. By these chemical actions and reactions, distillations and condensations, the alteration of the shale by heat becomes progressive and cyclic. This heat, under great pressure, and through subsidences and orogenic movements, is intensified. Chemical reactions are augmented through pressure and hydrothermal action. The hydrocarbons and other substances are distilled and condensed, dissociated and united a multiplicity of times. Mineral springs, ranging in temperature from that of the earth, from which they issue, to the boiling point, are frequent in the neighborhood of these burnt shales. They contain, in notable quantities, sulphate of sodium, magnesium, calcium, aluminum, carbonate of sodium, calcium and magnesium, chloride of sodium, potassium and calcium, silica, and, in excess, carbonic acid, sulphuretted hydrogen, carburetted hydrogen, and traces of arsenic, sulphuric acid and iron. Besides the visible phenomena, the warmth of these springs shows that this metamorphic action is still in progress.

Deposits of bitumen, in California, are found in sedimentary rocks of all ages, and principally in three different ways: 1. In superficial detritus. 2. In veins. 3. In porous or seamed strata. The bitumen may be of any consistency — gaseous, fluid, viscous, or solid.

Bitumen often occurs in superficial detritus or alluvium. This character of deposit is usually the overflow of tar springs, into which the detritus from the surrounding country has been washed or blown, and, in case where mineral tar is sufficiently liquid, it has percolated into the underlying earth. The detritus, when saturated with mineral tar, most always has been repeatedly burned, leaving black, vesicular clinkers, which are frequently refilled, where there has been a flow of tar after the fire.

Cooking utensils of the Indians are found in the vicinity of these tar springs, showing that they were used for purposes of fuel. After this they were set on fire by shepherds, so that the fleeces of their sheep would not be injured, or lambs suffocated by going into the sticky mass. Where the bituminized detritus has escaped the fires, and exists below the clinkers, it contains about five per cent of brown and friable bitumen, having a specific gravity less than water. It consists of seventy per cent of asphaltene and thirty per cent of petrolene. The accompanying map section and view show the Buena Vista Oil Spring, at Asphalto, Kern County, California. (Fig. 7.)

The mineral tar reaches the surface at the place marked "tar spring." At the place marked "tar springs," a number of trenches have been cut, and a tank erected, so as to intercept and save the tar. These trenches are cut in clinkers and comminuted shale, which is often saturated with tar. Thin layers of detritus, impregnated with mineral tar, lying nearly horizontal, are intercalated with thicker layers of detritus, which contain no bitumen. These beds are formed by flows from the tar springs and deposition of detritus denuded from the adjoining hills. Before this deposit was worked, mineral tar had flowed from the springs over the surface of the clinkers, until it had reached a thickness of from one to six inches; it had evaporated and oxidized, becoming stiff. It was of different degrees of purity, sometimes absolutely pure; at other times, it contained as high as eighty per cent of detritus. In 1891 and 1892 this concreted tar was being mined and refined. At the present time there are but a few tons, which are scattered over the ground, in small pieces. It has been estimated that, in recent years, these springs afforded about 160 barrels annually. There are a large number of other tar springs in California, nearly all of which have been burned in the manner described above.

In the shaft sunk at the Hancock deposit, lying northwest of the city of Los Angeles, in the center of what appeared to be an old

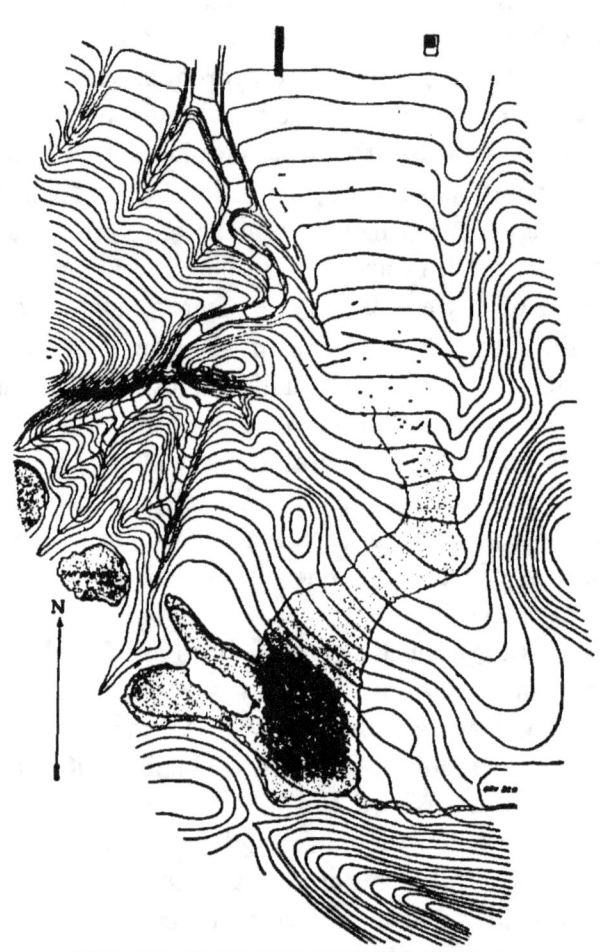

BUENA VISTA OIL SPRING, ASPHALTO, KERN COUNTY, CAL.

FIG. 7.

tar spring, on the surface were found the bones of domestic animals— horses, cattle, sheep, etc.; at a greater depth the bones of the bear, elk, and other wild animals, and, resting on the shale beneath the bituminized sand, at a depth of about thirty feet, were the bones of the *Elephas Americanus*. During the latter part of the summer season, in California, the natural grasses dry up, but, owing to the slight amount of water which ascends with the bitumen, as it does at the present time, many years ago, during the dry season, there grew, surrounding these asphalt springs, green and succulent herbage.

This herbage was the bait of the trap which tempted these herbivorous animals to their death. With everything else dry, these green herbs were an irresistible temptation. In struggling to get to them, the animals became mired in the tar lake and suffocated, their bones gradually sinking to the bottom.

The *Elephas Americanus* seemed to be the first one that met this fate, as his bones rest upon the underlying shale, below the remainder of the fossils. These springs must be very old, as it is many years since the American elephant fed upon the plains adjacent to the spring.

Veins of asphaltum, being rents, seams, and fissures filled with asphaltum, occur usually vertical, or not far from vertical. An innumerable number of small faults, contortions, and breaks occur in the formation in which these veins of asphaltum exist. These veins of asphaltum occur in the vicinity of red shales, or shales that have been burned and contracted by chemical heat; the contraction of the shales opened, and open cracks and seams in the adjoining formation permitted the ascent of the asphaltum. These cracks and seams extend to such depths as to reach deposits of petroleum oil.

From observations made at Asphalto, Mount Solomon, La Patera, and other places, it would appear that these fissures and rents and faults are filled with bitumen in the following manner: Heavy petroleum oil at various depths below the surface of the earth, in porous or seamed strata, is urged upwards, principally by rock pressure. The subsidence of the formation, in which the asphalt veins occur, is caused by the contraction of the shales by heat. In consequence of the subsidence, the shales and sandstones are very much broken and contorted. Around the edges of these subsidences, these veins of asphaltum are found.

The specific gravity of the asphaltum is too great for it to be buoyed up by water, and the formation in which the veins exist is too broken for gas to exert a pressure, but hydrostatic and gas pressure do, in some instances, urge the bitumen upwards. When ascending in porous or seamed strata, the petroleum oil is partly evaporated, the evaporated portion forming gas, which reaches the surface through passages which are closed to the viscous oil. The oil is readily oxidized on account of its divided state, and, if sulphuretted hydrogen or sulphur dioxide is present, the oil is resinified, forming asphaltum. By sulphuration, oxidation, and evaporation, the petroleum oil is finally hardened. When nearing the surface, it is indurated to such an extent that it rarely ever shows itself. Polymerization of the bitumen by the photochemical action of the light also assists in the induration of the bitumen when nearing the surface.

Adjoining the selvage of the veins, when near the surface of the earth, the asphaltum is brown and friable, fully two-thirds of which is asphaltene. In the interior of the vein it is black and shining, and, when cold, breaking with a conchoidal fracture, fully two-thirds of which is petrolene. The proportional amount of petrolene and asphaltene existing in the asphaltum is very variable. When at a sufficient depth, so as to be removed to a considerable degree from atmospheric influences, the brown margin of asphaltum does not exist. As a general rule, the further from the atmospheric influences, the more liquid the asphaltum will be. Bivalvular shells, filled with petroleum, are frequently found in these asphalt veins, the bitumen on the shells being much more liquid than the surrounding asphalt, and of a lighter color. The asphalt veins are rudely parallel to the periphery of the red shale deposits, and exist along the margin of the subsidences, the seams for the reception of the asphalt having been opened by the movement of the formation towards the contracting shales, permitting the ascent of the asphalt, forced up by rock

FIG. 8. — ASPHALT VEINS, SANTA MARIA, CALIFORNIA.

pressure. The parallelism of the veins to the periphery of the burnt shales is frequently changed by the varying hardness of different strata, or by orogenic contortions.

When the hanging wall consists of hard and close-grained strata, the bitumen will accumulate below the same, which, on account of its being conformable with the plane of bedding of the adjoining strata, gives it the appearance of having been deposited at the same time as the other strata. When these shales are burning, large amounts of sulphur vapors are disengaged. It is reasonable to believe that these vapors sulphurized the bitumen in their neighborhood, forming asphaltum. Mineral waters, containing sulphuretted hydrogen in large quantities, usually accompany the ascent of the asphaltum. The earth adjoining the veins is usually charged with mineral matter, deposited from infiltrated mineral water. If these openings and sulphur vapors had not been made and generated, through the contraction and burning of the shales by chemical heat, the asphaltum would not have been formed, nor could it have reached the surface of the earth.

The physical characteristics of the asphaltum filling these veins are extremely variable. In consistency, it gradates from a hard rock to a viscous condition. The asphaltum contains, intimately mixed, from one per cent to seventy per cent of impurities. The impurities hardly ever exceed seventy-five per cent. The presence of more than seventy-five per cent of impurities makes the bitumen so stiff that it cannot be forced through the cracks and seams. In the vicinity of the injected veins there are beds of shales containing from ten per cent to fifteen per cent of bitumen, but existing rock pressure is unable to move them.

The impurity in some of these asphalts is infusorial earth, in others, sand, while in others, finely ground shales and angular fragments of shales and fossil shells, or all of these impurities may be present in the same deposit. Shale preponderates as an impurity; in fact, the impurities are derived from the rock which the asphaltum encountered during its ascent. Most of the time the impurities are very fine and light, making the refining of the asphalt, by any known process, difficult.

The thickness of these veins of asphaltum is variable; sometimes they are twenty feet in thickness, decreasing until they become the thickness of a knife blade. In exploring for asphalts, these small seams are followed, and often lead to thicker parts. The asphaltum breaks with a conchoidal fracture, some of these conchiform pieces being ten feet in diameter.   The viscous asphaltum cannot enter the interstices or pores of the rocks of the formation; especially is this true where the pores or interstices of the rocks are filled with quarry water, consequently, when it is urged upward and forward by rock, hydrostatic or gas pressure, it exerts a pressure in all directions (similar in action to a hydraulic press), and through this pressure pushes the rocks asunder, making room for itself.

The pressure required is not great when the shale is contracted on one or both sides, the asphaltum being constantly alert to take advantage of any movement of the earth.

Next to these bituminous veins, the formation is of a bluish color, owing to the presence of a slight amount of bitumen.   These veins entered the formation subsequent to its subsidence, as the veins are not faulted; although frequently very tortuous, they are continuous. These veins sometimes bisect the brown bituminized sand, showing that they were injected subsequently to the filling of these sands with bitumen, and even after they had become brown and indurated through long exposure.   These veins of asphaltum frequently contain fossil shells and sharks' teeth, which must have been brought up from lower strata, as no similar fossils occur in the adjoining walls of the vein.   These fossil shells are filled with bitumen, mixed with the silt and sand, that entered them when they died.

The bitumen in the shells must have been much more liquid when it entered them than at the present time.   At the present time the bitumen is too stiff to enter the shells.   Sometimes the bitumen in the shells is much softer than that in the veins, and sometimes it is of a yellow color.   Slickensides, on both sides of these bituminous veins, show that the material has moved upwards.   Fossil shells, embedded in the hanging wall of these veins, have plowed grooves in the asphaltum as it ascended.   At the Waldorf Mine a groove was formed in this manner, ten feet in length.

These veins of asphaltum, in several instances, are found injected into the detritus, which has descended from the adjoining hills. Tunnels and shafts, excavated in the formation containing these veins, have been partly filled with the ascending bitumen.   At the La Patera Mine, in sinking upon some asphaltum, it was discovered that the excavation was being made in an old shaft, which, outside of the earth that had fallen into it, was filled with asphaltum.   At a depth of fifteen feet, an old-fashioned pick and sledge-hammer were found, which must have been buried for a long number of years.

Enquiries were made, but it could not be ascertained who sunk the shaft. At the La Patera, limpid sea water is inclosed in cavities in the asphaltum, its limpidity showing that sea water has no effect upon the asphaltum. In the Santa Barbara Channel, below high-water mark, near the La Patera Mine, veins of this mineral occur. Owing to the plastic nature of this character of asphalt, and the broken condition of the formation in which it usually occurs, it is very difficult to mine. When inclines, tunnels, shafts and other excavations are made in asphalt, or near it, they are hard to maintain, as the asphaltum lying above or near the excavations, even below them, commences to move through rock pressure, so that the excavations are soon destroyed.

Deposits near excavations give evidence of their existence by the earth bulging into the excavation. Wedges, picks and other tools, used in mining the asphaltum, become sharp, instead of dulling.

When mined and relieved from pressure, occluded gas expands in the asphaltum, making it very vesicular.

In Fig. 9, A A portrays unaltered shale, greatly distorted by subsidence, caused by the contraction of the red shale, D D. B B B B, veins of asphaltum squeezed and forced up by rock pressure, exerted by the shale A A, owing to their broken and contorted condition.

C C, shales filled with condensed hydrocarbons, which were vaporized by hydrothermal heat in the red shales.

Near or adjoining the red shales, the shales are filled with liquid asphaltum; in some places the shale and the liquid asphaltum have formed a mud, which is forced outwards and upwards by the weight of the superincumbent shale, through cracks and seams, to the surface of the earth. Further away from the periphery of the red shale, the shale has a smoky appearance, owing to the condensation of different vapors, generated by hydrothermal heat in the red shales, D D. D D, red shales which have contracted through the action of hydrothermal heat, such contraction causing subsidences in the overlying shales, A A—such subsidences opening fissures and cracks, B B B B, permitting the ascent of the bituminous mud, urged upwards by rock pressure, and, maybe, gas and hydrostatic pressure. Underlying the red shales is serpentine, being shales metamorphosed to a greater extent than the red shales. Engraving Fig. 9 on the following page gives a view of a formation into which veins of asphaltum have been injected.

Reservoirs for oil and asphaltum are created as follows: The porosity of limestone is created by chemical action, the changing of limestone into dolomite; the porosity of sandstone, by the solvent

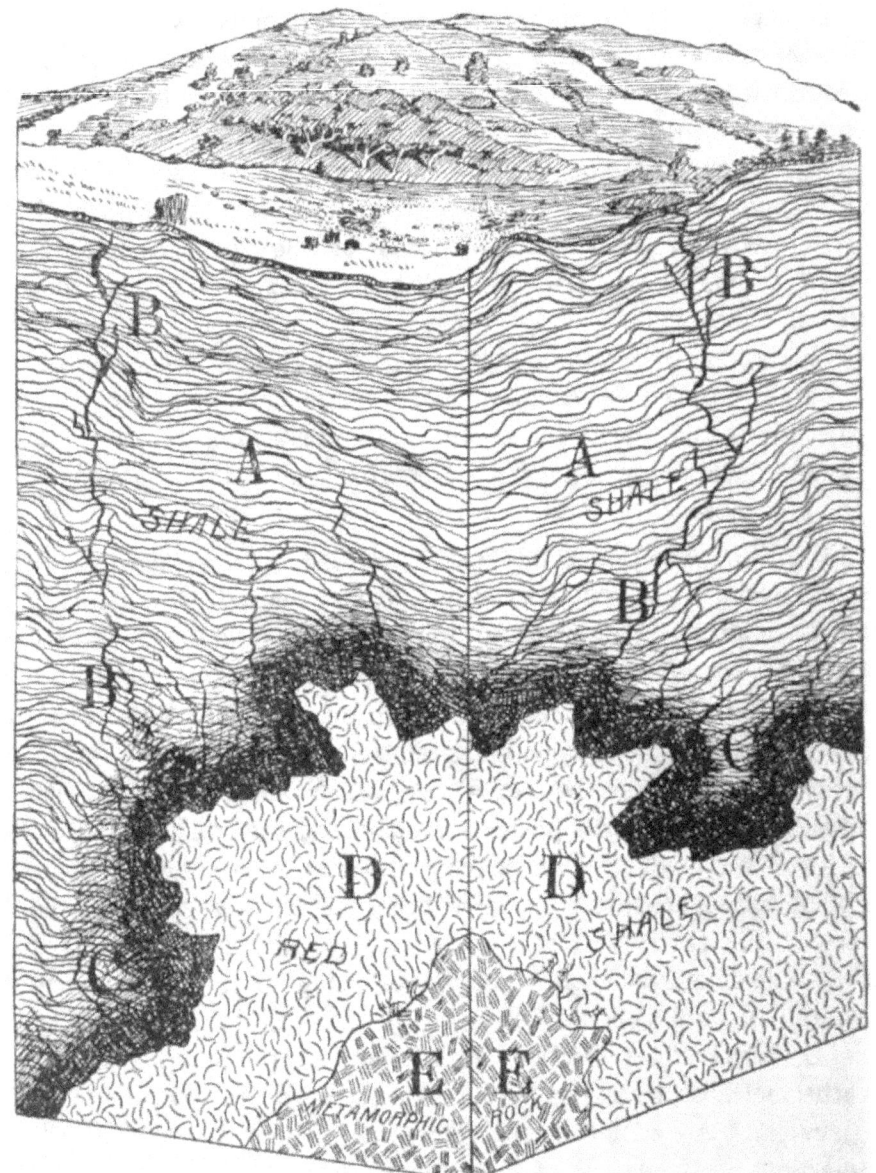

FIG. 9. — PLANO-SECTION SHOWING ASPHALT VEINS.

action of water leaching out the cementing material, such as lime, silica, iron, etc.; the capacity of shale for holding oil, or bitumen, by the mechanical bending and cracking of the strata, the cracks affording storage room.

LIMESTONE.—The Trenton limestone is very productive under certain circumstances. In its normal condition it is a compact rock, and then it contains neither gas nor oil; but, over large areas, limestone has been dolomitized, and so transformed into a porous and cavernous rock, in which the gas and oil are contained.

The dolomitization of the Trenton limestone is probably occasioned by the removal of carbonate of lime by the solvent action of water, charged with certain minerals, and as the Trenton limestone contains originally a small percentage of magnesia, it gradually becomes dolomitic in character, and, on account of its reduced bulk and crystallization, porous and cavernous.

When water has taken possession of shale in the shape of quarry water, or the shale is saturated with water, it is nearly impossible for oil to eject the water and enter the shale; the reverse is also true, for, when oil has taken possession of the shale, it is nearly impossible for water to enter the shale. This is undoubtedly owing to capillary attraction of the fluids in the shale. When the surface of a capillary tube is greased, it exerts but little capillary attraction upon the water, and when a capillary tube is moistened, it exerts but little capillary attraction on oil. Other rocks act the same as shale; the finer the grain of the rock the greater the capillary attraction, and the more difficult it will be for oil to replace water, or for water to replace oil.

Shale, occupied by water, makes a good encasement for oil and asphaltum.

The retention of petroleum and pissasphalt, in the porous and seamed rocks, cannot be effected without the accumulations or reservoirs having a cover or impervious encasement. This impervious encasement usually consists of unfractured shale, or other close-textured rocks, or porous and fractured rocks, cemented and sealed with indurated bitumen or other minerals.

When the outcrop of bituminized sand is exposed to the atmosphere for a long time, the bitumen contained in it loses its volatile parts by evaporation and oxidation, turns brown, and is easily pulverized between the fingers. The sand separates from the bitumen, and the bitumen is easily ground to an impalpable powder. This brown asphaltum extends to but a short distance below the surface of the stratum. Beneath the brown coating of bituminized sand, the deposits receive a coating of hard asphaltum, made hard by evaporation and oxidation. The condition of the bitumen in the seamed and cracked shale resembles that which is in the sand. This concreted surface is impervious to the flow of pissasphalt and petroleum oil, and frequently sufficiently tight as to inclose natural gas. In fact, all porous or seamed rocks, when the base of the saturating petroleum is asphaltum, becomes water and petroleum-tight, by reason of the petroleum becoming concreted by oxidation and evaporation; the same as when a tree is wounded by a cut or puncture, the impassated sap soon closes the pores, so that little sap escapes. The surface of the bituminized sands and shales is hard, increasing in fluidity as it enters the deposit, or is removed from atmospheric influences.

In some deposits, a short distance from the surface will show a petroleum oil of 10° Beaume, decreasing at 1,000 feet to 32° Beaume.

If rather stiff maltha is melted and poured into a hole in a sheet of iron one-sixteenth of an inch in diameter, so that it will form a thickness one-sixteenth of an inch on each side of the sheet, the sheet being one-sixteenth of an inch thick, it cannot be removed with a pressure of water equal to fifty pounds to the square inch. The prodigious pressure necessary to force maltha through the interspaces of sand, or irregular seams of shales, for a distance of several hundred feet, can hardly be imagined. In fact, the salvation of most of the accumulations of petroleum oil in California is owing to this induration of petroleum by oxidation and evaporation.

This impervious coating also protects the oil reservoirs from the entrance of surface water.

The petroleum oil, in passing through the sand or shale, collects the silt and carries it forward. This, also, assists in forming a cover, filling up the places through which the liquid hydrocarbons attempt to escape.

Under great pressure, the petroleum oils are constantly alert to take possession of any space created by the uplifting, or other movements, of the earth; and if these oils are concreted into asphaltum, by oxidation and evaporation, they retain possession. The oil, occupying the surface of the water in a formation, has an advantageous position to perform this work, as the formation is more fractured in these parts than in the synclines and the dips of the anticlines.

The cut (Fig. 10) shows strata of bituminized sand on a ridge running north and south, between the Coja Creek and a branch of the Baldwin Creek, seven miles west of Santa Cruz, California. The bituminized sand lies nearly horizontal, and extends from canyon to canyon, through the ridge. The dip of the shales and sandstones of the surrounding country shows that this is the apex of a large dome, covering an area of some twelve square miles. Overlying these strata of bituminized sand is a close-textured shale, forty feet thick, and underlying the same is a porous and incoherent sand.

Fig. 10.—C. S. I. Co.'s Mine, Santa Cruz County, California.

The Coja Creek, lying immediately west of this deposit, is 200 feet deep, and must have taken many thousands of years to form. Redwood trees, proving by their concentric circles to be several hundred years of age, are growing in the bottom of the creek. The impregnation of the sand with bitumen must have occurred before the gulch on either side of the deposit commenced to form through denudation, otherwise the liquid bitumen would have run out of the porous sand. From the horizontality of the surface of the porous sand, which underlies the bituminous strata, it must have been filled with water, forming a horizontal plane, upon which the bitumen floated. This, also, must have occurred before the denudation of the gulch on either side. The petroleum must have been under considerable pressure, as it has thoroughly saturated the sand between the porous sand and the overlying shale, and, where porous places have existed in the shale, petroleum has been forced into them.

Notwithstanding the thousands of years which this bituminized stratum has been exposed to the elements, the bitumen in the interior parts of the deposit is at present liquid, its liquidity being preserved by the concretion of the bitumen on the top, bottom, and sides of the deposit, stopping evaporation, oxidation, and leakage. These strata of sand, when bituminized, contain gold in considerable quantity, whereas, in those portions which are not bituminized, but little gold exists.

It would seem that the gold in the bituminized sand was protected from the solvent action of mineral water, the presence of the bitumen in the sand stopping the percolation of mineral water; whereas, in the sands not bituminized, this percolation is permitted, and the gold is dissolved and carried away. But further examination will be required to be positive that such is the case.

Sand rock, sand, and sandstone are composed mainly of rounded or broken grains of quartz of varying form, color, and fineness. The material cementing the grains is either argillaceous, bituminous, silicious, or calcareous, or a mixture of any of these four substances. Some of these sands contain cementing material, some in such a small quantity that they are friable. When found beneath the earth's surface, they are seldom found in an incoherent state. When they are porous, they are occupied by either natural gas, petroleum oil, or water (generally of a mineral character), or both of these fluids and gas.

The buoyancy of the oil in associated water is the force which impels the oil upwards. The oil is carried so far upwards that it sometimes escapes at the surface in the form of tar springs, or seepages, and is lost, or it accumulates in porous or seamed strata beneath the surface of the earth. For the reception of the petroleum, oil, or gas, the sandstone strata are at first made porous by the solvent power of water, which removes the calcareous and silicious cementing material.

Frequently these bituminized sands are jointed, and, although this adhesive and plastic material, when excavated and thrown into a pile, will stick together so that it has to be again mined, it will separate readily at these joints when being taken from the deposit. The joints are often filled with mineral matter, such as carbonate of lime, deposited by circulation of waters subsequent to the bituminization of the sands. (Fig. 11.)

These joints are probably partly due to pressure, as they seem to have a trend, which appears to be at right angles with the line of the steepest inclination upon which bituminous deposits rest, but always nearly perpendicular to the plane of bedding. Some deposits, when jointed in this manner, appear like a row of books upon a shelf.

Fig. 12 shows bituminous strata on the Sisquoc River, in Santa Barbara County. By the movement of the formation, the joints, filled with lime, have been distorted so as to nearly form the letter "S," showing that there has been a considerable movement since the bitu-minization and jointing of the sand.

FIG. 11.—BITUMINIZED SAND SHOWING JOINTS.

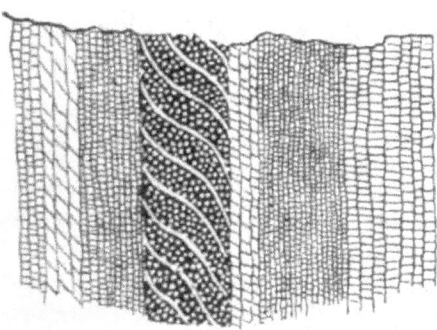

FIG. 12. — BITUMINIZED SAND SHOWING
CONTORTED JOINTS.

The strata are often faulted a few inches at these joints. In the folding of a formation, the shales will be contorted, and the bituminized sand faulted at its joints. The bendings of the bituminized sand are never very small and acute, whereas, in shale they are small and acute.

This makes a formation often appear unconformable, but it is the nonconformity made by bending and faulting, and not of deposition.

The argillaceous material cannot be removed like the calcareous and silicious cementing matter; consequently, sands cemented with argillaceous material seldom contain bituminous accumulations to any great extent. When the calcareous and silicious cementing material is removed by circulating water, the oil, if present, occupies the sand. This leaching generally occurs along the line of faults and the summits of anticlines, as they more readily offer avenues for the egress of water, but may occur along anticlinal dips and in synclines; in fact, in any place where the intricate subterranean course of percolating water reaches the sandstone. When the circulating water has been copious, the calcareous and silicious cementing material has been carried away to the surface of the earth, sometimes forming deposits of sinter, tufa, limestone or dolomite. When the flow of these waters, charged with carbonate of lime and silica, has been feeble, the sandstone has again been cemented and made impervious to water, then the flow of water ceases, or finds some other path of escape.

Deposits of petroleum oil, resembling pools, occur in the sandstone stratum in which the circumjacent sandstone is cemented with calcareous and silicious material. The portion of the stratum now occupied by the bitumen was formerly occupied by the carbonate of lime or silica, the lime or silica having been removed by the solvent action of percolating and circulating waters. If the gas and oil are ever removed from these porous strata, they will, in all probability, be again cemented by lime or silica, if water again takes possession, and its flow is feeble.

Shale is laminated clay, more or less indurated, splitting into thin sheets along the original laminae of deposition. In California, the majority of shales are quite soft, being easily cut with a knife.

A large proportion of the oil obtained in California is taken from the cracks and recesses in shale. The strata are arranged around the axis of the anticline in concentric circles. During distortion, occasioned by the uplifting of the strata, there would be an elongation of these concentric strata. If they consisted of non-elastic shale, they would be cracked and seamed transversely to the seams of their bedding. This would occur to a greater extent in the strata farthest from the axis of the anticline. If there was a great weight of superincumbent earth, the cracks and seams would not be so large along the places of bedding of the shale. Being more acutely bent, the strata on the steep side of an unsymmetrical flexure would be more broken than the other side.

During its uplifting, the strata on the slopes the farthest away from the axis of the anticline would move slower than those nearer the axis; consequently, one stratum would move upon the other, grinding the shale into plastic mud, and luting seams and cracks, . which would assist in forming an incasement for water and oil. This grinding movement also keeps cracks and seams from occurring parallel with and along the planes of the bedding of the shale. When the shales are bituminized in their cracks and seams, one-tenth of the bitumen is in the seams that occur parallel with the planes of bedding, and nine-tenths in the cracks and seams that occur transverse to these planes. Where the shales, the cracks and seams of which are filled with bitumen, which resists the action of denudation better than the shales that do not contain it, are side by side, the latter are worn away more largely than the former, and a valley results, owing to denudation acting unequally. Many bluffs, and prominent peaks and ridges, owe their existence and stability to the bitumen in the cracks and seams of their shales.

When water takes possession of shale, the capillary attraction offers so great a resistance that oil, even under enormous pressure, is incapable of forcing an entrance and ejecting the water. When oil is in possession of the shale, the same resistance is offered to the entrance of water. Shale, saturated with water, is a far better covering for petroleum than dry shale. Where the distortion of the strata has been acute, the multiplicity of these cracks and seams makes the storing capacity of these fractured shales very large — in fact, many will be equal in this respect to porous sands. For reasons stated heretofore, the cracks and seams will be wider on the summits of anticlines than on their slopes or in synclines. Owing to the broken condition of the shale, the petroleum has ascended from strata to strata, and not for any great distance through any particular stratum or strata, until the concretion of the bitumen, by exposure to atmospheric actions. assisted by silicious waters, sealed strata from each other.

SILICA AND LIME.—The hot waters, created by chemical heat, held in solution large quantities of silica. When the hot silicious water approached the surface of the earth, it was cooled; the cooler the water became, the less capable it was to hold in solution this large amount of silica. Cooling of the water eliminated the silica, which was deposited in the interstices of the shales, increasing their solidity and imperviousness. There may also be an interchange between the silica dissolved in the water and certain constituents in the sandstone and shale — for instance, carbonate of lime — the silica, having a greater degree of hardness than the substance removed, would be deposited. This silicification is a very frequent phenomenon in these rocks.

If the silicious water circulated through particular strata, the silica eliminated by the cooling of the silicious water, by its superior gravity, sank to the bottom of the strata and cemented the same. Strata, adjoining the hot, silicious water, may be cooler than those in which the silicious water circulates; in that case, the cooler strata act as condensers. This action created the strata known as "shells"; the shells often occur on the top and bottom of the sands, and throughout the shales. This silicification, together with the concreting of the bitumen, creates impervious strata, which is capable of holding petroleum, oil, and natural gas. These silicifications, when formed in strata in which bitumen occurs, are colored black. Their color is destroyed by burning, proving that it is owing to organic material.

Silica, in a very fine condition, is frequently attracted to some organic or inorganic nucleus which has grown in successive layers or bands, often of different colors. In a similar manner the small silicious particles, separated from hot, silicious solutions, are attracted by the incasements of porous, or partly porous, and seamed strata.

Lime is controlled usually by the same conditions and laws as the silica. Large masses of shale and sandstones have been calcified and salicified near the bituminous deposits in California.

Deposits of bitumen and petroleum oil are controlled by the line of permanent water. Below the level of the ocean, all cracks, seams, fissures, and interspaces are permanently filled with water. Above the level of the ocean, and below the beds of the streams, the supply of water in these spaces is fairly permanent. Above the beds of the streams the supply of water is dependent upon the rainfall, and the degree of freedom with which it leaves the formation.

Pervious and impervious strata modify the above conditions. Permanent water may be replaced or occupied by a deposit or column of bitumen, petroleum, oil, or natural gas. When the underlying water, which supports the oil, is released by the uplifting of a formation above the permanent water by orogenic movements, the water leaves the formation, and the oil drains into the voids formerly occupied by the water, and, possibly, reaches the surface by the same avenues taken by the water, and is lost or resinified to asphaltum. Frequently the porous strata accumulations of bitumen have resulted through the drainage of oil from a higher and large area of porous rock into a lower porous stratum.

The bituminized sands on the Mesa deposit, on the Sisquoc Ranch (Fig. 13), show the drainage action. They are drained into a centrocline, the sides of the centrocline afterwards having been carried away by denudation; the water which buoyed up the oil escaped, and the oil slowly sought the bottom of the centrocline. The sands forming the uppermost edge of the centrocline have, to a great extent, lost their bitumen; as soon as the bitumen leaves the sand, it falls into an incoherent mass, and is rapidly washed away by rains. The existing bitumen, in consequence of the long distance traveled, has become viscous, principally through oxidation and evaporation.

This drainage is still slowly progressing, as in hot weather balls of nearly pure bitumen form on the surface of the sands in the lower part of the deposit.

South of Asphalto, in Kern County, the bitumen has reached the bottom of the

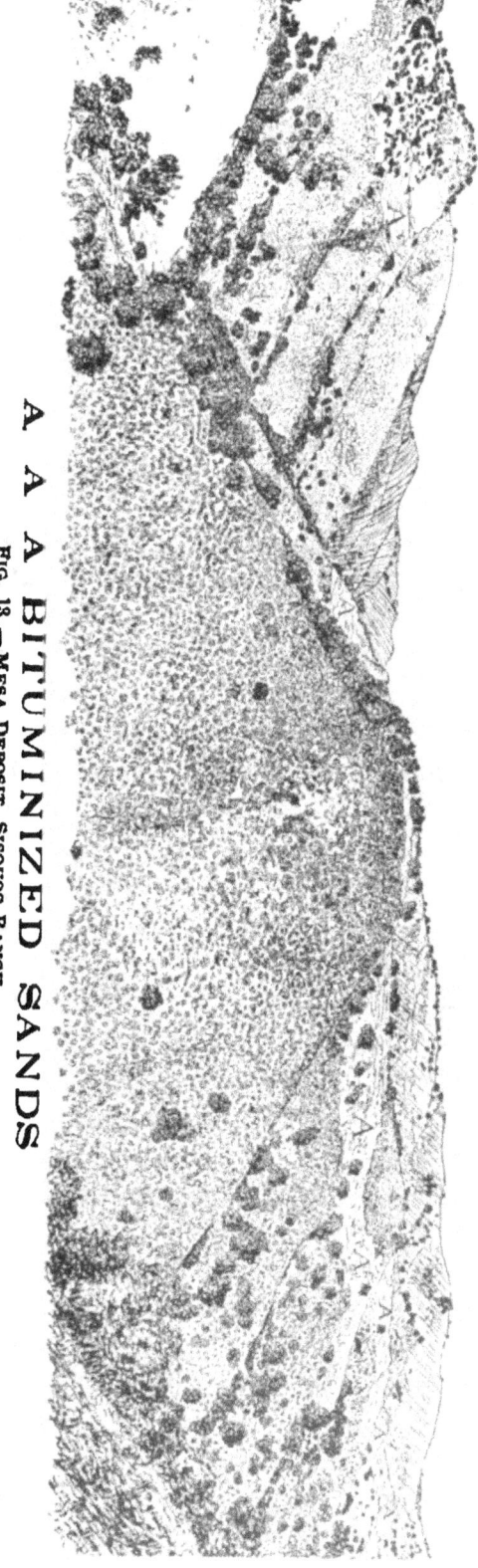

FIG. 18.—MESA DEPOSIT, SISQUOC RANCH. A A A BITUMINIZED SANDS

syncline by drainage, and is now gradually descending through the syncline, as through a ditch. Although this occurred ages ago, this drainage is still in process.

What evidence is there that water circulates and circulated through these formations? The presence of calcareous tufa at the outcrops of the strata; the presence of bitumen in the cracks and seams of the shales, and in the interspaces of the sands floated up by associated waters—if the bitumen was forced up exclusively by gas, it would not have been so evenly and generally disseminated throughout the shales and sands; springs of water, accompanied by natural gas and oil, issuing from the outcrop of strata, which dip towards the source of the water, and the absence of these springs from the outcrop of strata, which dip away from the source of the water; the seams of hard shale, called "shells," silicated by infiltrating silicious waters; nothing but molds and casts of fossil shells, the carbonate of lime having been removed in solution in water. Near the outcrops of these strata, in the creek and river beds, the sands and gravels are cemented together with carbonate of lime, forming conglomerates.

What is the effect on deposits and accumulations of bitumen by circulating waters, fresh and mineral? Petroleum oil, when exposed for a long time to water containing sulphuretted hydrodgen, is resinified by being sulphurized; especially is this true where the sulphur is liberated by decrease of pressure, or by the oxidation of the hydrogen. The sulphurizing of petroleum oils by sulphuretted hydrogen is a chemical combination, but, if only a mechanical combination, it is so intimate as to resemble a chemical combination.

Petroleum oil will be oxidized when exposed for a long time to water containing oxygen, or atmospheric air. Oxidation converts the oil into petrolene, and greatly increases its gravity. As slight quantities of petroleum oil are dissolved in fresh water, the lighter parts are dissolved; consequently, the effect of circulating fresh water is to carry away the lighter parts of the oil, leaving the heavier parts behind. Water, saturated with salts, has but little effect on petroleum oil. Sometimes the salts, through the agency of water, are mechanically and intimately mixed with the bitumen, so as to render their separation difficult.

Besides these chemical effects of water on the bitumen, water exerts the following described hydrostatic and hydraulic effects: In cases where strata are rendered leaky by denudation, and water in ascending to the surface through the same, the petroleum oil is floated out on the surface of the water and lost. Meteoric water, which falls on higher ground, penetrates the earth, sometimes to great depths, through inclined and porous strata, or through fissures, cracks, seams, and joints of rocks; and, after flowing a distance, sometimes the distance being very great, it must ascend through permeable strata to the surface, or, hidden, find its subterraneous way to the sea. The course of water flowing underground is not strictly analogous to that of a river on the surface, there being, in one case, a constant descent from a higher to a lower level, from the source of the stream to the sea; whereas, in the other, the water may at one time sink below the level of the ocean, and afterwards rise high above it, by hydrostatic pressure, due to the superior level at which the rain-water was received, and the encasement of permeable strata by impervious strata.

It must be borne in mind that the circulation of water through the rocks can be extremely slow. On account of the broken condition of the rock on the anticlines, caused by the acute curvature of the anticlinal arch, and its greater exposure to denuding agents, far greater quantities of rock have yielded to erosion than in the syncline, where the rocks have been hardened by lateral pressure, and cementing by infiltration of mineral waters. This broken condition of the backs of the anticlines permits water to enter them more freely than the synclines or the slopes of the anticlines, and also permits more readily the escape of petroleum, oil and natural gas. The accumulation of petroleum oil must have commenced when these rocks but slightly undulated. The former and ancient features and state of a formation should be taken into consideration, as well as those existing at the present time, in the examination of an oil field. If there is no opposing force or intervening obstacle, gas, petroleum, oil and water distribute themselves in porous or seamed strata, in accordance with the difference of their gravity. The gas lies above the petroleum oil, and the oil floats upon the water. It must be remembered that, under ordinary pressure, oil and water do not mix, and that the gravity of the petroleum oil is less than that of water, but for which little would have been seen on the face of the earth. If petroleum oil is introduced into the bottom of a vessel filled with water, it will rise to the top of the water, and if water is placed on the surface of oil, it will sink to the bottom of the oil.

The pressure of water is exerted below the petroleum oil, and the pressure of natural gas above. Porous strata, incased in nearly im-

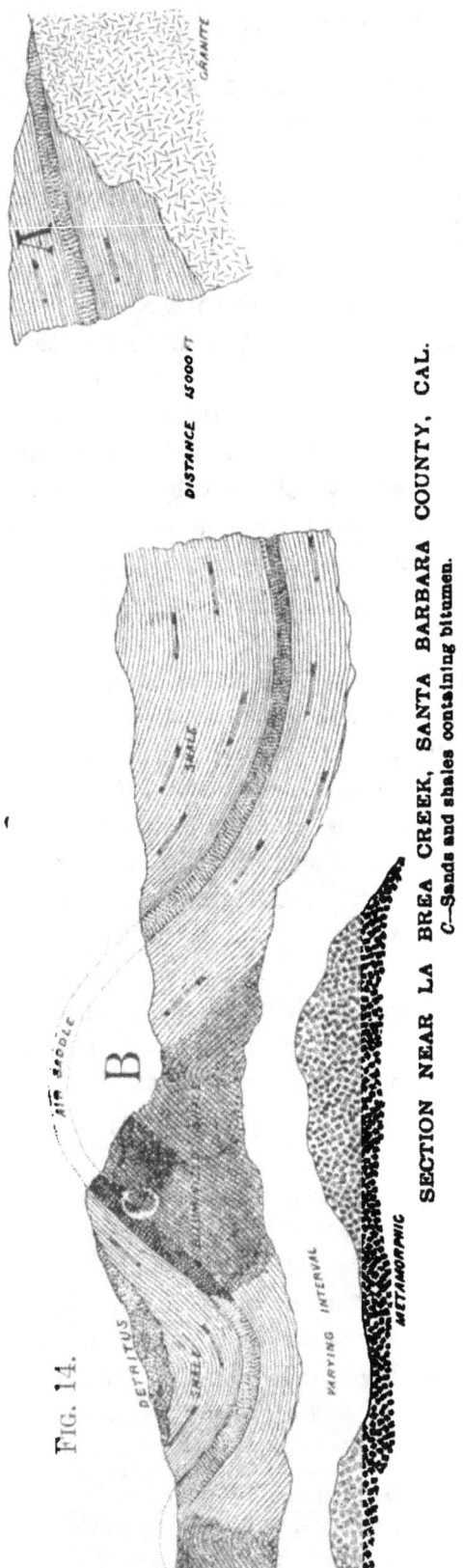

FIG. 14.

SECTION NEAR LA BREA CREEK, SANTA BARBARA COUNTY, CAL.

C—Sands and shales containing bitumen.

pervious strata, must be considered as conduits for the fluids, petroleum, oil and water. When petroleum, oil and water flow in the same direction, the porous and seamed strata are liable to be barren of oil. Where the oil and water flow in opposite directions, the porous or seamed strata are liable to be fruitful with oil.

Many years ago, before the denudation occurred, these strata were incased in an impervious cover; the right-hand dip of the anticline was towards the mountains; meteoric water, falling on the mountains, entered the porous strata at A (Fig. 14), and finally flowed up the right-hand dip of the anticline, B, as shown by the arrows, being impelled forward by hydrostatic pressure in the higher levels of the mountains, and the associated oil which entered these strata by its inferior gravity, when compared with water, flowed in the same direction as the water; therefore, this dip of the anticline is barren of bitumen. In the left-hand dip of the anticline, B, the water flowed downward, impelled by gravity, and, probably, hydrostatic pressure, seeking an exit, as springs in the valleys, or, unseen, found its way to the sea; the oil, by its inferior gravity, was buoyed up by the water, or ascended in a contrary way to the flow of the water; consequently, the pores and seams in this dip are filled with bitumen. These same conditions exist in the Zaca anticline, Las Pozitas anticline, lying west of Santa Barbara, and at Summerland and many other places.

The following is an exception to the accumulation of oil in the dip, remote from the source of circulating water. Petroleum oil is nearly always accompanied by natural gas. If the porous or seamed strata, serving for conduits for water or oil, have vertical curves or summits, and such summits are sufficiently tight as to hold the gas, in time the gas will accumulate in such summits and occupy a considerable part of the sectional area, and it will continue to accumulate until the velocity of the water, or oil, is sufficient to carry the gas forward, and down the incline. If the pressure never reaches such a point as to effect the removal of the gas, the flow of the water and oil will be more or less obstructed; and, finally, if the gas is not removed, or escapes, causes the flow of water or oil to cease. In this case, oil will be found in the dip of the anticline, which is towards the head, from which the water emanates.

Fig. 15 (see next page) shows alternating beds of different sands, all of which are bituminized.

Strata CCC are formed of coarse quartz sand, containing round pebbles of hard rocks, such as quartz. They do not contain many fossils.

Strata BBB are formed of fine, muddy sand, containing lenticular pebbles of shale. These pebbles have been silicated, forming chert, but their surfaces still retain the appearance of shale, and they have the lamination of shale.

This silification must have occurred before these sands were bituminized. This silification was, probably, effected by the infiltration of hot silicious water. The creation of this water was, probably, through the agency of metamorphism, which preceded the distillation of petroleum from carbonaceous matter by the heat of metamorphism.

In the figure, A represents fine shale with few fossil shells; in places it contains fossil fish-bones. The cracks and joints at right angles with the planes of bedding are frequently filled with bitumen.

No cracks at right angles with the plane of bedding could have existed in these shales until they were contorted; therefore, these cracks were bituminized after the folding of the formation had commenced.   Neither could bitumen have ascended through these shales before they were cracked or jointed, for when these fine shales contain quarry water, they are impervious to oil.

These different strata, BBB and CCC, are conformable, and do not pass into one another by gradation; the lines between them are clearly marked.   The muddy, fine sand does not weather as rapidly as the coarse sands; consequently, their faces of exposure are nearer verticality than the coarse sand.

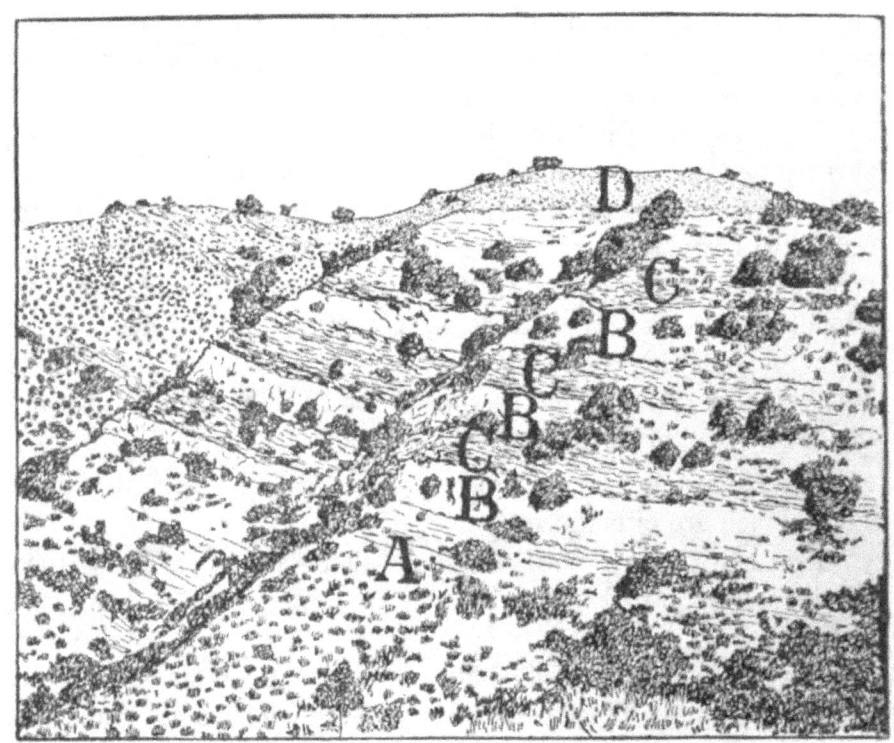

FIG. 15. — ALTERNATING BEDS OF DIFFERENT SANDSTONES.

The different strata must have been derived from two formations, one being composed of altered rocks, and the other of unaltered rocks, and the changes in the derivation of the sediments composing this formation, from the altered to the unaltered rocks, were quickly made, either by a change in ocean currents or the sudden uplifting of the land.

These strata of sand, described above, are situated on the Tinaquaic Rancho, in Santa Barbara County.   The character of the bituminized sands near Santa Cruz is very different.   They contain few round pebbles of hard rock, and no lenticular pebbles of shale.   They are clean quartz sands, of varying fineness, and were, probably, formed by the disintegration of granite. They do not contain many fossils, and contain gold in notable quantities.

That sudden upliftings did occur, is shown by the terrace structure at the Santa Cruz bituminous deposits, there being three terraces clearly defined. The uprise, or face, of the first and second terraces was partly cut from the bituminous sands, showing that the sands were contorted and bituminized before these terraces were made by the sea.

FIG. 16. — BITUMINOUS SAND DIPPING TOWARDS THE SEA.

When these bituminized strata rose above the line of permanent water, the flow of the bitumen was, and is at the present, down the dip of the strata towards the sea, and the sands in the upper part of the dome, which were vacated by the bitumen, are calcified, the lime probably being derived from overlying calcareous beds which have been removed at this place, but which exist further towards the east.

The following is a description of the only explored submarine oil fields in the world:

In Santa Barbara County, California, the Summerland oil-bearing strata consist of a fine-grained sand, encased in strata of clay or clay slate.   When near the oil-bearing strata, the clay, or slate, is of a bluish color, owing to its being slightly impregnated with bitumen. The sands and shales form an elongated dome, the longest axis of the dome being from east to west, running nearly parallel with the coast and parallel with the trend of the Santa Ynez Mountains. Near the eastern end of the dome the formation dips about S. 20° E., at an angle of about 50°, whereas, on the western end, the sand dips about S. 10° W., at an angle of about 60°. The general dip of the sand is southerly, at an angle of about 40° or 50°.

The first discovery of the hydrocarbons in this field was made on the south slope of the anticline.   At this place there was a fumarole, from which warm carburetted and sulphuretted hydrogen gas escaped.   No vegetation grew on this place, owing to the sulphur fumes; it was some twenty feet in diameter.   The Spaniards had a legend that a man was killed there, which, according to them, accounted for the fact that nothing grew upon it.   A pipe was sunk in the fumarole and capped, and a two-inch pipe inserted in the cap, when the gas was permitted to flow through the pipe.   It did so with considerable pressure, and, when lighted, gave a flame ten feet in length.   On a line nearly east and west with this fumarole, other wells have been bored, which have yielded gas.   The gas has been employed for domestic purposes.

South of the gas wells, on a nearly east and west line, are a line of oil wells; they are from 130 to 250 feet deep.   The first oil obtained in the field was from a well dug ninety feet in depth, which produced three or four barrels daily.   The oil is black or dark green, and is of a very heavy gravity, being 11° to 16° Beaume.

Judging from other oil fields, the northern dip of the sands of this anticline will be barren. Meteoric water, falling on the higher ground of the Santa Ynez Mountains, which at places reach an altitude of 3,600 feet, penetrates the earth through inclined and porous strata, or through fissures, cracks, seams, and joints.   After flowing through subterranean passages, it must ascend through permeable strata to the surface, or, hidden, find its way to the sea. As the Summerland anticline forms a barrier between the Santa Ynez Mountains and the sea to the passage of the water, it is forced by hydrostatic pressure to ascend through the north dip of the sand of this anticline.   Owing to its inferior gravity, the petroleum oil is floated upwards by the water, and is lost on the surface of the earth, or is carried over to the south dip of this anticline.   With the south-

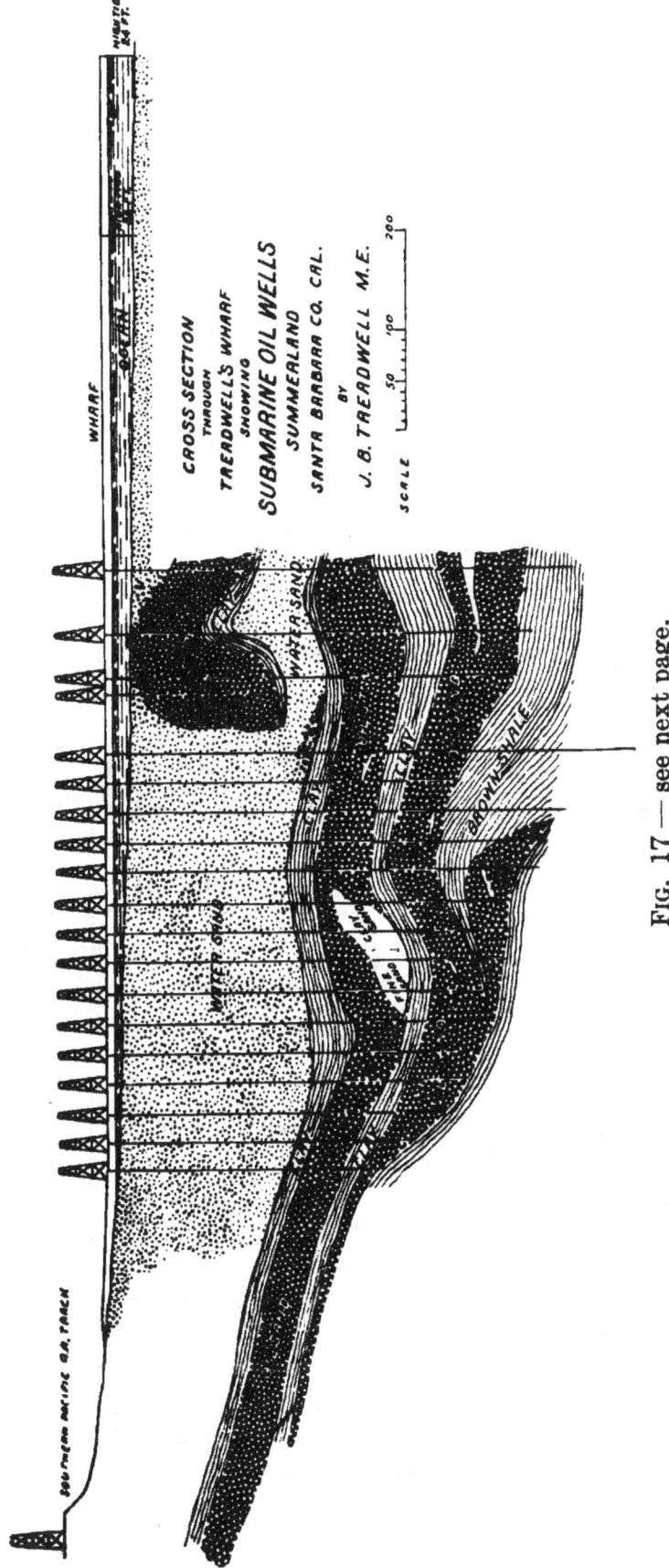

CROSS SECTION
THROUGH
TREADWELL'S WHARF
SHOWING
SUBMARINE OIL WELLS
SUMMERLAND
SANTA BARBARA CO. CAL.
BY
J. B. TREADWELL M.E.

SCALE

FIG. 17 — see next page.

ern dip of this anticline it is different; the flow of water is downwards, the oil remaining on top of the water by its buoyancy.

Owing to the large amount of organic matter in the shales underlying the Summerland oil field, if any iron was present during their deposition, it must have been in the form of ferrous carbonate. The carbonate of iron imparts a bluish or greenish color to the deposit. When the shales, in which carbonate of iron exist, are turned red, it is caused by chemical heat. The presence of red shales below the Summerland oil strata, as is shown by a well drilled to the depth of 1,000 feet, and the high temperature of the natural gas, shows that chemical changes are in active operation at present beneath this field. It is probable that sulphur compounds, liberated by chemical heat in the shales, have resinified the petroleum oils of Summerland, which will account for their great gravity.

A wharf has been extended into the sea towards the south, and at nearly right angles with the trend of the shore. From this, productive wells are drilled into the bottom of the ocean, yielding a petroleum oil somewhat lighter than the numerous wells upon the shore.

Fig. 17 is a profile made by J. B. Treadwell, M. E., showing the character and structure of the rocks encountered, and the number of wells which have been drilled.

The illustration Fig. 18 shows a formation lying east of Zaca Creek, in the county of Santa Barbara, California. A A A is bituminized sand, forming the south dip of an anticline; underlying the bituminous sand are bleached shales, B B B, and below the shales are metamorphic rocks, quartzite and serpentine. At one point the metamorphic rock has closely approached the bituminized sand. At this point the bitumen has been removed from the sand, and the sand is calcified and silicified. From the attending phenomena, it would seem that these sands were bituminized before metamorphism reached them.

This sandstone, when not reduced by erosion, is nearly 300 feet in thickness, and, like a mantel, covers a large part of the territory lying between the Santa Maria and Santa Ynez rivers, the Pacific Ocean and the Alamo Pintado Creek, in Santa Barbara County, some 600 square miles. In the summit of the domes, and in the dip of the anticlines which are farthest away from the mountains, which are higher than the summits of these anticlines, the sands are sometimes bituminized. Extending toward the north-west, from the place shown

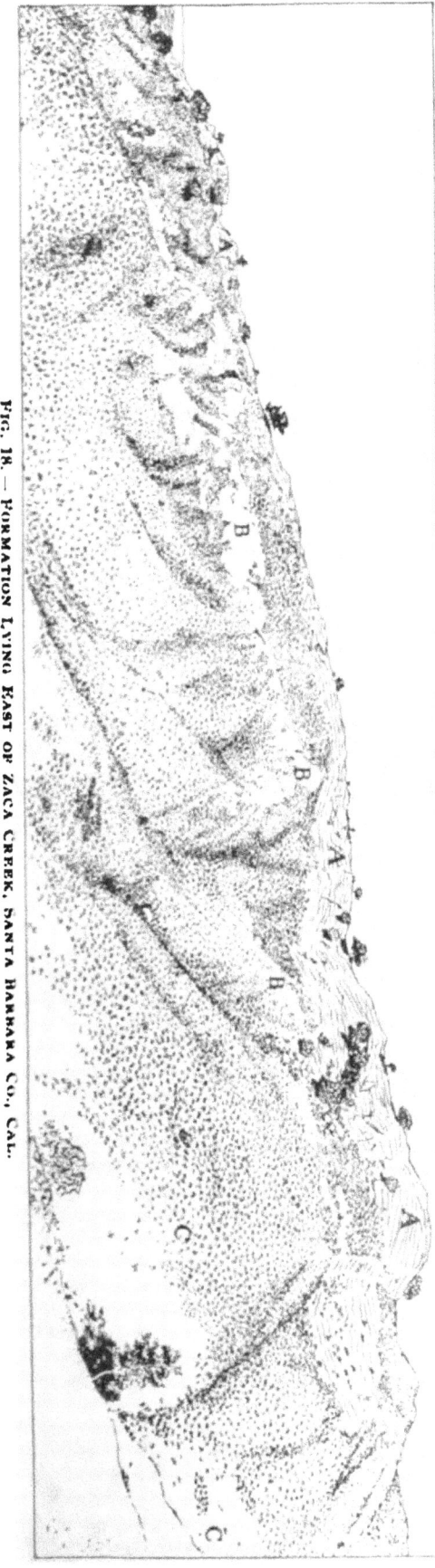

FIG. 18.— FORMATION LYING EAST OF ZACA CREEK, SANTA BARBARA CO., CAL.

in the cut, are sands containing millions of tons of bitumen. These bituminized sands are very prominent, forming high bluffs. Throughout the area, which this upper sand covers, there are places where these sands are silicified, in others calcified.

On the south slope of the Santa Maria Valley, and in some other places, these sands are uncemented, and have been formed into sandhills through the shifting of the sands by water and the winds. Also, scattered throughout this area, there are a number of sulphur blows, mineral springs, and places where natural gas escapes in large quantities. On the slopes of Mount Salamon, asphaltum is injected into the formation surrounding subsidences. Veins of asphaltum also occur on the Jonata Rancho. White, leached shales, and shales burned to various tints of red, occur in large masses. The shales and sandstones in this area are fairly conformable.

On the Sisquoc River there is another exposed sand which is bituminized, and which is, geologically, about 1,000 feet lower than this upper sand, and is separated from it by a bed of shale. How thick this lower sand is cannot be positively determined, but, judging from its exposures, it must be over 300 feet thick.   If this sand covers the same area as the upper sand, which it is reasonable to suppose it does, there is no reason why it should not contain in its domes, and the dips of its anticlines, millions of tons of bitumen, which, on account of being excluded from the atmosphere, should be in the form of petroleum oil, and which could be obtained by the sinking of wells.

If it be true that the bitumens are derived from terrestrial and marine vegetation, deposited in sedimentary strata, and then changed to carbonaceous matter, which was afterwards distilled by the heat of metamorphism, then we may expect to find petroleum oil or other bitumens in unaltered rocks lying above the metamorphic rocks, irrespective of the age of the unaltered rocks.

A number of facts that have been presented in the preceding pages tend to prove that this is the origin of bituminous accumulations in California.

A conclusive determination of the origin of the bitumens is of great importance, for if the origin is as set forth in this monograph, explorations can be continued to such depths so as to reach the metamorphic rock, and these explorations may be successful, especially so if the bitumens are found near the surface; but if the bitumens are indigenous to the rocks in which they are found, the depth to which they may extend is uncertain.

Fineness cf California Gold.

BY

FRANK A. LEACH,

SUPT. U. S. MINT AT SAN FRANCISCO.

O Mr. Chas. G. Yale, statistician of the Mint, I am indebted for valuable assistance in complying with the request of Mr. Benjamin that the Superintendent of the U. S. Mint should contribute a paper on the fineness of the gold product of California by localities.

The records of the U. S. Mint, at San Francisco, give some interesting facts relative to this subject. The following table is compiled from the averages of fineness of deposits made from the various mining counties of the State, covering a period of several months, embracing both quartz and placer gold.

	FINENESS.	VALUE.
Nevada County	855	$17 67
Siskiyou	852	17 61
Kern	754	15 58
El Dorado	868	17 94
Placer	792	16 37
Tuolumne	804	16 62
San Diego	803	16 59
Shasta	845	17 46
San Bernardino	705	14 57
Inyo	770	15 91
Plumas	851	17 59
Calaveras	835	17 26
Tehama	882	18 23
Amador	836	17 28
Sierra	858	17 73
Mariposa	805	16 64

	FINENESS.	VALUE.
Trinity	850	17 57
Yuba	881	18 2!
Los Angeles	789	16 31
Humboldt	688	14 22
Stanislaus	895	18 50
Butte	878	18 14
Fresno	805	16 64
Sacramento	898	18 56
Lassen	890	18 39
Mono	550	11 36
Merced	813	16 80
Madera	847	17 50

The above table credits Sacramento and Stanislaus counties with producing gold of the highest fineness. The majority of deposits from these counties, in fact, over four-fifths of the gold brought to the Mint from these locations, run over 900 fine. This gold is almost entirely from placer sources. It is but fair to say that the average for Mono County was derived from a small number of deposits, and that an average covering a longer period might raise the fineness quoted.

The highest grade of gold received at the Mint, for several years, was from San Bernardino County, in a deposit of placer gold. It returned a fineness of 980, or a value of $20.25 per ounce. Tuolumne County has the record of yielding even a higher grade of precious metal than this from San Bernardino. It came from a quartz mine, now said to be idle, which produced gold almost pure, running, at times, over 990.

The first attempt to compile a table of fineness of California gold, was made by A. P. Molitor, an assayer of San Francisco, who published his table in 1859. Some years later, J. S. Hittell compiled a table by mines and counties, dividing it into placer and quartz, and this was published in the State Mineralogist's Report, of 1884. Since Mr. Hittell's figures were published, no one has compiled a similar table until now.

It is worthy of note that, while the mines in the same locality usually have about the same average fineness of gold, there is sometimes a great variation; and often in one mine the gold will vary in value, to the extent of a dollar an ounce, in some cases.

There were thirty-one counties which produced gold in California in 1898. In the first table in this article, twenty-eight counties are named. Making an average of the fineness and value of gold from the counties named in the table, it is seen that the average fineness of California gold is 817.8, and the average value about $16.90 per ounce. The highest average county is Sacramento, where the gold averages 898, or $18.56 per ounce. Mono has the lowest average gold, 550 fine, or $11.36 per ounce.

The finest grade of gold produced in California, and probably in the world, is from the San Guiseppe Mine, half a mile from Sonora, Tuolumne County. It is a quartz mine, but the gold has uniformly run from 982, or $20.29, to 987, or $20.40 per ounce. Some years ago, the gold, from a run of ninety tons of ore, went 998, or $20.6305 per ounce. When it is remembered that perfectly fine gold is valued at $20.67, it will be seen how very pure this San Guiseppe gold is.

No mine in the world has such a record, the only approach to it being the Mount Morgan, in Queensland. The Neale Mine, near the San Guiseppe, has gold running 950½, or $19.64 per ounce, and it is also a quartz mine. The Table Mountain and Bald Mountain placers, of Tuolumne County, have yielded gold running 950, or $19.63 per ounce. The Spanish Hill placer, of El Dorado County, has yielded gold 987 fine, or $20.40. Some of the gold from the placers at Newhall, Los Angeles County, is worth $19.75, or 955½ fine.

Placer gold is usually of higher grade than that from quartz. The highest average of fineness in California is that of the gold from the placers around Folsom, Sacramento County, where it is worth over $20 per ounce, running from 974, or $20.13, to 978, or $20.21 per ounce. The average fineness of gold from that county last year was 898, or $18.56 per ounce, the highest in the State.

Gold of the lowest grade in the State comes from Mono County, where, at Bodie, it seldom goes over 580½, or $12 per ounce, and down to 411½, or $8.50 per ounce. The gold from the largest producing quartz mine at Bodie, in large quantity, averaged last year 484 fine, or $10.00 per ounce. Gold from some of the cyanide tailings plant in that camp ran only 130.5 fine, or $2.69 per ounce. At Benton, in the same county, the gold runs up to 919½, or $19 per ounce, but elsewhere it is of low grade. Average for the county is 550 fine, or $11.36 per ounce.

There is so much variation in localities that a county average does not tell the exact story. For example, in Calaveras County, gold from some mines at Angels runs 975, or $20.15 per ounce, and from Vallecito, 987, or \$20.42 (Fine Gold Mine); while one mine at Hodson yields gold of 627 fineness, or \$12.96 per ounce. There is a mine in Inyo County yielding gold worth \$20 per ounce, or $967\frac{1}{2}$ fine, and one in Kern County with same grade of gold. The largest producer in the latter county has gold worth \$16 per ounce, or 774 fine. A placer mine in Johnsville, Plumas County, also has gold worth \$20.36, or 985 fine.

The following tables of fineness and value of gold from California mines are compiled from returns sent in to the Statistician of the Mint, for 1898, while collecting figures of output of the mines for that year. Q means quartz mine: H, hydraulic; P, placer; R, river; T, tailings; B, beach-sand. Only a few mines in each locality are named, as a complete list would take too much space.

AMADOR COUNTY.

MINE.	LOCALITY.	FINENESS.	VALUE.
Q—Keystone	Amador City	790	\$16.33
Q—Jackson Exp Co.	Jackson	730	15.09
Q—Kennedy	"	820 to 830	16.95 to 17.15
Q—Kannakee	Pine Grove	772	15.95
Q—Burner Red Works	Sutter Creek	$822\frac{1}{2}$	17.00
Q—South Eureka	" "	$798\frac{1}{2}$	16.50
D—Gutten & Kepper	" "	$834\frac{1}{2}$	17.25
Q—Baliol	" "	850	17.57
Q—Wildman	" "	795	16.43
Q—Mahoney	" "	780	16.12
Q—Madrona & Harding	Volcano	790	16.33

BUTTE COUNTY.

Q—Poumerate	Berdan	760	15.71
P—Wild Yankee	Berry Creek	$822\frac{1}{2}$	17.00
P—Kentuck	Brush Creek	847	17.50
D—Princess	Chaparrel	887	18.33
H—King Bird	Clipper Mills	930	19.22
H—Scott & Stone	Concow	859	17.75
P—Farley	"	888	18.35
D—Pete Wood	Coutelenc	910	18.81
D—Sycamore Flat	John Adams	847	17.50
D—Emancipation	Lovelock	$943\frac{1}{2}$	19.50
Q—Matheson	Nimshew	837	17.30
P—Feather River Exp. (Dredge)	Oroville	922	19.05
H—Arno	"	871	18.00
D—Clark	Yankee Hill	900	18.60
Q—Tonkey Lease (Pocket)	" "	$834\frac{1}{2}$	17.25

CALAVERAS COUNTY

MINE.	LOCALITY.	FINENESS.	VALUE.
Q—Demarest	Angels	798½	$16.50
Q—San Justo (Gates)	"	960 to 975	19.84 to 20.15
Q—Hog Pen	"	880	18.19
Q—Lightner	"	885½	18.30
Q—Midland	"	850	17.57
Q—Gwin	Gwinmine	802	16.57
Q—Royal Cons.	Hodson	627	12.96
P—California	Milton	883	18.25
Q—Easy Bird	Mokelumne Hill	871	18.00
H—Green Mountain	" "	919½	19.00
H—J. Cuneo	Mt. Ranch	897½	18.55
Q—Legal Tender	Murphys	880	18.19
Q—Hinsdorf	"	756	15.62
D Hedrick	San Andreas	871	18.00
H—Fine Gold	Vallecito	940 to 987	19.43 to 20.40
Q—Keltz	West Point	750	15.50
Q—Oakland	" "	714	14.75

DEL NORTE COUNTY.

P—Wm. McKee	Crescent City	798½	16.50
P—Myrtle Creek	Gasquet	934	19.30

EL DORADO COUNTY.

Q—Larkin	Diamond Springs	825	17.05
Q—Crown Point	" "	774 to 871	16.00 to 18.00
Q—German	El Dorado	762	15.75
Q—Garden Valley M. Co.	Garden Valley	617	12.75
Q—Blue Rock	Georgetown	901	18.62
Q—Boardt	Greenwood	847	17.50
Q—Idlewild	"	699	14.44
Q—Oakland Cons.	"	847	17.50
D—Grizzly Flat	Grizzly Flat	871	18.00
Q—Grouse Gulch	" "	664½	13.73
Q—Lookout	Josephine	822½	17.00
H—Robert Bros.	Nashville	825	17.05
H—Snow Bros.	Newton	948½	19.60
P—Fernandes	Pilot Hill	900	18.60
P—Pilot Hill	" "	950	19.63
Q—Fisk	Placerville	847	17.50
Q—Grand Victory	"	706 to 776	14.59 to 16.04
Q—Zantgraf	"	570	11.78
Q—Friedman	Rescue	847 to 871	17.50 to 18.00

FRESNO COUNTY.

Q—Inyo	Auberry	701½	14.50
H—Little Joe	Pine Ridge	619½	12.80
P—Peck	Toll-House	919½	19.00
H—Kipp	Trimmer	664½	13.73

HUMBOLDT COUNTY.

MINE.	LOCALITY.	FINENESS.	VALUE.
H—Little Klondike	Blocksburg	834½	$17.25
B—China Flat	China Flat	822½	17.00
H—Trophy	" "	847	17.50
B—Pacific Beach	Dows Prairie	600	12.40
H—Mattah	Klamath	888	18.35
B—Union	Orick	934	19.30
H—Sarvorum	Orleans	834½	17.25
H—Rough & Ready	"	834½	17.25
H—Salstrom	"	847	17.50
H—Horseshoe Bend	"	818	16.90
H—Croton Bar	"	834½	17.25
H—Pearch	"	822½	17.00

INYO COUNTY.

MINE.	LOCALITY.	FINENESS.	VALUE.
Q—Roberts	Big Pine	677½	14.00
Q—Last Chance	Darwin	919½ to 967½	19.00 to 20.00
P—Gold Dollar	Independence	895	18.50
Q—Molus	Laws	677½	14.00
Q—St. John	Modoc	919½	19.00

KERN COUNTY.

MINE.	LOCALITY.	FINENESS.	VALUE.
Q—Amalie	Amalie	967½	20.00
Q—McKenny	Garlock	774	16.00
Q—Nugget Gulch	"	895	18.50
Q—Little Angel	Havilah	659 to 690	13.62 to 14.26
Q—Little Bonanza	Keyes	677½	14.00
D—Garnishee	"	629	13.00
Q—Juan Dosie	Piute	556½	11.50
Q—Rawhide	"	617	12.75
Q—Tip-top	"	613½	12.68
Q—Little Butte	Randsburg	725	14.98
Q—Napoleon Cons.	"	792	16.37
Q—Gold Coin	"	798½	16.50
Q—Yellow Aster	"	774	16.00
Q—Merton & Buckboard	"	810	16.74
Q—Butte	"	774	16.00
P—Ground Hog Placer	"	871	18.00
Q—G. B.	"	765	15.81
Q—Wedge	"	778 to 762½	16.08 to 15.76
Q—Sullivan & Black	Ricardo	871	18.00
Q—Glen Olive	Vaughns	613	12.67
Q—Pioneer	"	677½	14.00
Q—Polar Bear	"	689½	14.25
Q—Shoe String	"	700	14.47
Q—Yellow Jacket	"	667½	14.00
Q—Talc	Woody	726	15.00

LASSEN COUNTY.

MINE.	LOCALITY.	FINENESS.	VALUE.
Q—Brush Hill	Hayden Hill	653½	$13.50
Q—Golden Eagle	" "	653½	13.50
P—Casanova	Susanville	919	18.99

LOS ANGELES COUNTY.

Q—Red Rover	Acton	716 to 813	14.80 to 16.80
Q—Esperanza	Gorman	822½	17.00

MADERA COUNTY.

Q—Surprise	North Fork	745	15.40
Q—Lucky Louise	Raymond	653½	13.50
Q Ward	Zebra	774	16.00

MARIPOSA COUNTY.

Q—Independence	Bear Valley	851 to 864	17.59 to 17.86
Q—Merced	Coulterville	744	15.37
Q Virginia	"	783	16.18
Q—Castle Dome	"	871	18.00
Q—Mount Gaines	Hornitos	725 to 764	14.98 to 15.19
Q—Martinez	"	762	15.75
Q—Spencer	Mariposa	780	16.12
Q—Lindstrom	"	834½	17.25

MONO COUNTY.

Q—Cornucopia & Borasco	Benton	919½	19.00
Q—Standard Cons.	Bodie	484	10.00
T—Sunshine	"	130.5	2.69
Q—Jackson & Lakeview	Lundy	652	13.47
Q—Bonanza	"	654	13.51

NEVADA COUNTY.

D—Mary Jane	Eureka	885	18.29
H—Dockun	French Corral	938	19.39
D—Harrison French	" "	937½	19.37
H—Quong Wah	" "	847	17.50
Q—Graniteville	Graniteville	850	17.57
Q Cincinnati Hill	Grass Valley	840	17.36
Q—Brunswick	" "	812½	16.79
Q—Gold Flat	" "	795	16.43
Q—Norambagua	" "	790 to 810	16.33 to 16.74
Q—Slate Ledge	" "	725	14.98
Q—North Star	" "	850	17.57
Q—Pennsylvania	" "	852 to 856	17.61 to 17.69
Q—W. Y. O. D.	" "	840 to 846	17.36 to 17.48
Q—Columbus Tunnel	Nevada City	859	17.75
Q—Sneath & Clay	" "	750	15.50
Q—Bellefontaine	" "	726	15.00
Q—Chapman	" "	800	16.53

NEVADA COUNTY — Continued.

MINE.	LOCALITY.	FINENESS.	VALUE.
Q—Deadwood	Nevada City	791	$16.35
D—Harmony	" "	790	16.33
Q—Mountaineer	" "	822½	17.00
Q—Neversweat	" "	645	13.33
Q—Reward	" "	825	17.05
Q—Mayflower	" "	750	15.50
Q—Red Dog	" "	890	18.39
D—Reddick	" "	850	17.57
Q—Yellow Diamond	" "	871	18.00
D—Walkenshaw	North Bloomfield	906	18.72
D—Blue Lead	Relief	935	19.32
D—Eagle Drift	"	936½	19.35
H—Spanish	Rough-and-Ready	903½	18.67
P—Long Ravine	Spenceville	929	19.20
D—Fisk	Washington	880½	18.20
Q—Spanish	"	777	16.06
Q—Washington	"	822½	17.00

ORANGE COUNTY.

MINE.	LOCALITY.	FINENESS.	VALUE.
Q—Yaeger	Fullerton	967½	20.00

PLACER COUNTY.

MINE.	LOCALITY.	FINENESS.	VALUE.
Q—Alder Creek	Blue Canyon	921	19.03
D—Cedar Creek	" "	847	17.50
Q—Bazoca	Butchers' Ranch	780	16.12
Q—Christmas Hill	" "	821	16.97
H—Annie Laurie	Colfax	928	19.18
P—Tailings-Cambridge	"	890	18.39
D—Red Point	Damascus	927	19.16
D—Truro	Dutch Flat	871	18.00
P—Merrithew	Gold Run	871	18.00
D—Kotten	Iowa Hill	847	17.50
D—Morning Star	" "	900	18.60
D—Shelly	" "	822½	17.00
D—Big Dipper	" "	884	18.27
P—Shallow Placers	Michigan Bluff	822½	17.00
P—A. Dixon	" "	900	18.60
D—Hidden Treasure	" "	924 to 941	19.10 to 19.45
D—Basin Cons.	" "	924	19.10
D—Stewart & McGillivry	" "	900	18.60
Q—Dan'l Webster	" "	847	17.50
Q—Caledonia	" "	774	16.00
D—Sellier	" "	934	19.30
R—Mormon	" "	880½	18.20
Q—Hathaway	Newcastle	580½	12.00
Q—Eclipse	Ophir	871	18.00
Q—Boulder Cons.	"	650	13.43

PLACER COUNTY — CONTINUED.

MINE.	LOCALITY.	FINENESS.	VALUE.
Q—Gold Blossom	Ophir	647	$13.37
Q—Morning Star	"	605	12.50
Q—June Boy	"	900	18.60
Q - Central Quartz	Towle	893 to 900	18.45 to 18.60
Q—Pioneer	"	797 to 811	16.47 to 16.76
D—Glen	Westville	883	18.25
Q—Herman	"	846	17.48
D - Small Hope	Yankee Jim	925	19.12

PLUMAS COUNTY.

MINE.	LOCALITY.	FINENESS.	VALUE.
D—Star of Plumas	Buck	919½	19.00
Q—Morning Star	Box	750	15.50
Q—Grub Stake	"	837	17.30
D Cameron	Butte Valley	847	17.50
H—Arctic	" "	879	18.17
H—Jackson Creek	Cromberg	847	17.50
Q—Empire	Crescent Mills	785 to 794	16.22 to 16.41
Q—Jackson	" "	681½	14.08
Q—McGill & Standart	" "	700 to 750	14.47 to 15.50
H—Blue Nose	Eclipse	883	18.25
Q—Gruss	Genesee	800	16.53
H—Light Canyon	"	871	18.00
P—Gen'l Harrison	Johnsville	985	20.36
Q—Jamison	"	852	17.61
T—Alturas Tailings	La Porte	890	18.60
H—Clark Bros	" "	800	16.53
H - Hop Sing	" "	900 to 940	18.60 to 19.43
H—Yankee Hill	" "	860	17.77
P—Dey	Longville	845	17.46
P—Simmons & Reed	"	847	17.50
P—Albion	Meadow Valley	950	19.63
H—Deadwood	" "	940	19.43
P—Hillas	" "	950	19.63
R—Minerva Bar	Nelson Point	871	18.00
P—Olson	" "	920	19.01
Q—Dunn Bros.	Prattville	851	17.59
P—Butterfly	Quincy	883	18.25
D—Highland Cliff	"	917½	18.96
H—Quincy M. Co	"	956	19.76
P—Shinn	"	871	18.00
D—Lava Bed	Spanish Ranch	928	19.18
P—Long Bar	" "	834½	17.25
H—Pine Leaf	" "	871	18.00
P—Smith Flat	" "	892	18.43
P—Rich Gulch	" "	774	16.00
H—Halsted	" "	880	18.19

PLUMAS COUNTY — Continued.

MINE.	LOCALITY.	FINENESS.	VALUE.
Q—Lucky S.	Taylorville	726	$15.00
P—Ruffa	"	871	18.00
Q—Bullion	Wash.	627½	12.97

RIVERSIDE COUNTY.

Q—Lost Horse	Banning	800	16.53
T—Santa Rosa	Perris	605	12.50
Q Riverside Gold	"	756	15.62
Q—Corn Springs	Salton	787	16.26

SACRAMENTO COUNTY.

P—Eckhardt	Folsom	974	20.13
P—Rogers	"	978	20.21
H—Amador & Sac. Canal Co.	Michigan Bar	900	18.60

SAN BERNARDINO COUNTY.

Q—Gold Bar & St. George	Vanderbilt	653½	13.50
Q—Opera	Victor	822½	17.00

SAN DIEGO COUNTY.

Q—Ella	Banner	898	18.56
Q—Escondido	Escondido	800	16.53
Q—Free Gold	Hedges	800	16.53
Q—High Peak	Julian	890	18.39
Q—Senator	Senator	822½	17.00

SHASTA COUNTY.

Q—National Cons.	Buckeye	865	17.88
Q—Sybil	French Gulch	680 to 754	14.05 to 15.58
Q—Washington	"　"	820	16.95
Q—Mocking Bird	Igo	750 to 800	16.12 to 16.53
Q—Midas	Knob	845 to 850	17.46 to 17.57
Q—Sharp	Ono	919½	19.00
H—Smith & Benedict	"	847	17.50
H—Gardner Bros	"	847	17.50

SIERRA COUNTY.

Q—Croesus	Alleghany	870	17.98
Q—Mariposa	"	869 to 876	17.96 to 18.10
Q—Sierra	"	860 to 900	17.77 to 18.60
P—J. J. Morton	Downieville	822½	17.00
D—Exchange	Forest	789	16.31
H—New York	Gold Lake	895	18.50
H—Eclipse	Gibsonville	919	18.99
D—Garnett	"	883	18.25
D—Tahoe	"	883	18.25
D—Morgan	Goodyears	798½	16.50
D—Helmet	"	847	17.50

SIERRA COUNTY — Continued.

MINE.	LOCALITY.	FINENESS.	VALUE.
D—Bope	Pike	774	$16.00
P—Ryan	"	847	17.50
H—Bedrock	"	786½	16.25
H—Morristown	Port Wine	948½	19.60
D—Treasure	" "	895	18.50
Q—Mountaineer	Sierra City	883	18.25
D—Carmen	" "	931½	19.25
Q—Cleveland	" "	784	16.20
H—Hayes & Steelman	" "	866	17.90
Q—American Exchange	" "	622	12.85
H—Yuba	" "	895	18.50
D—Bruckeman	Table Rock	847	17.50
D—Hidden Treasure	" "	860	17.77
Q—Strassner	" "	842	17.40

SISKIYOU COUNTY.

MINE.	LOCALITY.	FINENESS.	VALUE.
Q—Black Bear	Black Bear	825	17.05
H—Montezuma	Callahan	859	17.75
H—Littlefield	"	847	17.50
Q—Hathaway Group	"	750	15.30
D—Miller	"	847	17.50
Q—King Solomon	Cecilville	847 to 895	17.50 to 18.00
H—Orton Gulch No. 2	"	865½	17.89
P—Miller	Summerville	847	17.50
H—Nolan Bar	Cottage Grove	822½	17.00
H—Nugget Bar	Etna	822½	17.00
Q—Myers	Fort Jones	907½	18.75
H—R. H. Campbell	" "	844½	17.45
P—Kurschel	Forks of Salmon	822½	17.00
P—Picayune	" " "	822½	17.00
Q—Knownothing	Gilta	798½	16.50
H—Willard & Hickman	Hamburg	822½	17.00
H—Clausen	Happy Camp	822½	17.00
P—Indian Capitan	" "	822½	17.00
H—New Brighton	" "	895	18.50
H—Pennsylvania	" "	871	18.00
P—Brazil	Hawkinsville	837	17.30
Q—Hazel	Henley	925 to 950	19.12 to 19.63
P—American Bar	Hornbrook	950	19.63
H—Blue Lead	Oak Bar	827½	17.00
Q—Michigan Bar	" "	818	16.90
H—Wright & Eastlick	Oro Fino	774	16.00
Q—Humpback	Rollin	858	17.73
H—Gold Hill	Sawyers' Bar	882	18.23
D—Bigelow & Son	" "	847	17.50
Q—Midwinter	" "	883	18.25

SISKIYOU COUNTY — Continued.

MINE.	LOCALITY.	FINENESS.	VALUE.
Q—Yreka M. Co	Sawyers' Bar	885½	$18.30
Q—Secretary:	" "	822½	17.00
H—Mississippi	Seiad	902½	18.65
H—Clark & Channier	Somes Bar	827½	17.00
H—Ten Eyck	Ten Eyck	840 to 866	17.36 to 17.90
H—C. Jensen	Walker	798½	16.50
D—Blue Gravel	Yreka	842	17.40
Q—Golden Jubilee	"	718	14.84

TRINITY COUNTY.

MINE.	LOCALITY.	FINENESS.	VALUE.
D—North Fork Placer	Abrams	871	18.00
P—Horseshoe Bend	Burnt Ranch	834½	17.25
P—Hoboken	" "	895	18.50
Q—Black Warrior	Carrville	847	17.50
Q—Eleanor	"	847	17.50
Q—Gold Dollar	"	822½	17.00
P—Reindeer	"	774	16.00
Q—North Star	Coleridge	750	15.50
Q—McLisle	"	854	17.65
Q—Enterprise	"	785	16.22
Q—Hoodoo	"	777	16.06
P—Moore	"	847	17.50
H—Watrous	"	847	17.50
Q—Yellowstone Group	"	759	15.68
Q—Globe	Dedrick	822½	17.00
H—Mahoney & Wallace	"	863½	17.85
Q—Amy Balch	Deadwood	872	18 02
Q—Lila & Summit	"	757½	15 65
Q—Brown Bear	"	847	17.50
Q—Lappin	"	786½	16.25
P—Emigrant Gulch	Denny	847	17.50
Q—Hidden Treasure	"	847	17 50
Q—Mountain Boomer	"	880	18.19
H—Frates	Douglas City	859	17.75
P—Thomas	Francis	847	17.50
P—Bridge Gulch	Hay Fork	786½	16.25
H—Carrier Gulch	" "	786½	16.25
P—Gurley Gulch	" "	786½	16.25
H—Hilliard	" "	786½	16.25
H—Hang Bar	" "	786½	16.25
H—Mark Poland	" "	786½	16.25
H—Cie. Fse.	Junction City	910	18.81
H—Harn	" "	834½	17.25
H—Heurtevant Group	" "	900	18.60
H—Mammoth	" "	930	19.22

TRINITY COUNTY — Continued.

MINE.	LOCALITY.	FINENESS.	VALUE.
H—Red Hill	Lewiston	891	$18.41
H—Eastman	"	822⅓	17.00
R—Phillips Bros.	"	822½	17.00
P—Poker Bar	Lowden's Ranch	889½	18.38
H—Humboldt	Minersville	847	17.50
H—East Fork	Trinity Center	900	18.60
P—Moore	" "	888	18.35
P—Underground Treasure	" "	900	18.60
H—Joss	Weaverville	859	17.75
H—LaGrange	"	850	17.57
D—Testy	"	859	17.75

TULARE COUNTY.

Q—Clara Gibbons	White River	550 to 608	11.36 to 12.56
Q—Richelieu	" "	738	15.25

TUOLUMNE COUNTY.

Q—Dreisam	Arastraville	830	17.15
Q—Longfellow	Big Oak Flat	726 to 822½	15.00 to 17.00
Q—Cons. Eureka	Carters	772½	15.96
Q—Grizzly	"	730½	15.10
Q—Eagle-Shawmut	Chinese Camp	805	16.64
Q—Mt. Lily	Columbia	895	18.50
D—Pine Log	"	871	18.00
Q—Confidence	Confidence	823	17.01
Q—Kanaka Group	Groveland	822½	17.00
Q—Tarantuta	Jacksonville	850	17.57
Q—Dutch	Quartz	740	15.29
Q—Jumper	Stent	878	18.14
Q—Santa Ysabel	"	750	15.50
Q—Atlas Dev. Co.	Sonora	775	16.02
Q—Equitable M. Co	Tuttletown	678 to 697	14.01 to 14.40
Q—Norwegian	"	900 to 911	18.60 to 18.83
Q—Strell	"	852	17.61

YUBA COUNTY.

Q—Cabbage Patch	Browns Valley	919½	19.00
D—Old Pittsburg	Camptonville	893	18.45
H—Horse Valley	"	906	18.72
Q—Honeycomb	"	898	18.56
H—Joubert	"	917	18.95
Q—Good Title	Dobbins	919½	19.00
T—Dredger Gold	Smartsville	900	18.60
P—Eagle Gulch	Strawberry Valley	909	18.79
P—Onken	" "	895	18.50

Petroleum in California.

BY

W. L. WATTS,

ASSISTANT IN THE FIELD TO A. S. COOPER, STATE MINERALOGIST.

[The following paper consists of extracts from the report of Mr. W. L. Watts, which will shortly be issued as a bulletin by the California State Mining Bureau. We are indebted to the State Mineralogist for allowing Mr. Watts to furnish us with this paper.—ED.]

CALIFORNIA'S mineral wealth consists not only in those minerals from which metals are obtained, but also in numerous other mineral substances, which become in greater demand as our manufacturing interests expand, and as our civilization advances.

The most important of the latter class of minerals, which, in a commercial sense, may be regarded as non-metallic, are the hydro-carbons; and of these, petroleum, in the form of asphaltum and oil and natural gas, is of the greatest value. This paper is confined to the last two of these items.

In California, the question of petroleum as fuel assumes a special importance, owing to the fact that the deposits of coal, which have thus far been discovered in our State, are inadequate to the steadily increasing demand for fuel.

The object of this paper is to summarize the leading facts concerning the occurrence of petroleum in California. This mineral, in the form of natural gas, oil and asphaltum, is found at various places in the Coast Range. The greatest showing of petroleum, and the greatest development of the petroleum industry are south of San Francisco. Oil and gas are found in Kern County, in the foothills of the Sierras, and natural gas is found in the great central valley of California. It is said that this gas has also been observed in other counties, including the foothills of the Sierras.

The geological formations yielding petroleum in California range from the lower cretaceous to the quaternary; and, in different localities, the geological horizon of the productive strata differs in point of vertical range. In the Puente hills, and at Los Angeles, the oil-yielding rocks are of Neocene age; these formations were first classed as Pliocene, on account of the numerous Pliocene fossils found in them.

On the south side of the valley of the Santa Clara River, in Ventura and Los Angeles counties, the principal oil-yielding formations probably range from the Neocene to the Miocene. The writer has not yet made a detailed examination of these oil fields.

On the north side of the valley of the Santa Clara River, in Ventura County, there is evidence of petroleum in rocks ranging from the upper Neocene to the lower Eocene formations; the productive formations ranging from the Miocene to the uppermost portion of the Eocene.

In the foothills west of Bakersfield, in Kern County, petroleum is found in formations ranging from the Eocene to the Neocene, and heretofore classed as Pliocene; but the oil-yielding formations, which have been tested by drilling, are supposed to be of Miocene age.

Natural gas, and some oil, have also been obtained in the foothills of the Sierras, east of Bakersfield, the formation being either of Pliocene or of Neocene age.

In Fresno County, and Kings County, there are exudations of petroleum from rocks of Miocene age; but the petroleum-yielding formations near Coalinga, in Fresno County, which have recently proved very remunerative, appear to be of Eocene age, formerly called cretaceous B.

The geological horizon of the oil-yielding rocks at Moody Gulch, Santa Clara County, has never been determined.

Some oil has also been obtained on the Tunitas and the Purissima creeks, in San Mateo County, from wells which penetrate strata which are, probably, of Eocene age.

North of San Francisco, petroleum-yielding formations crop out along the coast at Bolinas Bay and at Point Arena; at these places the exposed rocks are either of Pliocene or of Neocene age.

In Humboldt County, several wells have been drilled, from which some oil has been obtained, the rocks penetrated being either of Pliocene or of Neocene age.

On Bear Creek, in Colusa County, gas and oil are found in rocks of cretaceous age.

It is reported that in some places petroleum is found permeating eruptive or other crystalline rocks.

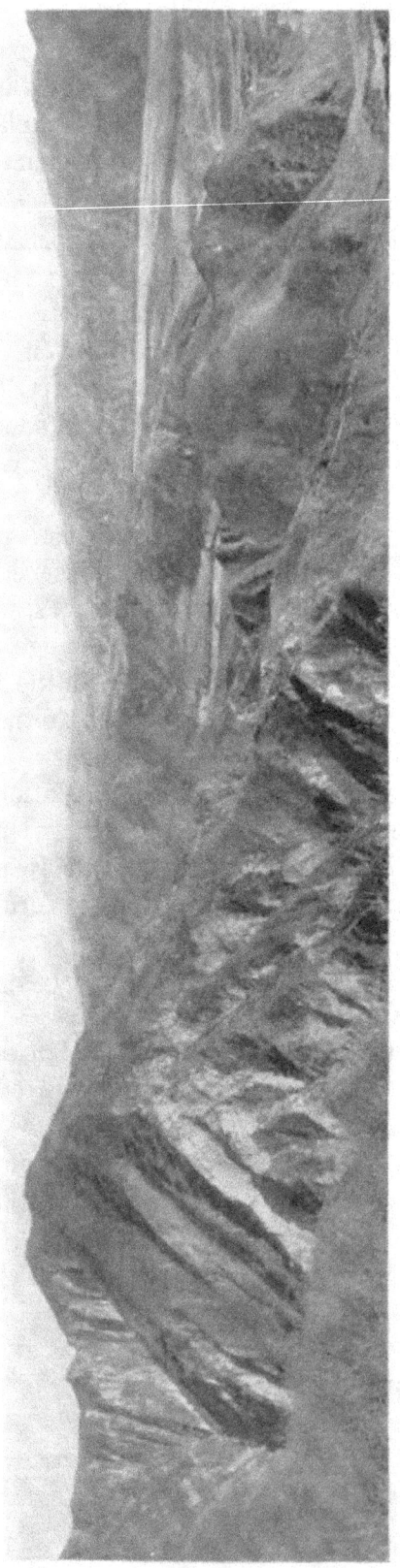

OIL YIELDING TERRITORY WEST OF PIRU CREEK, VENTURA COUNTY, CAL.

At Stockton, in San Joaquin County, natural gas is obtained in remunerative quantities, from wells permeating strata of quaternary age.

At Marysville Buttes, in the Sacramento Valley, natural gas is found in rocks of Eocene age, formerly called cretaceous B. There are several places in Sacramento and San Joaquin valleys where wells are yielding sufficient natural gas to be of local value.

The relative position in point of vertical range of the formations, wherein remunerative oil wells have been obtained, in Ventura and Los Angeles counties, is demonstrated by an investigation of the country between the Piru and Sespi creeks, in Ventura County, where the following sequence of formations can be seen: At the Piru Creek, a conglomerate formation is seen containing Neocene fossils, Pliocene forms being the most numerous. In some places, the conglomerate is impregnated with petroleum; it rests on a shale formation, containing Neocene fossils. The lower portion of the shale is interstratified with sandstone, which, in many places, is impregnated with petroleum, forming an oil sand, the outcropping strata of which resemble the oil sands seen in the Puente Hills, in Los Angeles County. The shale rests on a whitish sandstone of Miocene age. This whitish sandstone contains remunerative oil-yielding strata.

The conglomerate, the shale, and the whitish sandstone in Ventura County constitute a group corresponding to a group of certain conglomerates, shales, and sandstones which the writer carefully studied in the Puente Hills, in Los Angeles County. These formations probably form a large portion of the rocks in the oil districts of Los Angeles and Ventura counties; and, in both of these counties, these formations are of similar character, and contain fossils of similar age.

As observed in the Puente Hills, in Los Angeles County, these whitish sandstones, shales, and conglomerates rest on one another in the order here named. The principal oil-yielding formations found in these hills, and in the city of Los Angeles, are certain oil sands which interstratify the lower portion of the shale formation, and, probably, constitute the uppermost strata of the underlying sandstones.

The whitish sandstone formation extends westward from the Piru Creek, in Ventura County, to the Sespi Oil District, also in Ventura County, the distance between the two places being about eight miles.

At the Sespi Oil District, the whitish sandstone rests on a shale formation, whitish and grayish at the top, but passing into a dark-colored shale, which is interbedded with numerous thin strata, or nodular masses of hard bituminous limestone. These shales contain Miocene and Eocene fossils, and rest on a drab-colored sandstone, of no great thickness. The drab-colored sandstone rests on a brown sandstone, locally known as the Sespi brown stone. This brown stone rests on white sandstone, and the latter on a buff-colored sandstone. The Sespi brown stone, the white sandstone, and the buff-colored sandstone all contain typical Eocene fossils. All these sandstones are more or less interbedded with shale. The principal oil-yielding formations in the Sespi District are the lowermost portion of the dark-colored shales, the drab sandstone, and the uppermost portion of the Sespi brown stone.

There are numerous seepages of petroleum in the hard, buff-colored Eocene sandstones; but no remunerative oil-wells have as yet been obtained by drilling in these rocks.

Between the Piru Creek and the Sespi District, no marked nonconformity was observed by the writer, the variations of dip being referable rather to local geological disturbance than to non-conformability. Still, it by no means follows, that the formations actually rest conformably on one another. It is generally believed that, in California, the Miocene formations rest non-comformably on the Eocene. Observations in Los Angeles and Orange counties lead to the conclusion that the Neocene shales overlap the whitish sandstones; and there are some reasons for believing that the conglomerate rests non-conformably on the Neocene shales.

Since the Eocene period, there has been not only epochs of unusual geologic disturbance, but also disturbance of a chronic nature, which has been co-temporaneous with the deposition of the tertiary and quaternary formations. Similar disturbances continue to this day. It appears that, in many instances, these disturbances were of a local character.

The formations penetrated by remunerative oil wells, in such portions of Los Angeles, Orange, and Ventura counties, as have been examined by the writer, are as follows: .

In the territory extending between the Santa Ana River, in Orange County, and the ocean, at Santa Monica, in Los Angeles County, remunerative oil sands have been found in the lower portion of the Neocene shales, and, probably, in the upper portion of the Miocene sandstone. In one instance, oil was found in the overlying conglomerate.

At the Modello oil wells, near Piru, in Ventura County, remunerative oil sands have been found in the whitish sandstones of Miocene age.

In the Sespi District, remunerative oil sands have been found in the upper portion of the Eocene sandstones, and in shales, which appear to occupy a position between the Miocene and Eocene formations.

It is probable that these oil-yielding localities have their counterpart in many other places in California, besides those which have been mentioned, for the geological formations, constituting these oil measures, extend along the Coast Range, from San Diego County to Humboldt County. We have not yet obtained sufficient geological evidence to warrant the expression of anything more than a tentative opinion, concerning the actual geological horizon of the petroleum formations in the following oil fields: The oil fields on the south side of the Santa Clara River, those north of Santa Paula, in Ventura

OIL WELLS IN LOS ANGELES CITY.

County, and those of Santa Barbara, Kern, Santa Clara, and San Mateo counties. But the data already accumulated warrants the assertion that the oil measures, in the localities referred to, are of tertiary age.

From the description of the oil-yielding formations, the geological position of which has been definitely determined, it appears, that the productive strata are sandstones, underlying or interstratifying bodies of shale. In may be argued that these conditions indicate natural distillation as the chief cause of the accumulation of petroleum in the oil measures.

In many parts of California we find petroleum-yielding formations associated with shales, which show signs of having been subjected to metamorphic action. It is reasonable to infer that the petroleum, having been elaborated in the shale, may have been driven out of it by natural distillation, into enclosing or interstratifying beds of sandstone

It must also be borne in mind that, in other instances, petroleum may have been collected, by upward distillation, from sedimentary strata underlying the shale.

A very small percentage of petroleum, originally distributed through a great thickness of strata, might be driven into different zones by natural distillation; and, at certain temperature and pressure, it would pass readily through sandstone. The upward course of the petroleum might be impeded by strata of shale, and, when the temperature decreased, the petroleum might condense in any rocks sufficiently porous to afford it storage. If the shale were only partially impervious to the petroleum, the former would be, more or less, permeated by the latter, and fractures in the shale would give the petroleum access to overlying formations.

A modification of such processes, by gas or hydrostatic pressure, would be quite sufficient to bring about a re-distribution of the petroleum, and the formation of secondary deposits of that mineral.

The oil-yielding formations of California, in common with the other rocks of the Coast Range, show great geological disturbance; and, the complex structure resulting therefrom, has given rise to somewhat difficult geological problems in regard to the oil fields of this State.

In a general way, it may be said that the oil lines, or lines along which remunerative wells may be found, follow the strike of the axes of folds in the rocks, or the course of faults which have isolated blocks of strata inclosing the oil-yielding rocks. The tracing of oil lines in this State, and the development of oil fields, necessitate a competent knowledge of structural geology, without which the risks of oil mining would be greatly increased. The type of folds most likely to be met with in the oil fields of California, and the effect of such folds on oil lines, have been discussed in a paper by the writer in the "Mining and Scientific Press," of August 5th, 1898.

The counties in which productive oil fields have been developed, are :

NAME OF COUNTY	PRODUCT IN 1897	PRODUCT IN 1898
Fresno	70,140 Bbls.	154,000 Bbls.
Kern		10,000 "
Los Angeles	1,327,011 "	1,462,871 "
Orange	12,000 "	60,000 "
Santa Barbara	130,136 "	132,217 "
Santa Clara	4,000 "	3,000 "
Ventura	368,282 "	427,000 "
Total product	1,911,569 Bbls.	2,249,088 Bbls.
Total value	$1,918,269	$2,876,420

The growth of the oil industry of California is shown by the following statistical returns for the last ten years :

1888	690,333 Bbls.		1894	783,078 Bbls.
1889	303,220 "		1895	1,245,339 "
1890	307,360 "		1896	1,257,780 "
1891	323,600 "		1897	1,911,569 "
1892	385,049 "		1898	2,249,088 "
1893	470,179 "			

The depth of oil wells depends on the angle at which the oil sand dips, and the distance the wells are from the outcrop of the oil sand, or from the axis of the fold on which the wells are situated.

As a general statement, it may be said that the most productive wells are about 1,000 feet in depth, some being much deeper.

The "life" and yield of such wells are naturally varied. In some instances, they "started off" with a yield of 100 barrels, or more, a day; but, in the course of five or six years, the yield diminished to ten barrels, or less, a day. In other instances, the first yield was less than 100 barrels a day, but the rate of production was better sustained during the "life" of the well.

In some oil fields the wells are considerably less than 1,000 feet in depth, but, as a rule, their yield is not so great as that of the deeper wells.

The most remarkable features in the recent history of the petroleum industry in California have been the development of the Los Angeles oil field, of the Summerland oil field, in Santa Barbara County, and of the oil field near Coalinga, in Fresno County. In the Los Angeles oil field, which is in the residence portion of the city, between 1,000 and 1,100 wells have been drilled within an area of about two and a quarter miles in length, and less than a quarter of a mile in width. The depth of these wells ranges from 600 feet to more than 1,200 feet.

Naturally, the "life" of the wells, drilled so close together in the Los Angeles oil field, has not been of great duration. Their yield, during the first six months of their "life," averaged about twelve barrels of oil a day. In the Los Angeles oil field there are two strata of oil sand; and the "life" of the wells, which penetrate only the first stratum of sand, has been about three years. The wells which penetrate both the first and the second strata of oil sand, however, are of longer life; some have been producing oil for more than four years.

In June, 1898, 647 wells were being pumped in the Los Angeles oil field, which produced about 2,200 barrels a day. The gravity of the oil ranges from 12° B. to 17° B.

The Summerland oil field is situated about four miles east of Santa Barbara. The first oil well in this locality was drilled in 1890, by Mr. H. L. Williams. Subsequently, wells were sunk, remarkable for the amount of gas they yielded at a shallow depth. (See 7th, 10th and 13th Reports of the California State Mining Bureau.)

In 1895, there were twenty-eight productive oil wells at Summerland, as stated in Bulletin 11, issued by the Mining Bureau in 1896.

The following quotations, from Bulletin 11, describes the situation at Summerland, so far as it was indicated by the exposed rocks and the wells drilled at that date :

"These wells are situated a short distance from the ocean, along a line running N. 85 W. This direction corresponds to the strike of the oil-yielding formations. If this line were extended westward, it would pass a little south of the 455-foot well, drilled in 1890 by Mr. Williams, at the summit of Ortega Hill. There is a reasonable probability, therefore, of obtaining productive wells between this line

STREAM OF MALTHA IN SISAR VALLEY.

SEEPAGES OF HEAVY OIL FROM EOCENE SANDSTONE, NORTH OF THE SILVERTHREAD WELLS.

and the ocean; and the wells drilled by Mr. Fischer, at Loon Point, show that the oil-yielding formations extend eastward from Summerland for more than a mile. It is also evident that the oil-yielding formations extend south into the ocean; for not only are oil-yielding strata penetrated by the Fischer wells, but, at low tide, springs of gas and oil are uncovered on the sea shore."

How far subsequent events have substantiated these remarks, may be seen from the description of the Summerland oil field by another writer in this volume.

The Coalinga oil field, in Fresno County, has been under process of development during the last two years; and, by all accounts, it bids fair to rival, and, some say, to excel the Los Angeles oil field, in the quantity, as well as the quality, of its product. It is said that 60,000 barrels of oil were shipped from the Coalinga oil field during the month of June, 1899. In 1894, the writer made a reconnaissance of portions of the foothills on the south side of the San Joaquin Valley; and, as described in Bulletin 4, of the California State Mining Bureau, prospects of petroleum were found in many places.

A sample of Coalinga oil, obtained by the writer in 1893, from an old well, showed a gravity of about 34° B. A sample from another well, in the same oil field, was examined by Dr. Salathe in 1896.

It is not within the scope of this paper to discuss the vexed question as to the origin of petroleum, which has been treated exhaustively by many competent writers.

An able resume of the subject may be found in a paper by Mr. A. S. Cooper, State Mineralogist, in the "Mining and Scientific Press," for February 4, 1899.

He certainly makes out a very strong case for the theory of the vegetable origin of petroleum. This, of course, refers principally to marine vegetation. There is no doubt that the ancient seas contained accumulations of fucoids, similar to that in the Sargassa Sea, which, at the present day, is hundreds of thousands of square miles in extent, "and in places is dense enough to impede the progress of vessels."

With the exception of a sample of oil, from the cretaceous formations of Colusa County, all the samples of oil, which have been examined by the writer, showed an asphaltic base, i. e., the residuum, after the distillation of the lighter hydrocarbons, was an asphalt, or a heavy tar of asphaltic character. These asphaltic oils form asphaltum, on exposure to the atmosphere.

The residuum from the distillation of the Colusa County oil is not an asphalt; physically, it resembles the residuum from eastern asphaltum. The Colusa County oil does not form asphaltum, on exposure to the air.

Two samples of the Colusa County petroleum were distilled by the writer, and their distillates compare with distillates from a sample of asphaltic oil, as follows:

SAMPLE (A) FROM COLUSA COUNTY.

	BY VOLUME.	SPECIFIC GRAVITY.
Crude oil..		0.982 about 12° B.
Distillate below 250.......................................	1 per cent.	
" between 250 and 325 C.......................	60 " "	0.950 " 17° B.

Nearly all the distillates came over at 300° centigrade.

SAMPLE (B) FROM COLUSA COUNTY.

This sample contained b. s. (sludge).

	BY VOLUME.	SPECIFIC GRAVITY.
Crude oil..		0.9835 about 11° B.
Distillate below 280 C...............................	Traces.	
" " 300 C...............................	16.250 per cent	0.9111 " 24° B.
" " 350 C...............................	3.122 " "	0.9600 " 16° B.
At a somewhat higher temperature..............	43.750 " "	0.9788 " 13° B.

SAMPLE OF OIL FROM LOS ANGELES.

	BY VOLUME.	SPECIFIC GRAVITY.
Crude oil..		0.9534 about 17° B.
Distillate below 150 C..................................	Traces.	
" " 200 C..................................	"	
" " 250 C..................................	8 per cent	0.8330 about 38° B.
" " 300 C..................................	13.6 " "	0.8653 " 32° B.
" " 350 C..................................	3 " "	

The ultimate analyses of samples of oil from California and the Eastern States compare as follows:

Locality where Oil was Obtained.	Specific Gravity.	Nearest Degree B	C.	H.	O.	N.	S.	By whom Analyzed.
Oil Creek, Pa....	0.730	62° B.	82.	14.8	3.2	....	....	Deville.
West Virginia...	0.840	36° B.	84.3	14.1	1.6	....	....	Deville.
California........	....	....	86.934	11.817	....	1.1095	....	Peckham.
"	0.920	22° B.	84.	12.7	1.2	1.7	0.4	Salathe.

An examination of the foregoing table shows that the California oils contain more carbon than do the oils from the Eastern States. Concerning crude oils from Los Angeles and Ventura County, Dr. Salathe says:

"These crude oils, which all carry asphalt, held in combination with the high boiling members of the hydrocarbon series, are of a very complex constitution, which makes their refining exceedingly difficult. By a series of chemical reactions and fractional distillations, I have succeeded in isolating various hydrocarbons, which define clearly the presence of the following hydrocarbon series:

"(a) Hydrocarbons of the paraffine, or fatty series.

(b) Hydrides, or hydron, additional products of the benzole series, and homologous hydrocarbons.

(c) Pyridin and chinolin series.

(d) Isomeres of the terpene series.

(e) Sulphurette hydrocarbons.

(See Resume of Original Researches and Analysis and Refining Methods of Petroleum, mainly from the Southern Counties of California, by F. Salathe, Ph. D., in Bulletin 11, of the California State Mining Bureau.)"

Several years ago a careful examination was made, by Dr. C. P. Williams, of certain of the lighter distillates from southern California petroleum. His experiments showed that the samples tested were composed of the following hydrocarbons:

NAME OF HYDROCARBON.	Approximate amount contained in sample.
Paraffine	25 per cent.
Olefine	30 " "
Aromatic hydrocarbons	20 " "
Naphthalene	25 " "

As is well known, the petroleum of the Eastern States is composed principally of hydrocarbons of the paraffine series.

As previously mentioned, by far the greater portion of the California oil is used as fuel, and that in a crude state. A portion is used for fluxing asphaltum, and for the manufacture of illuminating gas, and a portion is refined. The portion refined yields: crude naphtha, illuminating oil, gas distillate, lubricating oil and asphaltum.

The naphtha, distilled from the California oils, is of special value for use in gasoline engines. Those who have made comparative tests of California and Eastern gasoline, in gasoline engines, claim a superiority for the California product.

As might be expected from the foregoing statements, concerning the relative composition of petroleum from the Eastern States and the asphaltic oil of California, illuminating oil manufactured in this State contains more carbon and less hydrogen than does illuminating oil manufactured from Eastern petroleums. The result is, that when burned under similar conditions, California illuminating oil gives a more smoky flame than does oil manufactured from Eastern petroleum. This is due to the fact that it requires more oxygen to effect the complete combustion of carbon than it does to consume hydrogen.

As previously stated, the petroleum obtained from the cretaceous formations of Colusa County, in California, is not an asphaltic oil. Should the petroleum from this county prove to be a paraffine, and be obtained in sufficient quantities, it might yield distillates which would blend with the illuminating oil manufactured from our asphaltic petroleum, and offset the excess of carbon which it contains.

A comparison of the fractional distillations of the Colusa County oil, with that of the asphaltic oil from Los Angeles, shows a marked discrepancy in the boiling point, and the specific gravity of the distillates; the excessive gravity, and high boiling point of the Colusa County oil, indicate that it is a valuable lubricant.

There is no doubt that, as time goes on, more use will be made of the constituents of our asphaltic oils in chemical manufacture. One use was pointed out by Dr. Salathe, who says: "The occurrence of pyridin and chinolin bases in California crude oils opens up a new resource for these products, which are largely used for the synthetical production of alkaloids, dyes, etc., and in a large measure for denaturalizing alcohol in Europe."

As previously mentioned, the greater part of California oil is used as fuel. It is the general opinion of those who use oil as fuel, that weight for weight there is not much difference between the fuel value of oils of different specific gravities, provided the oils are clean, or a suitable allowance is made for water and other foreign substances which they may contain.

In 1896, the writer made calorimetric experiments on the fuel value of California petroleum, as stated in Bulletin 11, Part 4, Chap. III, of the California State Mining Bureau. In the publication referred to, the fuel value of the petroleum, as determined by the calorimetric experiments, is compared with the fuel value of Nanaimo coal; also, with the fuel value of petroleum, as computed from practical working tests in locomotives on the Southern California R. R. In Bulletin 11, there is also a record of calorimetric tests of the fuel value of petroleum, made by Prof. H. Stillman, in the laboratory of the S. P. R. R. In 1898, calorimetric tests were made on samples, by Messrs. Jaffa and Colby, of the University of California: the samples tested were of a heavy grade of petroleum from Summerland, Santa Barbara County. The fuel values, determined by these different estimates, compared as follows:

FUEL VALUES OF PETROLEUM COMPARED.

	Available heat units in one kilogramme.	Available heat units in one ton, calculated as 909 kilogrammes.	One ton of 2,000 lbs. of Nanaimo coal and equivalent in bbls. of petroleum.
Nanaimo coal................................	6,684	6,075,756	
Sample of petroleum, 15° B., from practical working test in locomotives on Southern California Railroad............	....	9,886,585	3.87
Sample of crude petroleum, 16.5° B., tested by Professor Stillman............	9,800	8,908,200	3.49
Sample of lubricating oil, 16° B. to 17° B., tested by Professor Stillman........	10,788	9,796,192	
Sample of Los Angeles oil, 13° B., tested by W. L. Watts............................	10,203	9,274,527	3.63
Maximum fuel value obtained in calorimetric tests by W. L. Watts............	10,381	9,436,329	3.69
Minimum fuel value obtained in calorimetric tests by W. L. Watts............	9,991	9,081,819	3.55
Sample of Summerland oil (crude) tested by Messrs. Jaffa & Colby................	9,688	8,806,392	3.45
Sample of Summerland oil extracted by naphtha by Messrs. Jaffa & Colby......	10,242	9,309,978	3.64

It will be observed that the practical tests in locomotives on the railroad gave a higher fuel value to the petroleum than did the calorimetric tests in the laboratory. This is due to the fact that, in a furnace, a more complete combustion of petroleum can be secured than it is possible to obtain of coal.

In the calorimetric tests made by the writer, the petroleum was cut with gasoline, and the fuel value of the gasoline was deducted from the total calorific value. By this method, an estimate was obtained, which corresponds to that by "the gasoline cut," in common use among oil dealers for determining the amount of foreign matter in petroleum.

"The gasoline cut" consists in mixing, in a graduated glass, equal volumes of crude oil and gasoline. The water and foreign matter

SKETCH MAP,

Roughly showing the Areas over which the Oil-Yielding Formations of California extend. These Formations reach North into
Unexplored Territory, not shown on the Map.

Parallel Horizontal Lines represent Unaltered Cretaceous and Tertiary Rocks.
Parallel Vertical Lines show Unaltered Quaternary and Recent Rocks.
Crosses indicate Producing Fields.

Map Reproduced from *San Francisco Chronicle*.

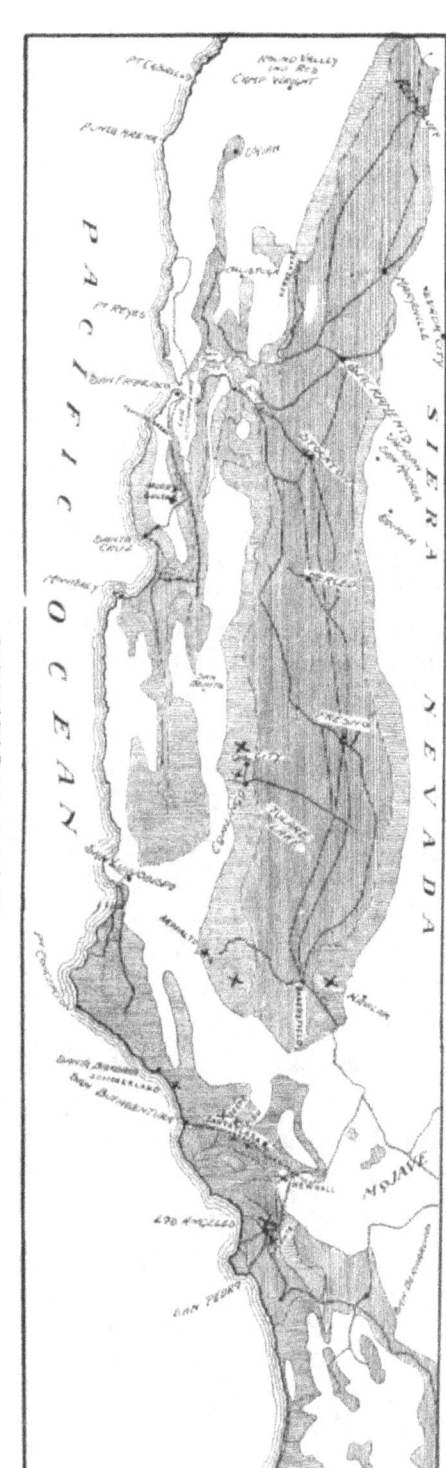

sink to the bottom of the oil; and the relative amounts of oil and foreign matter may be noted by reading the scale on the side of the graduated glass, at the point of contact between the oil and the residuum. The residuum at the bottom of the glass consists of earthy matter, water and sludge, or B. S., as it is known to the trade. In many instances, the sludge, or B. S., constitutes several per cent of the sample. It is usually a brown flocculent precipitate, heavier than oil and lighter than water; it is said to be an emulsion of petroleum and water. The calorific value of the sludge was estimated, by Messrs. Jaffa and Colby, at 4149 kilogramme calorics, or a little more than forty per cent of the fuel value of the sample of oil, which was dissolved in naphtha.

The relative fuel value of coal and Los Angeles oil as shown by combustion in furnaces, is as follows:

The heating furnaces of Los Angeles Steel and Iron Company, one ton Wellington coal equals 2.50 bbls. of oil; for steam purposes, one ton Wellington coal equals 3.00 bbls. of oil.

Los Angeles Consolidated Electric R. R. Company, steam purposes, one ton Wellington coal equals 3.62 bbls. of oil.

Los Angeles Court House, steam purposes, one ton good coal equals 3.10 bbls. of oil.

Southern California R. R. Company, steam purposes, one ton Nanaimo coal equals 4.00 bbls. of oil.

The only place where natural gas assumes sufficient commercial importance to be treated as a factor in the mineral statistics of the State, is at Stockton, in San Joaquin County. The yield of natural gas at that city, during the last two years, has been:

For 1897—63,920,000 cubic feet......................Value, $62,657.00
" 1898—74,424,650 "   " .... " $74,424.00

In 1893, the writer made a careful investigation as to the fuel value of the natural gas at Stockton. Its fuel value, as compared with that of coke and of Nanaimo coal, showed as follows:

2,000 lbs. of coke, carrying 10% ash......Equals 42,500 cu. ft. of gas.
2,000 " " Nanaimo coal.................. "   38,800 " " " "

As stated in Bulletin 4, of the California State Mining Bureau, the absolute value of natural gas is considerably in excess of its calorific value. In the Eastern States, the use of natural gas, instead of solid fuel, has been found to effect a saving of nearly fifty per cent, in addition to that arising from the greater cheapness of gas as compared with coal. This economy results from a saving in labor, and wear and tear of plant, and from the fact that a more uniform temperature can be secured by the use of gas than by the use of solid fuel.

The Copper Resources of California.

BY

M. M. O'SHAUGHNESSY.

HE copper-producing areas of California are confined, at present, to two districts: (1) The great Shasta belt, at the northern end of the Sacramento Valley, in Shasta County, crescent-shaped in form, swinging from Slick Rock Creek, west of the Sacramento River, with the Mountain Copper Company property as its southern apex; thence extending north-easterly, crossing Spring Creek, Squaw Creek, the two Backbone creeks, and the Sacramento River; thence easterly, across the McCloud River to Copper City, whence it appears to turn in a southerly direction, and terminate at Furnaceville; the axis of this crescent is about four miles east of the town of Redding, with a general radius of twenty miles. (2) The Sierra Nevada Mother Lode belt, reaching from Nevada County to Madera, a length of 140 miles. This belt varies in width, but stretches north-westerly, paralleling the Mother Lode and the granitic mountain chain of the Sierra Nevada, and is situated in the westerly foothills thereof, sloping toward the Sacramento Valley.

GENERAL GEOLOGY OF THE SHASTA BELT.—The county of Shasta lies at the head of the Sacramento Valley, and is bisected by the Sacramento River, which flows southerly, discharging into the bay of San Francisco. About three miles north of Kennet, the Pitt River flows into the Sacramento from the east, with about ten times its volume of water. Why the early inhabitants did not call the Pitt the Sacramento, and vice versa, is one of those peculiarities in nomenclature, for which it is difficult to make an explanation. The Pitt River is fed by the McCloud and various other tributaries having their sources in the southern slope of Mount Shasta, which towers to the north, with an altitude of 14,511 feet.

The Sacramento River is navigable to Red Bluff, and could be made so to Redding, by making channels at points which are at present dangerous to navigation. From Redding (elevation 550 feet) northerly, the rise of the river is, on an average, fifteen feet to the mile to Pitt River, whence it rises very rapidly.

To the east of the Sacramento River is a volcanic formation, running north-westerly, from the south-east corner of the county to within a short distance of the Sacramento and Pitt rivers, and having, as its most conspicuous feature, Lassen Butte (10,600 feet), Burney Butte, and Round Mountain. Where this lava flow has been eroded, the underlying Jurassie slate is exposed. Lying west of this is a metamorphic formation, merging into porphyry, in which all the copper ores, so far discovered, have been found.

West of this porphyritic formation is one of slate, dipping to the east, and underlaid and bounded on the west by the granitic rock of the Trinity Mountains. Occasional isolated beds of limestone are found near the contact between the slate and porphyry formations, such as the Squaw Creek-Backbone Creek body, west of the Sacramento River, and the McCloud limestone belt, east of that river and north of the Pitt, which, from its glistening color, forms a distinct landmark for a distance of fifty miles.

MINES OF THE SHASTA BELT.—I. The Mountain Copper Mine, commonly known as the "Iron Mountain," is at present the only big producer of copper on the Pacific Coast, and the first to be systematically developed in this belt. It was originally discovered by a land surveyor, named Magee, who acquired it on the theory that it was valuable for its iron ores. It was relocated by James Salee, who thought it valuable for its silver ores. They sold the property, in 1886, to some parties, who put up a twenty-stamp mill, and attempted to work the surface gossan ores as a free milling proposition. After spending $100,000 in this experiment, they discovered the ore was too base for successful treatment by that process, and forfeited their payments on the property with its improvements. Until 1895, the property was hawked all over this country and London, in search of a purchaser, and condemned by many alleged experts, until it was brought to the attention of C. F. Fielding, of New York, by Hugh McDonnell, the well-known promoter. After an examination by Mr. Alexander Hill, of London, the well-known mining engineer, the property passed into the hands of the Mountain Mines Co. (Limited), who disposed of it January 1, 1897, to its present owners, the Mountain Copper Co. (Limited), of London, an English corporation, with a capital of $6,750,000. The latter company, at the same time, acquired the New Jersey Metal Refining Works, Elizabeth, N. J.

The mine is situated about twelve miles north-west of Redding, in a mountain spur, about 7,000 feet wide, formed by Boulder Creek, on the north, and Slick Rock Creek, on the south, both tributaries of Spring Creek, which flows south-easterly into the Sacramento River. The altitude of the creeks at the mine is about 2,300 feet, and of the summit of the dividing ridge, 3,300 feet. The most pronounced ore-cropping is at the southerly end at Slick Rock Creek, where an immense mass of oxidized ore, commonly called gossan, 300 feet wide, traverses north-easterly the mountain slope to the summit of the ridge. From this point, north-easterly to Boulder Creek, the croppings branch out into two or three undulating belts, each nowhere exceeding 100 feet in width.

It is a peculiar fact, that the best copper sulphide ores are not always found under the biggest gossan croppings. Very often no sulphides have been found for a length of 300 feet—as shown by the Peck Tunnel, of this property—under the richest and most inviting surface croppings; and, frequently, big masses of sulphides are found underlying porphyry, with no indication whatever of surface mineral. The main ore bodies of this property are lenticular shaped masses, composed of gossan ores on the top, with the heavy, unaltered sulphides underneath, the division between the two classes of ore being well defined; the gossan ores rarely going beyond a depth of 125 feet from the surface. There is no doubt but, originally, they were the same composition as the sulphides, but the leaching process, continued for years, has abstracted all the copper contents, and left them proportionately richer in the precious metals. They will average $1 in gold, $3 in silver per ton, with practically no copper, while the sulphide ores, varying from one and one-half per cent to twenty-one per cent, with an average of seven and one-half per cent copper, carry only $0.70 in gold, with $1.50 in silver value per ton. A longitudinal section of the country over the outcrop, showing the explored masses in heavy lines, would be as follows:

The mass marked "A" on plan, near the south-western extremity of the ore body, is the present source of ore supply for the smelters. It is, roughly speaking, 800 feet long, 300 feet wide, and varying in depth from 100 to 300 feet at its lowest point. The walls at the bottom converge, so as to make it wedge-shaped in section, and of

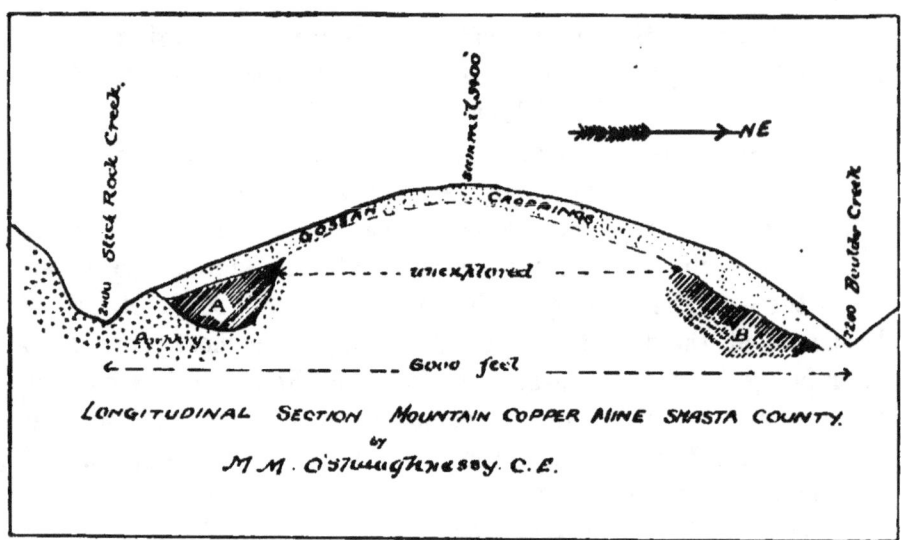

LONGITUDINAL SECTION MOUNTAIN COPPER MINE SHASTA COUNTY.
by
M. M. O'Shaughnessy. C. E.

about 1,700,000 tons in extent. The property was marketed on proof of this mass alone; since then, exploration work has been prosecuted with the diamond drill, demonstrating the existence of a large low-grade mass "B" at the Hornet, or Slick Rock end, while recent work shows a marked improvement with depth in the grade of this ore.

The three problems of greatest interest to the engineer, in connection with the development of a new property, are : (1) The mining, or extraction, of the ore; (2) the transportation; (3) the reduction. This property furnished a new and interesting field for the efforts of the specialist in each line. Mr. Alexander Hill introduced a very simple method for extracting the ore, by replacing quarry rock from the surface of the adjacent ground, in the cavities formed by the extracted ore, thus reducing the cost of timbering to a minimum; and making the extraction of such large masses of ore absolutely safe, in preventing any caves in the surface ground. The writer of this article was employed as engineer on the transportation problem, and had to meet the following conditions :

(a) To design a system for transporting 1,000 tons of ore per day, from an altitude at the mine of 2,300 feet, to one of 600 feet at the smelter site, now Keswick, at Spring Creek, about one mile west from the Sacramento River.

(b) To do so at a minimum of cost, and in the shortest time consistent with good work.

(c) To keep in mind facilities for calcining the ore, procuring fuel, and such other contingencies as might arise in the development of a property of such magnitude.

To anyone acquainted with the topography of this portion of Shasta County—whose surface is exceeedingly steep—the inability to procure only five or six acres of a level site for a smelter will be at once apparent. Such a spot was to be had only at Keswick, the site of the present smelters, with an altitude of 620 feet. It presented the additional advantage of a good supply of water, and abundant dumping ground for slag from the furnaces, without interfering with any vested rights in the adjacent lands or streams. I decided that a three-foot gauge railway would best fill the conditions, by laying an additional rail, for a mile and a quarter, from Keswick to the main line of the California and Oregon Railway, so that the broad-gauge freight cars could be hauled up to the smelter yards, and unload without transhipment. After many preliminary investigations of the intervening ground, which was very rough, I determined on a maximum grade of 200 feet to the mile, and to use as the sharpest curve a 34°, or one of 169.13 feet radius. Construction work on the railway was commenced in August, 1895, and the line was opened for traffic, February, 1896. In order to keep within the maximum gradient, it was necessary to build one loop around Tin Can Point, which adds to the scenic attractions of the line. Between reverse curves, sixty feet of tangent was always introduced, and on all curves the grades were lightened to compensate for curvature. At the mine end, the line terminates in a tunnel, from which upraises have been made to the ore bunkers overhead, to facilitate the loading of the ore cars. At the Keswick end, the cars dump from a trestle, elevated about twenty feet above the surface of the smelter yard. The locomotives used on the road have a short-wheel base, and were designed and constructed by H. K. Porter & Co., of Pittsburg. The smoke from the smelters and burning roast-heaps has a tendency to lubricate the rails and cause slipping of the wheels, and this has been overcome by using the sand-box freely; and, although the curves are sharp and the gradients steep, the road has been singularly free from accidents, having yet to record its first fatality.

The method of reducing the ore caused the management of the company much worry and expense before a practical system was adopted. The pyritic process was first attempted under the direction of Mr. Herbert Lang—who has written a book on the subject—but the results were far from satisfactory, and caused the company the loss of considerable valuable time, before it was abandoned as impracticable. Mr. H. A. Keller, then superintendent of the Parrot

Smelting Works, at Butte, was next retained to inaugurate a practical method of smelting the ore   He did so by building roasting stalls to burn the excess of sulphur out of the ore, and by erecting two ordinary Montana blast furnaces to smelt the reduced ore, with the addition of silicious ores and the necessary fluxes and coke.   The plant, at the present time, consists of three water-jacket furnaces, each 42 x 150 inches at the tuyeres, with a distance of nine feet from center of tuyeres to the charge floor.   They are charged, on an average, 160 times in each twelve-hour shift, the composition of each charge being as follows: roasted sulphide ore, 1,300 lbs.; raw sulphides, 400 lbs.; silica, 400 lbs.; limestone, 100 lbs.; total, 2,200 lbs.; coke, 170 lbs.   The resulting copper mattes carry forty-five per cent to fifty per cent copper.   The fluid slag was first shotted in running water, and carried away in a flume to the dump; owing to the low altitude at which the furnaces were placed, the available area of dump was quickly filled, so that now mechanical appliances have to be used to elevate and remove the slag.   The capacity of each of those furnaces is 200 tons every twenty-four hours.   Special calcining furnaces, built by the Parke & Lacey Company, of San Francisco, for the smelting of the fines, have recently been built, and have a capacity of about 250 tons per day.

Several tiers of roasting kilns were constructed of heavy masonry, six feet deep, seven feet wide, and fourteen feet long, with side and end drafts, each capable of holding a thirty-five-ton charge, and all connected with a big smoke-stack, leading up the side of the mountain, to carry away the sulphur fumes.   The concentration of the smoke from all those kilns proved so annoying to the workmen, that a simpler process is now in practice of burning the ores in various heaps, distributed at isolated and suitable points along the railroad toward the mine.   The ore is received in bunkers from the mine cars, and piled around flues made of wood, on the natural ground.   After the piles are properly shaped, and proper precautions taken for providing a suitable draft, the wood is ignited, and the piles slowly burn for sixty days, when the ore, less the excess of sulphur, is again loaded on the railway cars and conveyed to the smelter.

This system is much cheaper than the kiln burning, as the great heat of the burning ores in the latter causes breaking and distortion, which involve a constant expense in repairing them.   It is also more agreeable to the laborers, as the isolation of the heaps keeps the smoke nuisance down to a minimum.

The necessary power for supplying the smelting plant is furnished by water pressure in the winter time, and by steam from wood in the summer season, when the water runs short. The phenomenal drought of 1898 caused the erection of a pumping plant in the Sacramento River, near Spring Creek railroad bridge, which forces 2,000,000 gallons daily, against a head of 150 feet, through eighteen-inch and sixteen-inch pipe, to the smelting site. As the Sacramento River fluctuates in height as much as thirty feet, a portable railway has been provided, to pull the pumps out of the river bed in the flood seasons, when the water from this source is unnecessary.

During the year 1898, the quantity of ore taken out of the mine was 221,895 tons, averaging 8.42 per cent copper, or 740 tons per working day, against 165,060 tons, averaging 8.56 per cent copper, or 550 tons per day, in 1897; an increase of 56,835 tons, or 190 tons per day.

The quantity of ore smelted was 168,541 tons, producing 10,721 tons of copper, against 97,185 tons, producing 7,238 tons, in 1897. The quantity of finished copper marketed was 8,273 tons, against 6,025 tons, in 1897, an increase of 2,248 tons. The net profits, after deducting all costs and charges, amount to $815,000, against $315,000 in 1897. Seven and one-half per cent dividends, amounting to $468,750, were paid out of the profits, and the balance of $346,250 was placed to the credit of the reserve and depreciation fund. This would show a profit of $4.84 on each ton of ore smelted during 1898, when the price of copper was normal. With the splendid waterpower available for refining the copper by the electrolytic process, at Keswick, there is no reason why the profits could not be increased fifty per cent by producing fine copper at Keswick, and save the freight of $15 per ton on the sulphur in the matte now shipped to New Jersey, and the added extra expense of treating it at separate establishments there.

The starting of the Keswick smelters has proved of great advantage to the miners of Shasta County, who are able to dispose of their ores from the numerous quartz veins traversing the district, and have them smelted at a moderate cost. The silica of these ores is necessary for a flux, owing to the lack of this mineral in the Iron Mountain sulphide ores. The company has never favored the policy of procuring its own quartz for this purpose, preferring to see the individual enterprise of the small mine-owners encouraged, rather than attempt to make developments in this field.

THE BALAKLALA MINES. — This group consists of twenty-six claims, in Flat Creek Mining District, and is distant north-westerly from Redding sixteen miles, and north-easterly from Iron Mountain, four miles. The elevation of Kennet, on the Sacramento River, six miles distant in an eastern direction, is 670 feet; Squaw Creek, at the end of the wagon road below the mines, 1,600 feet, and of the ground covered by the mines, from 2,200 feet to 3,200 feet. They all lie on the south side of Squaw Creek, on the northerly slope of a mountain spur dividing it from Spring Creek, and half way up the mountain from the creek. The altitude of this spur varies from 3,200 to 3,600 feet.

They are in a porphyritic belt, from one and one-half to two and one-half miles wide, bounded on the west by a well-defined slate formation, which crops out and runs north-easterly across the country, and near which all the valuable copper ore bodies, so far discovered, are found.

The surface ore is an oxidized iron, carrying gold and silver, with slight traces of copper, and underlying this is found the heavy copper sulphide ore, averaging three per cent copper, $2 in gold value, and $2 in silver value to the ton. As the owners are people with moderate means, no heavy development work has been done. At one point, on the Huckleberry Claim, near the center of the property, one mass has been cross-cut, and a width of 135 feet of solid sulphide ores exposed. Various other tunnels and drifts prove a body of ore in sight of 500,000 tons, of the above grade. The surface of

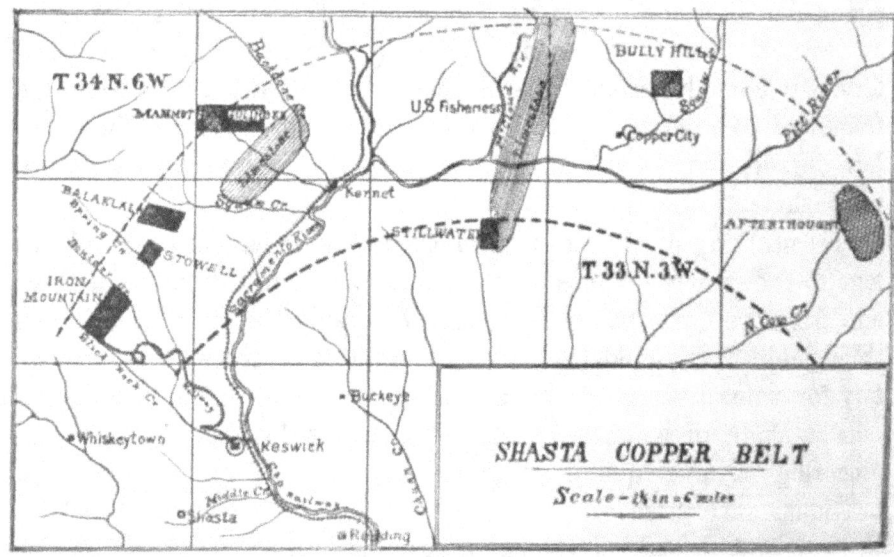

SHASTA COPPER BELT

Scale—⅛ in = 6 miles

the ore bodies has a general slope parallel with the ground surface, and dipping at an angle of 30° to the north. The walls, where exposed, strike N. 75° E. No deep workings or drill holes have been made to prove the depth of the ore mass, but from the surface showings made, and the occurrence of similar masses in different localities, I have little doubt that they will prove to have considerable depth.

A boring with a diamond drill, made in 1897, on the Mule claim, at a point about 1,500 feet east of the Huckleberry mass, showed, at a depth of 120 feet, eight and three-tenths per cent copper ore. Recent development work in a tunnel on the Mule, 600 feet north-east of the bore hole, showed two and seventy-nine hundredths per cent copper, $2.63 gold value, and $2.18 silver value per ton.

The whole country is covered with a growth of pine, so that abundant timber for mining purposes can be had at a moderate expense. All the future smelting sites for this base-ore country will be near the Sacramento River, which is a large permanent stream, and on the west side of which the California and Oregon system of the Southern Pacific Railroad runs. There are no agricultural lands in the vicinity whose products would be injured by sulphur smoke, but there are good quartz districts, both to the east and west of the Sacramento River, yielding gold-bearing ores, which can be purchased at reasonable prices, and thus increase the profit in the operation of the smelting process.

Although the Southern Pacific Co. has an exclusive monopoly of all the transportation of this section, its treatment of patrons in freight matters has been fair, and I am satisfied it will make reasonable concessions, in the way of making light rates, to all enterprises worthy of encouragement. There will be a lucrative field for an electrolytic plant, to treat all the copper matte, which will soon be produced here, and which must now be shipped east, at a cost of $15 per ton, to be refined. The metallic copper can then be shipped for one-half the present freight expense. Abundance of water power may be found for power purposes in the Sacramento, McCloud, and their many tributaries, and installed at a cost of $225 per horsepower. Steam power, as at present developed from wood and coal, costs $110 per horse-power per year; so the economy of electric power installation is at once apparent. The limestone cropping north of Squaw Creek, and along the two Backbone Creeks, will furnish a practicably inexhaustible field for procuring this fluxing material, necessary for those ores.

SURFACE ROASTING — KESWICK SMELTERS.

The Stowell Mines consist of seven patented claims, immediately south-west of the Balakalala. They have the usual gossan croppings on top, but only limited development work to show the extent of sulphides. The Mammoth group, on Backbone Creek, has some recent development work, which indicates an extensive body of sulphide ore, showing at one point a length of 120 feet of ore along the lode, which has been cross-cut for eighty feet without showing the wall. It is the property of R. M. Saeltzer, of Redding, and consists of eighteen claims, with 640 acres of patented land. The Goslinsky Mine lies east of the Mammoth, and has been bonded to Lewisohn Brothers, of New York, who have, for the past six months, been doing development work on it. Considerable copper indications are shown on Stillwater Creek, but the largest deposit has been proven to exist at Bully Hill Mine, near Copper City. This property lies twenty-five miles northeast of Redding, and is north of Pitt River, and has been recently purchased, with adjacent claims, for the sum of $225,000, by Captain De LaMar, the Utah mine-owner. The formation differs from the west side belt, in that the ores occur in thin veins, but are much higher in grade, carrying thirty ounces of silver and $9 in gold value to the ton. Two long tunnels, over a 1,000 feet each, have been driven into the hill, intersecting numerous veins of ore. The proverbial energy of the new owner gives promise that, within the coming year, the copper and bullion from this mine will add substantially to the mineral output of the State. With the preliminary work now being done, 1 have little doubt that in less than five years the copper from this section will exceed in quantity that now produced from Anaconda, where the output has recently fallen off heavily.

The ores east of the Sacramento River, in the Copper City and Cow Creek districts, are much more difficult to smelt than the Iron Mountain ores, as they are composed of heavy spar, zinc, antimony, sulphur, copper, arsenic, iron and sulphate of lime. William Kemp, the well known metallurgist, of Redding, has succeeded in smelting the most refractory of the Cow Creek ores, some of which contained over six and ten per cent of zinc. Lack of capital prevented his process being commercially applied to the reduction of those ores in bulk. Nearly all the owners of those properties are people of moderate means, unable, for lack of money, to erect plants on their own account, and, for this reason, a profitable field exists in this section for the investment of capital under intelligent supervision.

East of Cow Creek, the copper zone seems to be buried under the

Lassen lava flow, not appearing again till it shows up in Mountain Meadows, Lassen County, and Big Meadows, Plumas County, where it appears in similar strata to the Shasta surroundings.

The great altitude, and long distance from a main line of railway, will tend to discourage, for the present, active investments in this field.

[There is a large deposit of copper ore on Eel River, in the southwestern part of Trinity County. The above is a photograph of a huge boulder, or solid mass of chalcopyrite ore, which, from measurement, is calculated to contain 6,400 tons above the ground. It carries ten per cent copper, and $7.35 per ton in gold and silver.

This huge mass apparently slipped down from the ledge, situated some 300 yards above, where there is some 10,000 tons of the same ore in place, and lays on a little flat, with several thousand tons of float in smaller form. The property is known as the Island Mountain Copper Mines, and is partly owned and controlled by Frank A. Leach, Superintendent of U. S. Mint.]

THE SIERRA, OR COPPEROPOLIS BELT.—The veins in this region, in which copper occurs, are essentially fissure veins, and entirely different from the lenticular mass formation of the Shasta copper belt. They are all found in the metamorphic slates which lie west of, and are parallel to, the Mother Lode. The Spenceville mines, in Nevada County, are in the northern end of this belt, and have been worked, with moderate success, for a number of years. They claim to have an ore mass forty feet wide, carrying four per cent copper. The Calaveras County mines, including the Campo Seco and Copperopolis, have produced a good deal of copper in the past. The former contain both gold and silver, and the latter none. The better ores were smelted, and black metal produced, and the poorer ores piled in large heaps, roasted in the open air, and leached. From the liquid thus formed, cement copper was produced, which was marketed for paint purposes. Those mines were actively worked in the '60s, when copper was at a very high value, and practically abandoned in 1870, when the price dropped. Operations were resumed in 1887, and, through a combination of circumstances, ceased in 1893, when the copper product of the State dropped down to 120,156 lbs. for 1894. The shipping and refining charges, and the lack of concentration of the matte, are the apparent causes for the cessation of operations. El Dorado County has several veins of copper ore, from ten to seventeen feet wide. Copper has also been found in Amador, Fresno, Plumas, Madera, Del Norte, Humboldt, Marin, Mendocino, Riverside and San Bernardino counties, in small quantities, and considerable shipments of good copper ore, from many mines in those counties, are made to the Selby Smelting Works, at Vallejo Junction.

The records of the Mountain Copper Company, of Shasta County, show a product of 21,442,000 pounds for 1898, and of the State Mining Bureau, for the whole State, of 21,543,229; this would leave a product only of 101,229 pounds for all the mines outside of Shasta County.

COPPER OUTLOOK FOR CALIFORNIA.—I think the base-ore region of Shasta County will furnish an inexhaustible field for centuries in the copper mining industry of this State. The surface ores—except in the case of Iron Mountain—are unexplored. There is no reasons why the ores of this section should not go down to depths similar to those of Montana and Michigan. They are all of better quality than those of Rio Tinto, in Spain, which are now mined, and which were worked by the Phœnicians ages ago. The uses of copper for domestic utensils and electrical appliances are gradually extending, so that a demand for it, at a reasonable price, is always assured. Spasmodic inflations in the stock market will not tend to limit its production; neither will the attempt to substitute aluminum extensively be a commercial success, so that it is safe to predict a prosperous era in the development of the hitherto unexplored copper masses of this State.

The following table shows the number of pounds of copper produced in the whole State for the past ten years, and by the Mountain Copper Company, of Shasta County, for the past three years:

Year.	Mountain Copper Co. Pounds.	California. Pounds.
1888		1,570,000
1889		1,700,000
1890		1,600,000
1891		3,750,000
1892		3,200,000
1893		2,825,773
1894		120,156
1895		225,650
1896	1,820,000	1,971,545
1897	14,000,000	14,129,920
1898	21,442,000	21,543,229

The Use of Machine Drills in Stoping.

BY

B. L. THANE,

CLASS OF '99, COLLEGE OF MINING, UNIVERSITY OF CALIFORNIA.

HE mining industry has, within the past few years, taken a new impetus in all its branches. New mines are being opened every day, while old ones, which have been either working at a loss, or have been compelled to shut down, are now gradually being re-opened and placed on a paying basis.

This growth and new life is due, in a measure, to the discovery of new mining districts, but the most important cause of the progress is the wonderful advance that has been made in all branches of engineering.

By means of the important inventions and discoveries that have been made in mechanics, chemistry, and electricity, we are now able to work ores which, only a short time ago, would have been regarded as worthless. We find, for instance, the electrician eliminating two of our greatest difficulties, those of distance and superior elevation. While the mechanic has brought to the highest degree of perfection, not only ore-crushing and hoisting machinery, but all such devices as pertain to the mechanical handling of ores. At the same time, the metallurgist, with his combined chemical and mechanical skill, has helped us to extract the precious metals from the most refractory of ores.

But, when we turn to examine methods in actual use for actually mining or breaking ore, we are surprised at the small amount of progress that has been made in this direction. In the handling of the ore after it is broken, almost everything has been done to reduce the operation to its simplest and cheapest mechanical form; and every form of engineering skill has been brought to bear upon the problem. But the work of the engineer usually ends with the erection and installation of the machinery that he has designed. His interest rarely reaches to the details of breaking ore; this is supposed to be the peculiar province of hand labor. This branch of the business is usually under the immediate supervision of men risen directly from the ranks, who have learned to do exactly what was done before them, and, accepting without question that this is the only method, have bent their energies to bringing that method to the highest degree of perfection.

There is no intention of denying the value of such work, for it is done well, and the energy, the system, and the skill with which it is carried on are all of the highest order. But this does not prevent one, who has marked the revolution which mechanical devices have wrought in other branches of mining, from wishing to see the proper appliances brought to bear upon the problem of breaking ore.

The use of power drills in stoping is one of the methods that have been proposed to meet the problem. Although their use for this purpose is not new, the method has not, in my opinion, received the consideration that it merits. We find the big machine in the shafts, tunnels, and drifts a decided success, and in use almost everywhere; but when we look for "the little one" in the stopes and on the vein, we find instead the miner and his hammer pounding away, just as his fathers did before him. Perhaps I should make some exceptions here, and mention what has already been done in this line, but it has been so little, especially in this country, that it is hardly worth mentioning.

Attempts have been made to use machine drills in stoping, from time to time, by enterprising men. Most of these, from one reason or another, have resulted in failures; and failure once made is sure to be heralded far and wide, only to be lived down by the slow growth of success.

It was through the failure, or, rather, the repeated failure, and final success of one of these attempts, that I personally witnessed, that caused me to become personally interested in this subject, and, as it bears directly upon the question before us, I will endeavor to give a short outline of my experiences and their results.

While working underground in the Chief Mine, at Sumdum, Alaska, I had the good fortune of being placed as helper, or "chuck man," on one of these small machines, known as a "Baby Ingersoll." It was used for stoping and raising on the vein. It was, therefore, an excellent example of the problem before us, as the vein varied from one to three feet in width, the most unfavorable for machine work;

besides, the place of working was one of great difficulty, as it lay along the shaft, making any mistakes dangerous and costly. Thanks to my partner, who was a machine man *par excellence*, no such mistakes were made.

This was the fourth attempt to use the little drill at the mine, and the first to register a success. My partner, who was a late arrival from the Cœur d'Alenes, found the little drill thrown aside in disgrace, and covered with rubbish. But he knew what it could do, and begged to be allowed to give it one more trial.

Of course, it is out of the question to use the big machine, on account of its size and weight, for stoping in such narrow veins, even though in its own sphere it is already a success.

The cause of failure, in the use of the small drills, was due, in this mine, at least, to the fact that the men who had attempted to work them were used to the big machines, where the strength of stroke and larger size of the hole overcame many of the difficulties met with by the smaller ones. The latter naturally requires more care and skill, but this care and skill is not so great but that any man of ordinary ability can master it.

The drill used by us weighed 170 pounds when mounted on a tripod. We found the latter to be much better than a bar, as it allows greater freedom of motion, something absolutely necessary in following a hole, and takes much less time in moving from one place and setting up in another. With the machine that we used, the bolts are so arranged that a drill may be driven in any direction by a simple manipulation of the legs of the machine, which is easily acquired by practice.

The chief cause of trouble in machine drilling arises from the many "slips" and layers of alternately hard and soft rock, which are found running in every conceivable direction in vein formation. A drill once started, for instance, will run freely till it strikes one of these slips, which it will, naturally, have a tendency to follow; or, again, if driven through soft seams, it comes in contact with harder material, it will immediately begin to slip along the new surface. This tendency, if allowed to continue, will cause the drill to bind against the bend in the hole, and will not only prevent the drill from entering the hole, but will hold it fast and prevent its withdrawal. This not only causes much annoyance and loss of time, but it frequently causes the total loss of a hole.

With a big machine, where the power is sufficiently great, the drill is driven ahead and pulled out regardless of the slips and bends, and this difficulty is more easily surmounted ; but with the smaller one, constant watch must be kept, and the instant the drill starts to slip or bind, the machine must be re-adjusted to follow the hole, and the stroke shortened, as in starting.   To do this it is seldom necessary to stop, for a perfect understanding of the drill, and between the men working it, allows the chuck-tender to loosen the proper bolts while the machine is still in motion, and as soon as its position is changed enough to allow it to run smoothly, it is fastened again. From personal experience, I think it is not only unnecessary but unwise to attempt to drill holes too deep.   This practice is usually the cause of much trouble and loss of time.   From three to six feet is plenty deep enough for such work, generally speaking.   This is all the more the case, as the time required to move the machine from one place to another is very small indeed.

A tripod necessitates the use of a platform to work on, but as it is very easy to build one out of lagging, this is a matter of very small consideration, especially as the planks may be used over and over again ; and a platform once built saves time in the end, as one can move about freely and rapidly, and with less danger of accident.

So far the economical side of the question has not been mentioned, but simply the practicability of the small machine.   With regard to this new side of the question, and a most vital one, it is almost impossible to collect data of any general application, because of the small amount of work that has been done with the small drills up to within the last few years.

However, I am informed by Capt. Thos. Mein, recently superintendent of the Robinson Mine, in the South African Gold Fields, that in this great district the use of small machines, on narrow veins, is an established fact, and is carried on with great success.

In our own State, I find that within the last six months some three plants, of five small machines each, have been put in ; that in the North Star, at Grass Valley, is, perhaps, the best known.

My own experience with this work is about as follows :

With the "Baby Drill," two men, my partner and myself, were able to drill about forty feet a day in hard quartz, full of slips and seams.   This work was done between the hours of nine and half-past four. The rest of the shift was sufficient for us to do all our own timbering, to build our chutes and ladder-ways, to shovel away our broken rock, and to keep the place in ship-shape order.

Circumstances favored me, in the following way, with an opportunity to compare the machine and hand labor to advantage.   During the holidays, the compressor was shut down for a week or so, preventing the use of the machines ; during this time we worked in the same stope, using now the hammer to drive our holes.   The comparison is an excellent one, for the following reasons : All the conditions were the same in both methods; the rock was the same throughout; the size of the vein and the nature of the working places were the same for both methods, and as we were situated away from the other workmen, the quantity of rock broken daily was accurately measured.

The results of the comparison were all in favor of the drill, for we found that for an average of the week's work, we were able to break down just half the ground that we did before, in the same time; or, in short, that two men are able, under similar circumstances, to do twice as much work with the machine as they can do by hand. There is also another incidental advantage, that of ventilation.   One of the greatest difficulties of the superintendence is to see that the men are properly supplied with fresh air, particularly in making upraises. This difficulty is entirely eliminated when machine drills are used. It must be evident, from these considerations, that the small machine drill is practical and economical, as far as labor is concerned, when it is properly used.

There still remains the cost of drills, compressors, power and pipe line, etc.   It is unnecessary to go into these questions in detail, because each case must be settled by itself, with a full knowledge of the situation, source of power, size of plant, etc.   But with an abundance of water power, and the elimination of distance and position made possible through electricity, the running expense for power is very slight, and the first cost of plant is within the reach of any stable company capable of either opening or running a mine.   While, therefore, a certain increase of expense for power and plant is necessitated, the reduction in the expense for labor is just one-half as much more than offsets this outlay.

The reduction in the number of hands needed for breaking ore is one of the chief causes of the prejudice that many miners have against the machines.   But this prejudice is entirely unwarranted, for every cause that increases the output of the miner, and decreases the expense of mining, only tends to open new mines, and enables old ones to increase their force, thus giving plenty of work to the supposed surplus of miners.

On my return to the mine, I hope to secure further and fuller data on this subject, in order to convince others of what I firmly believe myself, for I think it is only a question of time before we will have the little machine everywhere high up in the stopes, pounding its way into prominence and success.

INGERSOLL SERGEANT "BABY" MACHINE DRILL.

NOTE.—Through the kindness of Captain Thomas Mein, I am able to add the following extract from a private letter of the well-known South African engineer, Mr. L. T. Seymour, to him. The letter is dated Johannesburg, March 15, 1899. The extract reads as follows:

"STOPING.— Hand stoping in medium ground, up to four feet wide, costs five to six shillings per ton, with boys (Kaffirs). We are, at the Heriot and Gold Deep, stoping with machine drills at that price, on six foot stopes. Each miner now runs two and three-quarters inch Ingersoll Sergeants, or three and one-eighth inch Little Giants. As a rule, a drill of this size costs (without dynamite) from £115 to £125 sterling a month to run, for all expenses, including depreciation and interest. The miners' wages come to from £58 to £63 out of this, or almost exactly half. Therefore, two or three drills, distributing the white wages over them, reduces this cost largely. I believe, in three years' time, we shall do practically all our stoping with small machines on the little stopes, and with large ones on everything over four feet. The two and one-half inch Little Giant works very well.

"At some mines we make the man in a boys' stope run a small machine as well. On this basis, on a four-foot six-inch stope, each small machine stopes 279 tons per month, at a total cost, including dynamite, of 5s. 9d. per ton, stoped; and with the large machines, as high as 960 tons per white man per month, when he runs two drills."

These figures tend to corroborate the position taken by Mr. Thane. Machine drills have been used in stoping at several mines in California for some time, notably the North Star and the Utica. Their advantages over hand labor as to cost will depend almost entirely on the cost of power. Where the latter is cheap, there can be no question as to their advantage. The development of the water-power resources of the State, that is now going on so rapidly, will be surely followed, in the near future, by the more extensive use of the machine drill in stoping.

S. B. CHRISTY.

On a New Form of Scorifier and the Results Obtainable by its Use.

BY

ERNEST H. SIMONDS.

THE relative merits of the scorification and the fusion assay of ores for gold and silver have recently been the subject of discussion by many trained metallurgists and chemists of the very highest reputation and lifelong experience, in our own Institute* as well as in other technical societies and publications. It will be conceded by almost everyone that the superiority of scorification lies largely in its simplicity, its wide range of application with little or no change of fluxes to adapt it to different ores, and its ease of execution; its great and fatal weakness as regards gold ores is that a single scorifier will not treat at one operation, and with the production of a lead button small enough for cupellation, as much ore as is necessary for a reliable result. Without entering into a discussion of disputed points, I wish to bring forward a slight modification of the scorification method which, without sacrificing any of its advantages, permits the use of from one-fourth to one-half of an assay ton of the sample, and thus, in most cases, does away with the serious weakness stated above.

The improvement lies solely in a change in the form of the interior of the scorifier, the whole idea being apparent to any assayer on simple inspection of the accompanying cuts, which show sections of the standard scorifiers with which we are all familiar and of scorifiers of the new shape.

The idea occurred to me while developing, before a class in assaying of the University of California, the theory of the scorification

---

* Trans. A. I. M. E., Vol. XXIV, pages 867–872.

assay of ores, as given in Goodyear's translation of Bodemann and Kerl's *Assaying*,* page 112, where it is shown that: "The chemical reactions in the scorification or slagging process depend partly on the roasting of the ore that takes place during the smelting, partly upon the peculiar power of the oxide of lead produced to form a slag with all earths and oxides of the base metals, and thus, if present 'in sufficient quantity, to act as a universal flux; and besides, also, very essentially upon the influence that litharge exercises upon metallic sulphides, and which commences actively even at a moderate red heat." I was showing before the furnace that, as is the case, if one-half A. T. of a pure galena, without other metallic sulphides, be

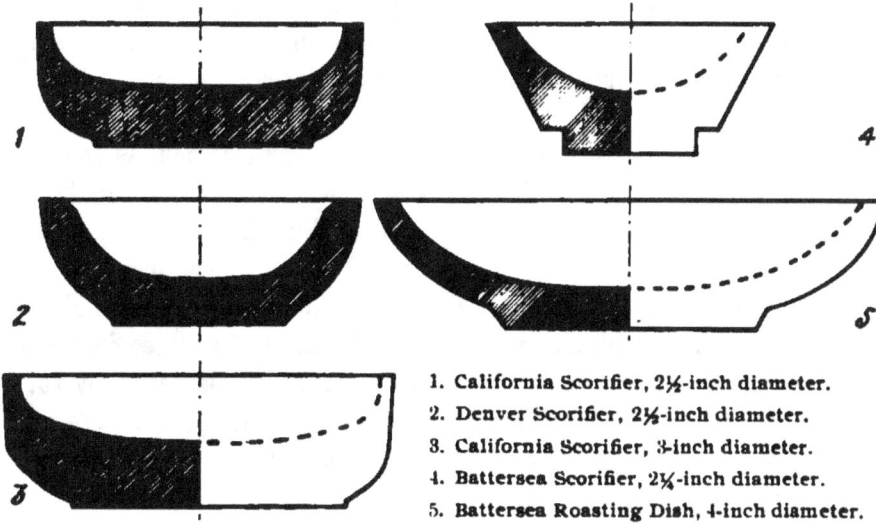

1. California Scorifier, 2½-inch diameter.
2. Denver Scorifier, 2½-inch diameter.
3. California Scorifier, 3-inch diameter.
4. Battersea Scorifier, 2¼-inch diameter.
5. Battersea Roasting Dish, 4-inch diameter.

THE CALIFORNIA SCORIFIER.

placed in a scorifier of the ordinary shape, without any lead or other fluxes, and roasted in the muffle, first at a medium red, and toward the end at a full yellow, the ore is perfectly scorified, with the production of a clear slag and a bright button, which, upon cupellation, yields the same result in ounces of silver per ton of ore as the usual charge for the scorification of galena ores of one-tenth A. T. of ore, fifteen to thirty grams of lead and one-half gram of borax glass.

The idea presented itself to me that the same effect might be obtained, to a greater or less extent, with pyrite and all other ores, by the use of some lead, if the scorifier were given a shape similar to a roasting dish; and investigation has proven this to be the case. By

* A Treatise on the Assaying of Lead, Copper, Silver, Gold and Mercury. Translated from the German of Th. Bodemann and Bruno Kerl by W. A. Goodyear. John Wiley & Sons. 1889.

making the bottom of the interior of the cup flatter, the charge is spread out, and more surface exposed to oxidation by the air; hence, the oxidation is done nearly all by the air, and the litharge formed has little burning to do, but need only slag the oxides produced, with the aid of the borax added. But a flat, shallow bath favors the oxidation of the lead as well as the oxidizable constituents of the ore, since the lead cannot be covered over until only a little remains. The result of this is that not only is the weight of the lead button to be cupelled greatly reduced, but a correspondingly large amount of litharge is added to the slag, which is very effective in fluxing the gangue of the ore and the base metal oxides produced by the roasting, and in completing the oxidation of the last traces of such elements as sulphur. Thus, the amount of work which the litharge has to do is dinimished at the same time that the quantity formed is increased.

The result of such a saving at both ends is that, as the results below show, one-half A. T. of ore, in general, can be perfectly scorified at one simple operation in a scorifier three inches in outside diameter to a fifteen to twenty gram button, by the use of a universal charge of seventy-five grams of lead and two and one-half grams of borax glass; while one-quarter A. T. can be scorified in a dish of this shape, two and one-half inches in outside diameter, by forty to sixty grams of lead and three grams of borax glass, with the production of a button of the same weight. The change is so simple, and the gain is so decided for gold ores, that it is rather surprising that no one should have hit upon the idea before.

The following extracts from notes of work done by the writer in the assaying laboratory of the University of California will serve to show the results obtainable:

Experiment 1. The sample consisted of rich gold ore containing seventy per cent of quartz, twenty-five per cent of chalcopyrite, a few per cent of galena, and much coarse gold. Ran a charge in a four-inch Battersea roasting dish, with the rim ground down to three and one-fourth inches outside diameter, made up as follows: Ore, one-half A. T.; gran. lead, seventy grams; borax glass, two grams. Charge placed in a muffle heated to a low yellow and melted for three minutes with the door closed. The door being opened and a good draft maintained in the muffle, a complete bull's-eye formed in fifteen minutes from the time of putting in, but effervescence continued a little longer. Poured in thirty-five minutes from the time of putting in, when the bull's-eye was one-half inch in diameter. Slag fluid and clear; button bright, soft, and malleable; dish lined with only a thin glaze of clear slag. Weight of button, nineteen grams; this could have been considerably reduced by allowing the scorification to go on five to ten minutes longer. Dore 11.60 mg., Au 9.03, Ag 2.57, Oz. Au 18.06, Oz. Ag 5.14. Mean of five scorification assays, each one-fifth A. T., by the regular method: Oz. Au 18.19, Oz. Ag 5.17. Absolute agreement cannot be expected on account of the large amount of coarse gold in the sample.

Numerous experiments, on various samples, show that a charge of one-half A. T. of ore, seventy-five grams of lead, and two and one-half of borax glass works well on any ore with a quartz gangue and from none to seventy per cent of pyrite, chalcopyrite, galena, or almost any metallic sulphide or mixture of them, being thus a universal charge for almost all gold and silver samples, except concentrates, metallurgical products, and very rich silver ores.

Experiment 2. The sample consisted of vanner concentrates from an important mine at Jackson, Amador County, California, and are typical vanner concentrates of the Mother Lode. It contained about eighty-five per cent pyrite, with small amounts of silica, arsenopyrite, chalcopyrite and galena; no amalgam or coarse gold.

Ran a charge as follows, in a scorifier of the new form, three inches in diameter, as shown in Fig. 3: Concentrates, one-third A. T.; gran. lead, seventy-five grams; borax glass, three grams. A bull's-eye formed in ten minutes; poured in thirty-five minutes from the time of putting in, when the bull's-eye was one-half inch in diameter. Slag fluid and perfectly oxidized, lead button was bright and malleable, and the scorification was in every way as perfect as any by the regular method. Result: Oz. Au 4.29, Oz. Ag 0.78. The combined parted gold from four charges, each one-tenth A. T. of concentrates, forty-five grams Pb and one-half gram of borax glass, in two-and-one-half inch Denver scorifiers gave: Oz. Au 4.29, Oz. Ag 0.67; two more of the same charges gave: Oz. Au 4.30, Oz. Ag 1.05.

When one-half A. T. of a sample containing over seventy per cent of pyrites is scorified in a flat dish with the charge given under Experiment 1, the slag is not clear, but the results for gold are little, if any, below those found when only one-third A. T. is used.

The last three experiments, now to be given, were made to submit the method in scorifiers of the new shape to severe tests as to its rigid accuracy compared with the results on the same samples given from scorifiers of the standard form. An outline of the plan adopted for the tests is as follows:

Three typical and rich samples were selected and each was assayed in duplicate by the standard method and singly by the new method, the three scorifications and cupellations for each ore being done together in the muffle at the same time, under as nearly as possible the same conditions. Charges one and three were done in scorifiers of the standard shape and size, using one-half gram of borax glass and the amounts of lead and ore indicated with each; charge two, with all three ores, was scorified in a four-inch Battersea roasting dish (the portion of this dish touched by the charge has the same shape as the flat three-inch scorifier shown in Fig. 3) with seventy-five grams of granulated lead and two and one-half grams of borax glass, the same universal charge given under Experiment 1. In all cases, two thirds of the lead was mixed with the ore, the rest of the lead and the borax glass scattered on top as a cover. The slag and cupel losses were determined for each assay of Experiments 4 and 5.

Experiment 3. The sample was a silver-lead ore, containing about equal parts of galena, quartz, and mixed sulphides, the latter largely pyrite.

	1	2	3
Ore	1-10 A. T.	½ A. T.	1-10 A. T.
Granulated lead	30 grams	75 grams	30 grams.
Dish	2¼ in. Batt. Scor		2¼ in. Batt. Scor.
Weight Pb button	12 grams	15 grams	13 grams.
Time in furnace	25 minutes	40 minutes	25 minutes.
Ounces silver per ton	270.0	270.80	269.0.

The lead button of two fell out separate and clean from the slag on breaking the latter. The cupel was not so heavily feathered as those of the one-tenths, and I took more pains to brighten the bead of this assay; it split slightly, and an examination of the bottom of the bead with a lens showed it to be pure and absolutely free from bone-ash.

Experiment 4. The sample was an oxidized ore, carrying its silver in ruby silver ore, and containing over sixty per cent malachite, the rest quartz and cerussite.

	1	2	3
Ore	1-10 A. T.	½ A. T.	1-10 A. T.
Granulated lead	45 grams	75 grams	45 grams.
Dish	2¼ in. Denver Scor		2¼ in. Denver Scor.
Weight Pb button	20 grams	17 grams	26 grams.

First cupellation:

Ounces gold................................0.3.....................0.40.....................0.45.
Ounces silver..........................320.2.................317.20.................318.4.

In slags:

Ounces gold............................None ....Trace ....Lost.
Ounces silver............................0.9.................4.64.................Lost.

In cupels:

Ounces gold............................Trace ....Trace ....Trace.
Ounces silver............................8.7.................6.50.................10.2.

Total ounces gold.....................0.3.................0.40.................0.45.
Total ounces silver.................329.8.................328.34.................329.5.

All three were poured in thirty-five minutes from the time of putting in. Button two was rather hard from copper. All three were driven alike in cupellation and nicely feathered; one was brightened well, part of the feather litharge being burned off; two brightened well and without difficulty, all the feather litharge being burned off, and the bead spit strongly but without loss; three was brightened too cold.

Experiment 5. The sample was an ore carrying about $5,000 per ton in free, fairly coarse gold, and consisting of a quartz gangue and large percentages of chalcopyrite and galena.

	1	2	3
Ore......................................	1-10 A. T............	½ A. T...............	1-20 A. T.
Granulated lead....................	45 grams............	75 grams...........	45 grams.
Dish..................................	2½ in. Denver Scor....................		2½ in. Denver Scor.
Time in furnace....................	35 minutes.........	35 minutes.........	35 minutes.

First cupellation:

Ounces gold............................250.0.................242.80.................235.6.
Ounces silver..........................400.0.................400.50.................392.4.

Mean of 1 and 3: Ounces gold, 242.80; ounces silver, 396.2.   Compare 2.

In slags:

Ounces gold............................0.1.................0.22.................Lost.
Ounces silver............................0.7.................2.32.................Lost.

In cupels:

Ounces gold............................0.7.................0.34.................1.4 (?)
Ounces silver............................9.3.................5.46.................12.6.
Total ounces gold....................250.8.................243.36.................237.1.
Total ounces silver.................410.0.................408.28.................405.7.

Mean of 1 and 3: Total ounces gold, 243.95; total ounces silver, 407.85. Compare with results of 2.

It will be noted that, in both experiments 4 and 5, the slag loss is greater by a few ounces in the large scorifiers but that this is a little more than compensated by a smaller loss on the cupel; so that the results of the uncorrected assay by either method agree very closely, the total losses being slightly less in the flat than in the standard scorifiers. It will be noted also that these experiments show that the volatilization losses, as well, are not greater by the new than by the standard method. The sample was not as finely ground as it should have been, having been ground through a number eighty screen, which probably caused the wide discrepance between 1 and 3 in Experiment 5. I have made no attempt to check the results by the flat scorifier directly against those by crucible assay; but my general experience is that results by the standard scorification and crucible assays agree very closely when both are well done.

Anyone wishing to test the method can do so by using the charges given in four-inch Battersea roasting dishes, as can be seen by comparing figures 3 and 5; the outside one-half inch of radius takes no part in the work and could be cut off. The writer found that the best results are obtained with rich ores by having the muffle a full yellow, with the draft cut off, when the charges are put in, and melting thoroughly with the muffle door closed for two or three minutes; then opening the door and flue damper and roasting at a red to orange, with a fair draft, until the "bull's-eye" has formed and roasting is fairly complete. The heat should then be raised to a full yellow, closing the door, to complete the re-action of the litharge with the ore, to collect any lead globules scattered through the slag in one button, and to set up strong convection currents in the slag to carry every particle of it against the lead. This method was followed in making the above check experiments.

So far, I have been unable to obtain quite as high results, in the first crude experiments, by the flat as with the standard scorifiers on copper matte, an ore rich in gold (thirty-six ounces) and tellurium, and on precipitate rich in silver sulphide from a leaching works; but believe that better results might be obtained by more experiment and modification of the details of working. This method can no more claim to be superior in all cases than any other method of chemical determination; but the above experiments show it to be accurate for gold ores of any richness and for silver ores up to 400 ounces per ton. It is certainly very simple, convenient and universal in application as compared with other methods, especially for ores having a high and varying reducing power.

No attempt will be made to secure a patent; indeed, the idea is probably not patentable.

I wish to propose for a scorifier of this shape the name of the "California Scorifier," for the double reason that it is the result of investigation carried out at the University of California; and that it is especially adapted to assaying the ores of California, on account of their non-uniform character.

The College of Mining, University of California.

BY

PROF. S. B. CHRISTY.

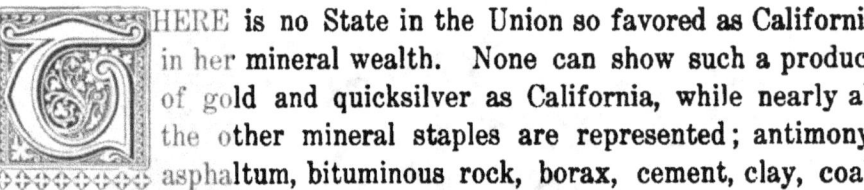

HERE is no State in the Union so favored as California in her mineral wealth. None can show such a product of gold and quicksilver as California, while nearly all the other mineral staples are represented; antimony, asphaltum, bituminous rock, borax, cement, clay, coal, copper, iron, lime, lead, mineral waters, petroleum, slate, and an admirable assortment of building stones, hardly yet appreciated. Any one of the long list is important enough to give employment to thousands of men, when the value of this kind of natural wealth is once properly appreciated by capital seeking investment.

For the proper utilization of this vast treasury of mineral wealth without waste and at the least outlay and loss of capital, it is necessary to bring to bear upon the problem all the resources which science and art have put at our disposition. The days are past in this State when the miner's pan and rocker would open up a fortune in every mountain stream, but wealth, greater than that already realized, lies before the men who bring the forces of nature to bear upon the raw material that she has hidden away in our mountains.

It was to train young men who should help to develop this natural wealth that the College of Mining was organized, and it has been its constant purpose to educate young men who should be in thorough sympathy with their work, and ready to meet conditions as they actually exist.

In getting together the material for a paper on "The Growth of the American Mining Schools, and their Relation to the Mining Industry," which was read before the American Institute of Mining Engineers, at their session at the World's Fair, at Chicago, in 1893, I became more than ever convinced that the existing relation between the mining schools and the mines was altogether too loose. This is largely due to the fact that in this country there exists no official relation between them, as there is in Germany and France, where both are under Government control. The result has been that each has got more or less out of sympathy with the other, to the detriment of both. The mining work too often gets into ruts, and rule-of-thumb methods prevail too much, while the graduates of the mining school too often show themselves utterly helpless before the first practical problems that come to them.

It seemed to me that there was no real necessity for this, and I determined that, with the help of the California miners, this condition of things should not exist in California's School of Mines.

The only way to meet it was to see that the mining student had the same kind of practical training as the practical miner gets, that is, actual work underground. There is no necessity of postponing this experience till after a student graduates. During the four years' course in college, the student has two vacation periods in each year, that, taken together and wisely utilized, can give him nearly a year of practical experience in several of the more important phases of mining work. In order to put this plan into operation, it was necessary to secure the co-operation of a large number of mine operators, and this has been done, largely through the aid and influence of the California Miners' Association, whose members have heartily approved of the plan and helped to make it a success. The plan has been very simple. It has consisted not merely in having the boys look on at other men working, but they have carried out Edward Everett Hale's motto, "Lend a Hand." And it has been good for them all, in many ways. In the first place, it gives the weaklings, and those naturally unfit for the rough and tumble of mining life, a chance to find it out before it is too late to seek some other occupation. In the next place, those who are naturally adapted to the work learn to love it and to feel the spirit of enterprise and the unconquerable determination to overcome difficulties that is characteristic of the true miner. This experience, had at intervals of their theoretical work, is of the greatest importance to them. It reacts on all their college work, and they come back to their studies with a new zest and an insight into the bearing of their studies that they never had before.

Not the least of the advantages of their experience is that they come back with a new standard of manhood.   They have learned to respect the brawn that hammers out the wealth of the world in the rock-ribbed mountains.   They have learned that a man is still a man in overalls and a flannel shirt.   They have learned the truth of the lines of Bobbie Burns:

"A man 's a man for a' that, for a' that."

There is no lesson that they can learn in college more valuable to them. Incidentally, many of them are able to earn enough to help pay their way through college, and there is a lesson in work for wages underground with the pick, the shovel, and the sledge that cannot be learned in books, and no man can hope to be a successful miner who has not been through that initiation himself.

It is our purpose to train a lot of young men who shall be ready, with heads or hands, for any problem that may come to them, without boasting and without fear.   If we do not always succeed, it will not be for lack of trying.   We could not have accomplished what we have if it had not been for the universal sympathy with our efforts, and the unfailing good will and spirit of co-operation that have been shown by the miners of California, from mine-owner, superintendent, and foreman down to the miners themselves.   For this I wish to thank them heartily, one and all.   It is a bad thing for a State when her young men feel themselves above work, and there is no surer guarantee of the success of a State than when her college men are proud to be able to use their brawn in productive labor.

I have been asked to write an account of our Mining College as it now exists and the opportunities the University of California now offers to mining students.   The following will give an idea of its organization and of its work, and the opportunities as they now exist.

THE FACULTY OF CALIFORNIA'S COLLEGE OF MINING.— President, Benjamin Ide Wheeler; Dean, Samuel B. Christy, Professor of Mining and Metallurgy; F. G. Hesse, Professor of Mechanical Engineering; Joseph Le Conte, Professor of Geology; Willard B. Rising, Professor of Chemistry; Frederick Slate, Professor of Physics; Frank Soulé, Professor of Civil Engineering; Irving Stringham, Professor of Mathematics; Sidney Cloman, Professor of Military Science; H. T. Ardley, Associate Professor of Free-hand Drawing; George C. Edwards, Associate Professor of Mathematics; M. W. Haskell, Associate Professor of Mathematics; Andrew C. Lawson, Associate Professor of Geology and Mineralogy; E. A. Hersam, Assistant Professor of Metallurgy; Hermann Kower, Assistant Professor of Mechanical Drawing; H. I. Randall, Assistant Professor of Civil Engineering.

In addition to the above professors, there are something over twenty-five instructors and assistants occupied with departmental work and instruction.

ENTRANCE REQUIREMENTS.—The entrance requirements of the Mining College have been carefully adjusted to the possibilities of the best public High Schools, and are being continuously raised as fast as the condition of the secondary schools warrants it. The present requirements may be divided into literary and scientific, as follows:

Literary.—1st. "A. Oral and written expression," the ability to read, write and speak the mother-tongue correctly, clearly and pertinently.

2nd. "English 1." The elementary English requirement which covers a critical knowledge of some eight selections from the works of standard authors.

3rd. Any one of the following subjects:

"Latin 6." Elements of Latin Grammar, Composition, and first four books of Cæsar's Gallic War.

"Greek 8." A similar requirement in Greek.

"English 14." A more advanced requirement than "English 1," involving a more extensive and critical knowledge of certain English authors.

"French 15a." The ability to read at sight simple French prose, and to translate simple English into French, and a knowledge of the principles of French Grammar.

"German 15b." A similar requirement for German to that for French.

4th. Government and History of the United States.

Scientific.—5th. "Algebra 3." Through Quadratic Equations.

6th. "Plane Geometry 4."

7th. Advanced Mathematics: "12A1," Solid and Spherical Geometry and "12A2," Plane Trigonometry.

8th. "Elementary Physics 11." The requirement implies at least a year's work daily with laboratory practice.

9th. "Elementary Chemistry 12B." This requirement implies at least one year's work with laboratory practice.

10th. "Free-hand Drawing 16." Representing not less than two years' work of not less than four hours per week. The study of light and shade and perspective, by drawing and shading from simple geometric models and from simple objects that are related to these in form.

All the above subjects are taught in any of our good High Schools at the present time. Examinations for admission are held annually, about the middle of August, in Berkeley.

THE ACCREDITING SYSTEM.—Such High Schools as may have been carefully examined by committees of the Academic Council, and whose work has been duly approved, are placed on what is called the Accredited List of Schools for the year for which they are examined. Graduates from these schools, provided they bring a recommendation from the principal, and his certificate that they have satisfactorily completed all the studies of the course named in the above list, are admitted to the College without examination. At the present time there are eighty-two such accredited High Schools scattered all over California. The records show that the students thus admitted by the accrediting system average a little higher than those who pass the regular examinations. The system has proved of the greatest service to the State of California in setting up a high and uniform standard, and has brought the High Schools and the University into a close and helpful relation that has been of mutual benefit. It has given the schools the influence of the University in favor of good and honest work, and it has shown the University people better than anything else the local needs and difficulties, and enabled them to raise their entrance requirements in such a manner as to help the schools rather than to hinder them.

EXPENSES OF STUDENTS.—Tuition in the colleges at Berkeley, including the use of the libraries, is free, to non-residents as well as to residents of the State. There are, however, the following incidental expenses to be met by the student at Berkeley :

Diploma Fee.—Each student, on entering any of the colleges at Berkeley, is required to deposit five dollars with the Secretary. This deposit, if the student graduates, will constitute the payment for his diploma. If he leaves the University without graduating, the money, less any unpaid fines or charges, will be returned to him on presentation of the Secretary's receipt for it.

Laboratory Fees.—In the laboratories a charge is made for materials actually used. This charge, for students in the elementary laboratories, amounts to from $10 to $30 per annum.

Military Uniforms.—Every able-bodied male undergraduate student is required to drill. A sum sufficient to cover the cost of the uniform, about $16, must be deposited with the Secretary immediately after admission.

Board and lodging may be obtained in private families in Berkeley and Oakland at from $18 to $30 a month. They may occasionally be had in return for various personal services in the household. The hours of recitation are such that many students reside in Oakland and San Francisco. The journey from San Francisco to the University requires from an hour to an hour and a quarter. The cost of board and lodging, in students' boarding clubs, ranges from $15 to $22 a month. A few students "board themselves" for as low as $10 a month, but this plan of living is not generally to be recommended.

There are no dormitories maintained by the University.

Other expenses are: Gymnasium locker and suit, about $4; books and stationery, from $18 to $25 per annum. A life membership in the Students' Co-operative Society costs $2.50, which amount is soon made up in the purchase of books and other supplies at prices below current retail rates.

The *dining hall* in the University grounds provides board at cost price.

SCHOLARSHIPS.— *The State of California Scholarships.*— In accordance with action taken March 9, 1897, the Regents of the University will set apart annually, out of the income furnished to the University by the State, the sum of $3,500, to be distributed equally among the (seven) Congressional Districts of the State, for the purpose of aiding poor and deserving students to attend the University. The scholarships so founded will be known as the State of California Scholarships, and will not exceed twenty-eight in number for any one year.

The Levi Strauss Scholarships.— At the same meeting of the Regents at which provision was made for the State of California Scholarships, Mr. Levi Strauss, of San Francisco, duplicated the action of the Board by providing for not to exceed twenty-eight additional scholarships, to be distributed and awarded on precisely the same terms as those prescribed for the State of California Scholarships. In formally accepting the magnanimous offer of Mr. Strauss, the Regents emphasize the fact that "the terms of his gift constitute a significant recognition that the purposes and beneficence of the University extend to every portion of the State alike."

Numerous scholarships have been founded by private citizens for special purposes and under special conditions.

UNDERGRADUATE CURRICULUM. — This curriculum includes only those studies which are absolutely essential to the efficiency of the mining engineer, and these studies are arranged with a view to concentration and close interdependence. The number of independent lines of study carried on at the same time is limited, as nearly as possible, to three. The intimate relation of practical application to theory is constantly impressed upon the student; all subjects of study are, from the beginning, illustrated and applied by exercises in the laboratory, the draughting-room, and the field; and the summer schools of surveying and practical mining held during the University vacations are organized for the purpose of affording the student a more extended application of his knowledge, and as an introduction to the practical work which he must undertake after graduation.

Students free from deficiencies may, in addition to the minimum course of study prescribed, elect four units per half-year from any courses given in the University. With a view to a greater breadth of view, students are advised to choose these Free Electives from the literary courses, such as English, French, German, Spanish, History or Political Economy. But they may follow out special lines, in Mathematics, Physics, Chemistry, Geology, Mineralogy, Petrography, Mechanics, Surveying, Electricity, Mining, or Metallurgy.

With the desire to secure "sound minds in sound bodies," all students are subjected to a careful physical examination on entering, and for two years are given a thorough course in Physical Culture. In securing a familiarity with the organization and management of large bodies of men, the course in Military Science, which extends through, is of the greatest value.

The requirements for graduation from this College, with the degree of B.S., are set forth in the following scheme. The studies are explained in detail in the description of the Courses of Instruction.

Freshman Year.	1st Half-year. (Units.)	2d Half-year. (Units.)
MATHEMATICS—(3A) Elements of Analysis (Algebra, Trigonometry, Analytic Geom.), with applications..	5	5
PHYSICS—(1) Elementary Course: Lectures and Laboratory.........	3	3
CHEMISTRY—(1) (2) Inorganic: Lectures................................	3	3
(3) Laboratory Experiments, and (4) Qualitative Analysis....................................	2	2
DRAWING—(1) *Instrumental, and (2A) Descriptive Geometry......	2	2
MILITARY SCIENCE—(1) Two exercises each week......................	$\frac{1}{2}$	$\frac{1}{2}$
PHYSICAL CULTURE—(1) Three exercises each week...................	...	...
Totals..	$15\frac{1}{2}$	$15\frac{1}{2}$

* Until, but not including, the academic year 1900-1901, Freshmen who are not proficient in the elements of Free-hand drawing will also be required to take Course 1 in Decorative and Industrial Art.

	1st Half-year. (Units.)	2d Half-year. (Units.)
Sophomore Year.		
MATHEMATICS—(3B) Elements of Analysis (Differential and Integral Calculus), with applications...	3	3
PHYSICS—(2A) General Course: Lectures..................................	3	3
(3) Laboratory..	2	2
CHEMISTRY— (5) (6) Quantitative Analysis: Laboratory..............	3	3
MINERALOGY—(1) Laboratory...	1	1
SURVEYING—(1A) Lectures, with (1B) Field Practice and Mapping	3	3
(3) Summer School, four weeks...........................	...	...
MILITARY SCIENCE—(1) Two exercises each week......................	½	½
PHYSICAL CULTURE—(2) Three exercises each week..................	...	...
Totals..	15½	15½
Junior Year.		
GEOLOGY—(1A) General Course...	3	...
MINERALOGY—(2A) Crystallography, (2C) Physical Properties......	2	1
(3) Descriptive, or GEOLOGY (2A)* Economic.......	...	2
METALLURGY—(5) Structural Metals, Fuels.............................	2	...
(7A) (7B) Assaying.....................................	2	2
MINING—(3) Lectures and Laboratory................................	...	2
(4) Summer School, one month...............................	...	...
MECHANICS—(1) Analytical..	4	4
CIVIL ENGINEERING—(8A) Strength of Materials.......................	...	4
DRAWING—(5) Graphostatics...	2	...
MILITARY SCIENCE—(1) Two exercises each week......................	½	½
Totals...	15½	15½
Senior Year.		
GEOLOGY—(2) Field Work..	2	2
(3) (4) Petrography...	2	3
(2A) Economic, or MINERALOGY (3)* Descriptive.....	...	2
METALLURGY— (6) Ore crushing, sampling and fluxing................	2	...
(8) (10) Gold, Silver, Quicksilver.........................	2	2
(11) Lead and Copper...............................	...	2
MINING—(1) (2) Lectures..	4	4
Thesis..	...	...
MECHANICS—(2) Hydrodynamics......................................	3	...
MILITARY SCIENCE—(2)...	1	1
Totals...	16	16

Each unit in the above curriculum involves three hours' work, including the time spent in preparation.

---

* Mineralogy (3) and Geology (2 A) are given in alternate years to Juniors and Seniors together.

GRADUATE COURSES.—Students desiring to pursue advanced or special work after graduation will be afforded every facility that the libraries, laboratories, and collections of the University offer.

A candidate for the degree of Mining Engineer must be a graduate of the College of Mining of this University, or must have successfully completed a course of study equivalent to the regular undergraduate course of that college, and must pass a satisfactory examination in the following subjects: Mining, ore dressing, petrography, economic geology, the elements of thermodynamics, construction of mining machinery, and political economy. The applicant must have had at least one year of actual practice in the field in the course chosen, and must show, by an original memoir upon some subject bearing upon this profession, power to apply to practice the knowledge acquired. The degree will not be given earlier than three years after completion of the undergraduate work.

A candidate for the degree of Metallurgical Engineer must pass a satisfactory examination in the following subjects: Metallurgy, ore dressing, assaying and analysis, the elements of thermodynamics, construction of furnaces and metallurgical machinery, and political economy. In all other respects the conditions are the same as those required for the degree of Mining Engineer.

Special Students.—The regular undergraduate course is recommended to beginners in preference to any other. But in cases where it is impossible to follow it throughout, students of mature years may concentrate their entire attention upon mining, metallurgy, and assaying, together with the subjects directly related, provided they have the necessary preparation for the courses they elect.

Special students in the College of Mining must have a preparation equivalent to the regular entrance requirements in either (*a*) Mathematics: Algebra (3) and Plane Geometry (4), or (*b*) Physics (11). or (*c*) Chemistry (12*b*).

Since all of these subjects are prerequisite to many of the mining courses, it is highly desirable for special students in mining to be prepared in all three of these subjects. In order to take advanced work in mining, special students must have the same preparation for the courses they elect as regular students. The prerequisites for each of the courses given are stated under Courses of Instruction.

Most of the courses begin in August of each year and continue throughout the academic year. It is, therefore, important for special students to begin their work in August. Those who enter later do so at a great disadvantage and will not usually be admitted unless they have advanced standing. It is impossible for a special student to complete the work of any course in less time than a regular student unless he enters with advanced preparation. A course of this sort lasting a year is the shortest from which any advantage can be expected.

The following descriptions, condensed from the University Register, give an idea of the opportunities for instruction which the University of California offers to mining students:

MUSEUMS. — The several collections composing the University Museum have, by action of the Regents, been more closely co-ordinated with the several departments to which they pertain than was formerly the case; and, owing to the extremely crowded condition of the University buildings, it is possible at present to place on public exhibition only a very small portion of the collections, comprising mainly the ethnological collections, in the upper corridor of South Hall, and the mounted mammals and birds, in East Hall.

The materials are obtained from many sources, chief among which may be named the following: (1) The State Geological Survey, which contributed not merely its extensive collections of minerals, of fossils, of marine and land shells, but especially that series of skins of California birds which were the type-specimens of the species described in its report on ornithology. This nucleus of the Museum was subsequently enlarged by a set of Wardian casts made up of selected types of the larger fossils. (2) The Pioche collection of shells, fossils, minerals, and ores illustrative of Pacific Coast forms, though principally from South America. (3) The collection of D. O. Mills, containing a large series of California land shells, and of native ores and rocks. (4) The collection of James R. Keene — a costly group of minerals. (5) The various expeditions of the Zoölogical Department. Recent additions have been numerous and valuable. Type-specimens of new species are placed in the Museum, as are also specimens of various species, genera, etc., illustrating interesting cases of variation, geographical distribution, and other facts in the natural history of the California marine and terrestrial fauna.

Besides the museums of botany, zoölogy, and ethnology, the following are of more especial interest to mining students:

Paleontology.—The collections of the Geological Survey, which have become the property of the University, contain either the types or representative specimens of nearly all the known California fossils.

In addition to this, the paleontology of the State is illustrated by a collection of splendidly preserved fossils, collected by C. D. Voy, and presented to the University by D. O. Mills, Esquire.

A large collection, purchased some years ago by legislative appropriation, represents fully the development of invertebrate life in North America. A carefully selected series of crinoids, from the celebrated locality near Crawfordsville, Indiana, is one of the most interesting features of the paleontological department of the Museum.

A number of valuable invertebrate and vertebrate fossils have been donated to the University during the past year.

Structural Geology.—A number of fine models of the most interesting geological regions, chiefly of the United States, and embodying the results of the researches of the United States Geological Survey, but partly of other countries. To these has recently been added an excellent relief map of the peninsula of San Francisco, from latitude 37° 30′ to the Golden Gate, on a scale of two inches to the mile, the map having been constructed by the Department of Geology.

Economic Geology.—Sets of specimens from numerous mines on the Pacific Coast—gold, silver, copper, quicksilver, iron, and coal—showing for each mine the ore minerals, veins-tones, wall-rocks, and other important features.

Mineralogy.—A very large collection, fully arranged, and supplied with ample case room. It completely illustrates the instruction in mineralogy, and offers inexhaustible material for investigation, facilities for which are freely placed at the disposal of the student.

Mineralogical Methods.—Deposited in the Mineralogical Museum is a collection of glass and wooden crystal models, the former illustrating fully the relations of holohedral, hemihedral, and tetartohedral forms.

Petrography.—The collection contains many hundred rock specimens from the eastern States and the Territories, from England and the European Continent, and a very large number of California rocks, collected by the corps of the State Geological Survey and by C. D. Voy. The collection of rock-sections for microscopic study contains over three thousand slides, numbered to correspond with the hand-specimens from which the slides were prepared. The California rocks are being determined and placed in the collection.

Machine Models.—A small but valuable collection of machine models, the basis of what will ultimately become an extensive and important collection; and a cabinet of mathematical models, for use with the classes in Descriptive Geometry and in the draughting-room.

Civil Engineering Models.—This department has a large and excellent assortment of models in wood of the various bonds used in masonry, and of joints and fastenings in carpentry. It possesses a number of models in wood, iron, and brass of the most noted roof-trusses; also, models of bridges, walls, arches, gateways, and domes. Diagrams of many noted European and American structures are used in the class-rooms. There is also an admirable supply of surveying and hypsometrical apparatus.

A large collection of photographic slides of engineering apparatus and structures in process of erection, of photographs, and of technical drawings, is in use by the Department.

Mathematical Models.—A collection of about three hundred models of mathematical curves and surfaces in plaster, thread, wire, wood, and celluloid, including the Brill collection and the Schröder models of Descriptive Geometry.

Donations to the museums of the University are gratefully received. Messrs. Wells, Fargo & Co. will transport such gifts to the University gratuitously, if the weight of the package does not exceed twenty pounds. Special instructions for collecting and forwarding any particular kind of material will be furnished to any who may desire them.

LABORATORIES.—The Chemical Laboraties are large and commodious, well lighted and well ventilated, and offer excellent facilities for the study of chemistry. They comprise the following: An Elementary Laboratory, for beginners; a Qualitative and a Quantitative Laboratory, each containing all the usual appliances; an Organic Laboratory, for special and advanced studies in organic chemistry, and two large Research Laboratories. Special rooms are devoted to volumetric analysis, gas analysis, spectrum analysis, and electrolysis. Ample facilities are provided for chemical analysis and for investigations in foods, drinking waters, mineral waters, poisons, etc. A chemical museum, with a large collection of chemical products and apparatus, is open daily for inspection and study.

The Physical Laboratory occupies the entire basement floor of South Hall, which affords favorable conditions as regards stability and evenness of temperature. There are set apart rooms for elementary and for advanced work, for photometry, for spectroscope work, for engine and dynamos, and for a workshop. The apparatus includes many instruments and standards for fundamental measurements from makers of the best reputation, and the laboratory employs a competent mechanician, in order to increase the equipment from original designs. It offers good facilities to students who wish to pursue the study of physics beyond the limits of the prescribed course, whether in connection with special work, like electrical engineering, the use of polarized light, and physical chemistry, or for the sake of physics itself. Such students may make special arrangements for using the laboratory.

The Mineralogical Laboratory is provided with a large collection of unlabeled minerals, which students determine by their physical properties and by blow-pipe analysis. The department possesses a large reflection goniometer and spectrometer, by Fuess, of Berlin, reading direct to ten seconds, and a Groth's universal apparatus, consisting of a polarization instrument for both parallel and converging polarized light, an apparatus for determining the angle of optical axes, and a small goniometer and spectrometer; also apparatus for cutting and grinding crystal sections. Special students of mineralogy will find ample facilities for investigation in optical mineralogy.

The Petrographical Laboratory contains a large collection of rocks, and several thousand thin sections of the same. These are at the disposal of students. For the preparation of thin rock-sections the laboratory possesses all needful apparatus. Seven petrographical microscopes, from Fuess, of Berlin, fitted with all the appliances for petrographical investigations, are at the disposal of students. The material to be investigated is practically inexhaustible.

Mechanical Laboratories. The completion of the Mechanical and Electrical Engineering Building, in 1893, made it possible for these laboratories to be much enlarged. With the additional space and apparatus thus provided, the equipment is unusually complete in all the departments of experimental engineering.

The Machine Shops afford excellent facilities for mechanical practice. They have a floor area of about 4,000 square feet, and comprise the following: (1) A main machine-room, fully equipped with metal-working machines and with bench and hand tools for various kinds of metal work; (2) a carpentry and pattern-room, containing an excellent assortment of carpentry tools and a full set of wood-working machines; (3) a blacksmith and foundry room, containing appliances for forging and casting; (4) a room for fine work, especially well supplied with fine machines and tools for delicate processes. Many pieces of apparatus in the laboratories are from original designs, and were constructed in the machine shops.

The laboratories for experimentation and investigation have a floor area of 5,000 square feet, and the covered court has an additional area of 6.300 square feet, and they are well equipped for experiments in electrical engineering, thermodynamics and steam engineering, hydraulics, and testing of materials used in machine construction.

The entire east wing of the building is set apart for the Electrical Engineering Laboratories. They comprise the main power-room and the dynamo, electrical testing, photometic, and standardizing laboratories.

The main power-room contains a Ball engine of 100 horse-power, which drives through a counter-shaft and friction clutches, a thirty Kilowatt 1,000-volt alternating-current generator, a twenty Kilowatt compound wound, constant potential generator, which may be used as a dynamo or motor, and a ten Kilowatt arc lighting dynamo. The installation in this room is a typical central station of its kind, and is arranged primarily for experimental work. Mains are run from the station switch-board in this room to the large laboratory switch-board in the dynamo laboratory, making a flexible system throughout.

The Dynamo Laboratory contains a fifty horse-power Straight Line engine, belted through counter-shafting and friction clutches to twenty-two dynamos and motors especially arranged for experimentation and investigation. When a variable speed is desired, current is taken from the main power-room, and the laboratory is driven by motors.

The machines are of different capacities, ranging from fifteen Kilowatts down, and represent, as far as possible, the best American practice in dynamo machine construction. The continuous current, constant potential, and constant current types and single and multiphase alternating current machines are all represented in this laboratory, including three experimental dynamos constructed by the students in the machine shop. A number of Brackett dynamometers, also made by the students, may be used in efficiency tests. The switch-board, containing over four hundred terminals, is connected with the machines and also with the instrument table, containing sixteen Weston alternating and direct current ammeters, voltmeters, and wattmeters. A bank of eight transformers and liquid, metal, arc, and incandescent lamp resistances also have connections with this switch-board. The testing and standardizing laboratories contain a full equipment of accurate scientific and commercial instruments, among which may be mentioned three Kelvin Electric Balances, four Kelvin Electrostatic Voltmeters, an Anthony Wheatstone Bridge, four Nalder Wheatstone Bridges, and a permeameter, magnetometer, and Ewing curve tracer for magnetic investigations. These two laboratories and the photometric laboratory have sub-switch-boards connected to the main switch-board in the dynamo room. In connection with the photometric and other experiments requiring an unvarying potential, a storage battery of sixty chloride cells is available. The rooms are all supplied with solid masonry piers for the mounting of sensitive instruments.

For experiments in thermodynamics and steam engineering, the 100 horse-power Babcock-Wilcox boiler, four steam engines and two gas engines of the department are used, they being erected with all accessories for investigation.

For the experiments in hydraulics, there are available the water-tanks, gauges, and meters, and various types of motors and turbines. There are also appliances for efficiency tests and determination of the resistance to rotating disks and cylinders in water.

The Testing Room contains machines for tension, compression, and torsion of different capacities, and a wire-testing machine for experiments on cables and ropes.

The Civil Engineering Laboratory has recently been established and fitted with apparatus of the best make, particularly designed for experimental tests and original investigations, especially as related to the materials used in civil engineering construction.

A latest improved Olsen automatic and autographic testing machine of 200,000 pounds capacity, a Riehlé cement-testing machine of 2,000 pounds capacity, and a new Thurston Torsion Machine are among the recent additions to this set of apparatus. The timbers, building-stones, cements, and bitumens of the Pacific Coast receive especial attention in this laboratory; and practical questions connected with water for domestic use, tests of macadam rock and of sanitary mechanism are considered.

Metallurgical Laboratories.—I. The Assaying Laboratory is equipped to give instruction by the most improved methods in the fire assays of gold, silver, lead, antimony, tin, iron, nickel, cobalt, and quicksilver ores, and furnace products. It occupies six rooms on the lower floor of the Mining Building.

The crushing and sampling room contains a Taylor sample-crusher, large iron mortars, and rubbers; a panning sink, with a full assortment of miners' pans, horns, bateas, and other devices for making vanning tests of ores; a complete assortment of sieves, and a large sampling table. In this room the small-scale sampling is done, and the sample is prepared for assaying. From here the sample goes to the fluxing-room. This is provided with eight Becker pulp-scales, and desks containing all the necessary fluxes; it also contains a Fairbanks' platform scale, graduated in kilogrammes, as well as pounds, for convenience in large-scale tests. The sample, after fluxing, goes to the furnace-rooms. These contain four crucible furnaces and three muffle furnaces, built after an improved design, and arranged to burn either coke or charcoal; also, two large crucible and muffle furnaces for burning soft coal, like those used in Freiburg, Przibram, and Colorado. All these furnaces have been carefully designed, built in the walls, and iron-clad in a substantial manner. Besides these, examples of eight portable charcoal and gasoline furnaces commonly used are also provided to familiarize students with their use. From the furnace-room the doré buttons go to the parting-hood and then to the balance-room, which is provided with eight Becker assay balances. A convenient store-room completes the assaying laboratory.

11. Research Laboratory. This laboratory occupies a set of four rooms on the floor above, covering the same ground area as the assaying laboratory. The largest of these contains two iron-clad crucible furnaces, with an air blast, in which cast-iron can be easily melted, and two iron-clad muffle furnaces; the electric current is available for all kinds of electrolytic determinations in electro-metallurgy, both in the wet way and by fusion; and small-scale apparatus, specially designed for demonstrating with exactness all the essential points in chlorination, cyanide, and other leaching tests, is provided. This room is also fitted with horizontal and vertical shafting, so that power can be used in every part of it, and a ten horse-power transmitting dynamometer is provided for such tests as require it. This room is also fitted up for the humid, or mint, assay of silver bullion and general volumetric work. A second large room is fitted up for analytical and special research work, and contains a distilled water and drying apparatus, a water-blast blow-pipe, gas muffle, and crucible furnaces, and a small Pelton water-motor, besides the appliances usual in an analytical laboratory. Two other rooms contain eleven

of the finest Oertling and Becker assaying and analytical balances and a small reference library.

The laboratory is provided with a complete equipment for measuring high temperature, such as an air thermometer, a Fischer calorimeter-pyrometer, a Siemens electric pyrometer; also the Orsat, Bunte, and Fischer apparatus for the analysis of furnace gases, and a Thomson calorimeter for determining the heating effect of fuels.

Mining Laboratories.—The mining laboratory originally constructed by legislative appropriations has grown entirely inadequate to the needs of the large classes now enrolled in the College of Mining, and through the generosity of Mrs. Phebe A. Hearst it has been enlarged to about four times its original size. The completed laboratory consists of the following divisions:

1. *Gold and Silver Mill.*—This laboratory is designed to illustrate on a working scale the methods in successful use in crushing, sampling, concentrating and amalgamating gold and silver ores. The first floor is fifty by thirty-four feet, the second floor forty-one by thirty-four feet. The floor of the first story is of concrete, and the floor above is double, for sampling ores.

At one extremity of the laboratory is the dry-crushing and sampling plant. A platform elevator lifts the ore by the car-load to the upper floor, where it is fed to Dodge jaw-crushers, provided with automatic samplers which select automatically a sample of the crushed ore. This remains on the upper floor to be quartered down for assaying. The main lot of crushed ore passes through the floor by chutes, at will, either to a pair of sixteen-inch Krom steel rolls especially designed for this laboratory, or else to a six-inch Sturtevant mill. After being crushed in either of these machines, the ore is elevated by a bucket elevator and delivered to a Krom revolving trommel, where it is classified into three sizes and the coarse particles are returned to the machine for further reduction. The whole plant is connected with an exhaust fan and a set of dust chambers, so that the operation is conducted without inconvenience or loss from dusting.

For reducing the assay samples, an adjustable Fraser & Chalmers Comet Crusher and an E. P. Allis Sample Grinder, and a Krom Crusher with a set of Krom shaking screens, together with the usual "bucking-plates," are provided.

The wet-crushing plant consists of a stamp battery with three 500-pound stamps; the mortar is especially designed for either single or double discharge, so that it can be used, if necessary, for either wet or dry crushing of silver ores, or as a deep mortar for gold ores.

For gold ores a high discharge is used in the mortar, and the pulp passes first over a set of amalgamated copper plates and then over a Frue concentrator, specially designed for either side or end motion, and altered to facilitate accuracy in making experimental runs. An automatic sampler, also of special design, takes, at intervals of three minutes, samples of the pulp simultaneously as it leaves the mortar, the plates, and the concentrator. The frequency of the sample-taking is also adjustable. An exact knowledge of the entire operation at each step is thus made possible. The tailings are all impounded in a concrete settling-tank and are finally sampled by quartering as a check. Owing to a scarcity of water and to avoid a loss in slimes, the clear water is returned by a centrifugal pump to the battery.

For silver ores the ore may be crushed either dry or wet, and the pulp, either raw or roasted, may be treated either by amalgamation or by leaching. For this purpose four amalgamating pans and two arrastras, and numerous leaching and precipitating boxes are provided.

The concentration of closely disseminated ores is provided for by the dry-crushing and sizing plant, and two Harz ore-jigs; while finely disseminated ores are crushed wet, sorted in spitzkasten and spitzlutten, and either jigged on an ore-bed or treated on vanners, according to fineness.

Students are afforded every opportunity to acquire practical familiarity with each detail of the best methods now in use. For this purpose, parcels of ore varying in amount from five hundred pounds to several tons are assigned to each member of the class. He is expected to take charge of the work, and, with the assistance of his fellow-students, to weigh, crush, sample and assay his lot of ore; to determine the best mode of treatment, and then to carry this plan into execution, determining the amount and nature of the losses at each step, and the best method of reducing them to a minimum. Thus each student, in turn, acts as workman and foreman in charge of work, and all acquire experience covering a wide range of methods.

II. Chlorination Room. A room twenty by thirty-four feet has been provided to contain a reverberatory roasting furnace and the necessary leaching vats to illustrate the extraction of gold ores by the chlorination process, as well as the chloridizing roasting of silver ores.

III. Cyanide and Hyposulphite Leaching Rooms. Two floors, each forty-three by fifty-eight feet, have been provided for experimental work in leaching gold and silver ores by these methods. The lower floor is of concrete and contains large sumps to contain the solutions. These rooms will be fitted up to contain solution tanks, pumps, precipitating boxes, etc., all on a working scale of from one to five tons.

IV. Forge and Drilling Rooms. For the purpose of familiarizing the mining students with the operations of the hardening and tempering of drill steel and of rock-drilling and blasting, a space of thirty-four by fifty-one feet has been provided to contain fourteen down-draft forges of the most improved pattern. It will also contain a full equipment of single and double hand drills, hammers and sledges, Ingersoll and Rand machine drills, with all the appliances for sharpening, hardening, and tempering drills of all kinds. A supply of quarry blocks of sandstone and Rocklin granite is also provided for students to familiarize themselves with the work of rock-drilling.

Blasting and the effect of explosives are illustrated during the term by actual work at the rock quarries near Berkeley, and in vacation by work in the mines.

V. Power and Work Shops. Power is provided for all large work by a twenty horse-power Babcock and Wilcox water tube boiler, and steam is conveyed in covered steam pipes to any point where it may be needed. Electric power is also available for light work, electric motors being provided wherever necessary.

For the construction and repair of apparatus used by the department, workshops have been provided, covering an area of thirty-four by forty-three feet, which contain a band saw, universal saw bench. a planer, a wood lathe, and a full set of wood-working tools; also a Hendy-Norton fourteen-inch swing engine lathe, and a small speed lathe, a Cincinnati universal milling machine, shaper, drill press, and a full line of machinists' tools. It is also equipped with tools for water, gas, and steam fitting.

VI. Lecture-Room and Museums. These are located on the third and fourth floors of the Mining Building, and are designed to give a complete illustration of the lectures by maps, drawings, and plans, and by collections of models and products. The arrangement of the lecture-room is such that lantern-slide illustrations may be projected upon a screen when needed. A large collection of lantern-slide illustrations is designed to illustrate the mining art as practiced in all parts of the world.

ATTENDANCE. — Table IV is taken from my paper on the "Growth of Mining Schools," and gives the attendance in some of the principal mining schools, giving a four years' course in 1887.

TABLE IV. — ATTENDANCE AT MINING SCHOOLS FOR 1887.

AMERICAN	EUROPEAN
Columbia....................87	*England* — Royal School of Mines 60
Lehigh..........................68	*Saxony* — Freiberg...............................163
Massachusetts Inst. Tech.....30	*Prussia* — Aachen.................... 43 ⎫
California........................23	Berlin ....104 ⎬ 243
Pennsylvania...................17	Clausthal................. 96 ⎭
Ohio 14	*Austria* — Leoben .... 29 ⎫
Lafayette.......................12	Klagenfurt.............. 8 ⎪
Michigan University........... 8	Mährisch-Ostrau .... 23 ⎬ 117
Washington 7	Dux...................... 17 ⎪
Illinois............................ 3	Prizbram .... 40 ⎭
Wisconsin...................... 3	*Sweden* — Stockholm 17
Total.....................272	*France* — École des Mines de Paris..111 ⎫ 191
	École des Mines, St. ⎬
	Etienne............... 80 ⎭
	Belgium — Liège.............................. ——
	Total............................. 791

The following figures are taken from a paper of Dr. M. E. Wadsworth, "On some Statistics of Engineering Education," Trans. A. Inst. M. E., July, 1899. They give the attendance at various mining schools for the year 1896-97.

University of Alabama...Mining Students,		4
Case School Applied Science .... " "		9
Colorado School of Mines (Civil, Elect., Mining and Chem. students combined)...		161
Columbia School of Mines...Mining Students,		88
Lafayette College... " "		3
University of California ....Mining Students only.		113
Massachusetts Institute of Technology...............................Mining Students,		24
Michigan College of Mines (includes Mining, Civil and Mechanical)...................		140
University of Minnesota ....Mining Students,		34
Ohio State University... " "		33
Pennsylvania State College....................................... " "		23
South Dakota School of Mines (includes Civil, Mining and Scientific)..................		51
Western University...Mining Students,		5
University of West Virginia (Civil and Mining together)..,..........................		75
University of Wyoming...Mining Students,		7

The rapid growth of the College of Mining, at the University of California, is shown by the following table.

At the University of California, the attendance of Mining Students only (not counting those in Civil and Mechanical Engineering) has been as follows :

Year 1887–88, Mining Students 24	Year 1893–94, Mining Students...... 33
" 1888–89, " " 24	" 1894–95, " " 42
" 1889–90, " " 26	" 1895–96, " " 65
" 1890–91, " " 30	" 1896–97, " " 113
" 1891–92, " " 33	" 1897–98, " " 164
" 1892–93, " " 25	" 1898–99, " " 174

This growth is all the more remarkable since it has taken place in the face of a very substantial increase in the entrance requirements.

Mining Debris Legislation.

BY

CHARLES G. YALE.

HEN some fifteen or more years ago the hydraulic mines operating in the drainage basin of the Sacramento and San Joaquin valleys in California were closed down by injunctions of the United States Courts, many persons not familiar with the subject supposed mining was to a large extent prohibited in this State. This impression has not yet been entirely dissipated. The facts were, however, that the mines were only prohibited from working where the "debris" or tailings from their extensive operations entered the streams and injured their navigability, or damaged the farming and orchard lands along the banks. Where no such damage or injury was done, the mines continued to work.

Now, under an Act of Congress, especially applying to the drainage basin referred to, all these mines are permitted to work under certain conditions and restrictions and under the supervision of a Board of Engineers appointed for the purpose. That this entire matter may be properly understood by those who may be interested, but not familiar with the details, the following history of the legislation connected with these hydraulic mines in California is submitted. The writer was, to some extent, closely connected with the efforts tending toward the rehabilitation of the hydraulic mining industry of the State, and has endeavored to state the facts as briefly as possible without entering into any of the controversial features of the subject, which can be of little moment to others than those directly interested :

"DAMAGE BY DEBRIS."—For many years the hydraulic mines operated on a very large scale, and at one time there was over $100,000,000 invested in them in this State. The extent of their operations, however, proved their undoing. The immense mass of

detritus or tailings, commonly called "debris," resulting from the
work, was allowed to go where it might, and, while much was deposited in the canyons or gulches, vast quantities entered the navigable
streams, injuring and shoaling them, and doing damage to the farming and horticultural lands along the banks. The bay of San Francisco, into which the main rivers empty, was also shoaled in the upper
portions. The amount of debris in some localities, notably about
Marysville, was almost beyond conception to those who have never
seen it. In some places it piled up in the streams twenty-five or

DEBRIS-RESTRAINING DAM BELOW THE RED DOG HYDRAULIC MINE,
NEVADA COUNTY.

thirty feet deep and caused overflows in the streams during the
spring, thus creating great damage. In many localities, orchards
and farming lands were almost entirely covered by debris and
thus destroyed. The miners, in many cases, paid damages where
direct, but the whole matter brought on numerous suits and much
controversy. Finally, in the test suit of Woodruff vs. The North
Bloomfield Mining Company, the miners were beaten, and the U. S.
Circuit Court gave a decision which resulted not only in closing
down the mine named, but all the principal hydraulic mines in the
central-northern part of the State. The mining company and its

agents and employés were perpetually enjoined and restrained from discharging and dumping into the Yuba River or any of its forks, ravines, or branches, or any stream tributary to the river, any tailings, boulders, cobblestones, gravel, sand, clay, debris, or refuse matter from the tracts of mineral lands or mines, and also from allowing others to use the water supply of their mines for washing such material into the rivers or streams.

THE LITIGATION.—It will be observed that the decree of the U. S. Court was not against hydraulic mining in name, but against dumping the debris into the streams, ravines, etc. On the large mines operations were suspended and many costly works were allowed to go to decay. Mining camps were deserted and large districts were depopulated. The miners, in many cases, persisted in continuing to work their mines, and the Anti-Debris Association, composed of farmers of the Sacramento River Valley, carried on an organized opposition to hydraulic mining. Long and costly litigation and bitter controversy between the farmers of the valley and the miners of the mountains resulted, and this continued for years. Meantime, mine after mine was closed down and the mining property, ditch systems, etc., became worthless. The closing down of these mines materially reduced the gold yield of the State. The water companies which sold the miners water were enjoined from such sale, and the ditch properties also became worthless. The great reservoir systems with ditches, flumes, pipe lines, etc., were partially abandoned and, in some cases, the dams were blown up and destroyed. Thousands of men were thrown out of employment, and whole camps and villages in the mining regions were deserted. This condition of affairs continued for some ten or twelve years.

ARRANGEMENTS FOR A COMPROMISE.—In the fall of 1891 a number of miners in Placer County, who had suffered by the closing down of these mines, agreed to meet together and try to see if some arrangement could be made by which the mines could again be started. At this meeting they decided to call a county mining convention to discuss the subject. Five representatives from each voting precinct in the county were called to attend the convention, which was held in the city of Auburn, Placer County. The principal idea of this county convention was to formulate a plan for a State miners' convention and memorialize Congress as to needed legislation for the mining industry of California.

It was the general sense of the convention that the public prosperity demanded a speedy and amicable settlement of the debris question, a settlement whereby the rights and interests of the miner and farmer shall be protected. An address to the people of the State of California was issued, setting forth the condition of the mining industry as affected by the court decisions, and a statement made as to the remedial measures proposed, by which the hydraulic mines might again be worked without the injuries formerly inflicted. With this address was also issued a call for a State mining convention, to be held in the city of San Francisco, and representatives from all the counties, both mining and farming, were invited to come and discuss the subject. In accordance with this call, representative miners came from all over the State, and the valley counties sent also delegations of farmers.

The proceedings of the State convention were harmonious beyond expectation. The remarks of the valley representatives were such as to re-assure the mining men that the valley people could be their friends as soon as assured that the miners intended in good faith to do nothing beyond the law. The questions with which the convention had to deal were of such a nature as had involved a bitter controversy extending over a period of many years. The property interests were valued into the millions of dollars. Yet mutual concessions on both sides, and sober judgment, enabled them to reach an amicable conclusion and agree upon a common plan.

BASIS OF AGREEMENT.—The basis of this agreement was the report of a Government commission of engineers. The Legislature of California, realizing that rehabilitation of the hydraulic mining industry would benefit the people of the State and whole nation, if it could be accomplished, passed a joint resolution, bringing the matter to the attention of Congress. In accordance with this resolution, Congress passed an Act appointing a commission of engineers for the purpose of ascertaining if some plan could be devised to adjust the conflict between the mining and farming sections, and the mining industry rehabilitated, and for examining the navigable rivers and tributaries, with a view to improvement and rectification of the rivers. This Commission finally made the investigations referred to and reported results. It was this report which formed the basis of agreement between the mountain and valley men of the State. It appears by this report that dams and other restraining works may be erected in many of the canyons, which will not only under certain conditions restrain the material producing the damage complained of in the past, but will also restrain debris now dislodged but still remaining in the canyons. The report of these engineers specified the places where these dams could be erected, their cost, the amounts of gravel which could be set free after their construction, and all necessary details.

The convention of miners asked Congress to accept and adopt the report of the Commission appointed for the purpose stated, and that it at once take steps to put into practical and effective operation the means suggested, in order that mining might be resumed in the manner indicated, without the injury complained of in the past. The memorialists further suggested that Congress, having appointed the Commission to determine the question, should accept and act upon its conclusions, which are of a nature to be acceptable to both parties to the controversy, in that they provide that mining may be again carried on under certain specified conditions, and also that the debris will be restrained from the rivers and farming lands.

The fact was recognized by the convention that until Congress took proper action for the erection of suitable works for the restraining of mining debris, hydraulic mining was absolutely restrained by the courts, and, as law-abiding citizens, they recognized that the decrees of the courts must be respected. They therefore urged immediate action upon Congress.

"THE ACT OF CONGRESS."—In March, 1893, Congress passed the so-called Caminetti Act, which permits these auriferous gravel mines to be operated by the hydraulic process under certain restrictions and conditions.

The essential features of the law are that all such mines, operated under this system, shall impound or restrain their debris or tailings, and prevent them from entering the navigable streams, or injuring the lands of other parties. Under the Act, the California Debris Commission, consisting of three officers of the Corps of Engineers, U. S. A., was appointed by the President. This Commission is empowered to issue licenses for mining by the hydraulic process under this Act, when it is satisfied that the debris dams or impounding works are sufficient to restrain the debris. The hydraulic miner must make application to the Commission for license to mine, and submit his plans of the proposed restraining works, which are subject to the approval of the Commission. Each separate application is advertised for a specified time, and a hearing is held before the Com-

ELECTRIC TRAM-WAY AT HIDDEN TREASURE DRIFT MINE, PLACER COUNTY.

mission, at which those who may be opposed to the issuance of a license may state their reasons. When the plans are approved, and the necessary works constructed, members of the Commission make a personal examination of them, and, if satisfied that the debris can be restrained, a license to mine by the hydraulic process is issued, and the mine may begin operations. If they see any reason to believe, however, that damage may be done to the rivers or to individuals by the operation of the mine, no license is granted, and the mine may not be legally worked. Moreover, even after the license is granted, if the debris, or water carrying too much of it, is for any reason permitted to enter the stream, the license may be recalled. Frequent examinations are made to see that the miners are complying with the laws.

It should be borne in mind that the miners themselves must bear the expense of the restraining works for their respective mines, and for this reason hundreds of the smaller ones have never been started up again, because their owners, having become impoverished by enforced cessation of operations for a series of years, have not the money to construct the necessary impounding works.

Since the passage of this law, several hundred of the hydraulic mines have built the works and received licenses to mine. The product, however, does not by any means come up to its former dimensions, since the debris must now be run into settling basins behind the dams and allowed to settle, so that a much smaller quantity of gravel can be handled than when the tailings did not have to be cared for.

WHERE THE LAW APPLIES.—It is proper to state and should be borne in mind that the law applies only to that section of the State drained by the Sacramento and San Joaquin rivers and their tributaries. In the north-western portion of California, in Siskiyou, Trinity, Humboldt, and Del Norte counties, where there are extensive deep gravel deposits, there is no restriction on hydraulic mining, and never has been. The hydraulic mines of the section dump their tailings into tributaries of the Klamath River, which has been officially declared a non-navigable stream. Moreover, there is very little cultivated land along that river, so no damage is done.

The hydraulic mines in the drainage V basins of the Sacramento and San Joaquin rivers have been gradually resuming work under the operations of this Federal law. Under this beneficent law this class of mines may be operated without doing any damage to the rivers or the lands along their banks, and without being involved in the endless and bitter litigation which formerly resulted. The output of these mines has materially increased the gold product of the State, and their renewal of operations has helped to rehabilitate sections of the State which were partially deserted and impoverished when the mines were closed down by injunction of the courts. Naturally, under the conditions requiring the work to be done behind debris dams, the output of the mines is restricted as compared with that when the amount of gravel washed was only limited by the water supply, since now the gravel must be washed into reservoirs and the muddy water there settled until it flows out practically clear of debris.

DEFINITION OF HYDRAULIC MINING.—It is to be noted that there is no law against hydraulic mining in California, and never has been, either Federal or State. The injunctions granted were because damage was done to individuals in certain localities or to certain navigable streams. Suits were not brought generally against the industry by name, but because of specific injury done by a certain mine, and that mine was sued. But the powerful Anti-Debris Association, composed of farmers in certain valley counties, prosecuted the miners wherever possible and secured numbers of injunctions.

The two sections of the Civil Code of California relating to this subject are very clear and explain the present legal status:

SEC. 1424. The business of hydraulic mining may be carried on within the State of California wherever and whenever the same can be carried on without material injury to the navigable streams or the lands adjacent thereto.

SEC. 1425. Hydraulic mining, within the meaning of this title, is mining by means of the application of water, under pressure, through a nozzle, against a natural bank.

Nevada County.

HE chief gold-producing county of California has been, and yet is, Nevada. From both historical and industrial points of view, its mining story, but briefly outlined here, is not equalled in interest and importance by that of any of its sister counties of this rich Commonwealth. From its vast beds of auriferous gravels, and from the rich and deep-reaching veins of its quartz-ribbed hills, has come, approximately, $212,000,000 worth of gold, or nearly one-sixth of the total gold product of the State.

Although Nevada suffered more than any other county in the State from the stoppage of hydraulic mining in the early '80s, by which its chief mining industry was paralyzed, it has, because of the richness and permanency of its quartz mines, preserved the leading place through all the years of depression in the industry, producing annually $2,000,000 or more in gold. It has also, until 1898, been the banner county in the combined production of all minerals, Shasta alone having passed it, because of its great copper mine. The Register of Mines of Nevada County, recently published by the State Mining Bureau, lists 234 quartz mines, and many placer and drift mines. There is, of course, a great number of promising mines not so developed as to call for description. The Mining Bureau's census of the gold mills of the State, published in 1896, credits Nevada County with sixty-two stamp mills, dropping 777 stamps, and eight patent mills and two arrastras. The Register mentioned lists sixty-seven stamp mills.

It was in this county that California quartz-mining and milling crudely began and saw its early development. It was in Nevada County that hydraulic mining and drift mining had their origin and chief development. It was here, too, that the first quartz-mining district in America was organized, and the first district laws regulating quartz mining adopted.

With such a record, sustained through a period of depression in the mining industry, this mineral region may be assumed to be one with a large mining future before it, and one worthy of the attention and confidence of mining investors, and this assumption is borne out by a knowledge of the conditions and indications to be found.

Nevada is one of the comparatively small, irregularly shaped counties of the State. It is a strip of mountainous and picturesque territory seventy-five miles long, reaching from the State line across the snowy Sierra Nevadas, at an elevation of 8,000 feet, down the slope toward the Sacramento Valley, where its western end lies but little above the sea. Its area is about 1,125 square miles. Its boundaries enclose a region of much diversified topography, soil, climate, and resources, the chief one of the last being here the theme. The county is part of the bounteously wooded and watered Sierra Nevada mining region, where the conditions present what has often been called the world's paradise of the miner.

Nevada County shares the credit of California's fame as a gold-producing State chiefly with the great Mother Lode region, the northern end of which is generally taken to be found twenty miles or so to the south in Placer County, which it bounds on the north. The gold-quartz formations of the county have distinguishing peculiarities, and its auriferous gravel deposits of the past and present are of great magnitude and importance. Although gold is very widely distributed over the Sierra Nevada region in veins and placers, there is no other part of it which shows such a concentration of valuable deposits within a narrowly circumscribed area as the main gold region of this county. The veins are not connected with the long, continuously-linked veins of the Mother Lode to the south.

Gold mining began in 1848, soon after Marshall's discovery, in El Dorado County, and in 1849 prospectors swarmed into its wild and virgin hills and canyons, spreading along the Middle Yuba, South Yuba, and Bear rivers and numerous tributaries, and establishing typical mining camps at the richest discoveries, two of which were the beginnings of Grass Valley and Nevada City. These surface placers yielded richly by the primitive methods then in use. During the first years the alluvial placer mines furnished the largest amount of gold. Very soon, however, the older tertiary hill gravels were discovered. The deposits of these were far richer and more abundant at Nevada City than in Grass Valley. Here drift mining began, and between 1856 and 1860 and between 1865 and 1870 the ancient channels on

EARLY PLACER MINING SCENE IN 1852, AT WHAT IS NOW NEVADA CITY.

the Alta, Towntalk and Independent hills were worked by the drifting process. As early as 1851 the hill gravels above Nevada City were discovered. Small shafts were sunk and low drifts run in different directions. Sluices, the first improvement on the pan, rocker, and long-tom, were first used, it is said, in the Coyoteville diggings at what is now Nevada City.

Ground sluicing first came into general use in 1851-2 in the "coyote" claims at Nevada City. A. Chabot, who was mining at Buckeye Hill in 1852, introduced about forty feet of hose into his claim, the water being conducted from the bank to the bottom of the diggings in a closed wooden box, strengthened by iron clamps to withstand the pressure. This was found to be very useful in sluicing off the dirt, but there was no nozzle attached, and the idea probably never occurred to Mr. Chabot that a stream of water under a high pressure, directed against the bank, would accomplish a great deal. In April, 1853, E. E. Matteson, who was then working a claim on American Hill, conceived the idea that by directing a stream of water against the bank he could facilitate matters. He accordingly attached a nozzle to a piece of hose, and found that this would accomplish the work of several men. This was the first attempt at "hydraulicking," and Mr. Matteson's simple appliance, which was exhibited at the California Mid-winter Exposition at San Francisco in 1894, was the forerunner of the immense steel and iron-pipe systems and the giant monitors of to-day, which, unluckily, are lying idle in the great hydraulic mines of the county at the present time.

It was in October, 1850, that quartz was discovered and quartz mining began. The discovery was made on Gold Hill, close to the present city of Grass Valley where, on the hill-slope, was found the broken top of a ledge very rich in gold. The occurrence of gold-bearing quartz was likely noted elsewhere at about the same time or previously, as has been variously asserted, as gold was found in California before Marshall saw his nugget; but the Gold Hill find was the first to attract fame, and it started gold milling in this State on its great career. The first gold was pounded out in hand mortars, etc., but picturesquely crude and inefficient mills were quickly put up. Although an excitement over quartz mining here and at the east ensued, the lack of knowledge, skill, and appliances, and the cost of milling, which made worthless rock assaying less than $50 a ton, prevented any early development of quartz mining on a notable scale.

For the first ten years after 1850 the growth of the quartz-mining industry was slow, for the further reason that the rich placers with which the county abounded still held out, maintaining their favor because of the less amount of labor, capital, and skill required to work them. In 1855, there were but seven producing quartz mines. As the richest of the surface placers began to fail, attention was turned to quartz mining, which has since flourished uninterruptedly.

COMPRESSED AIR POWER-HOUSE. NORTH STAR MINE, GRASS VALLEY.

The earliest important quartz mines were developed in the Grass Valley district, which has surpassed any other gold-producing district of a similar area in the world in its production of considerably over $100,000,000, and in the free-milling character of its ores and their demonstrated permanency with depth.

Nevada County, geologically, is divided into three distinct auriferous belts west of the Sierra Nevadas. These are the Washington gold belt, the Grass Valley gold belt, and the Meadow Lake gold belt. The rocks composing these auriferous belts are mainly jurassic. The formation in the south-western part of the county is metamorphic slate and schistose rock, not known to be auriferous or gold-bearing.

Copper, iron, magnesite, and lime are found in this belt. Masses of serpentine occur among these slates, and a large body of serpentine crops out on the surface, about a mile west of the west bank of Wolf Creek. About a mile west of Indian Springs is a very large mass of iron ore of excellent quality. The auriferous slates, schists, and metamorphic rocks join on the east of this iron deposit, and continue about twenty miles easterly to the serpentine belt near Washington, forming what is known as the Grass Valley gold belt. This belt is very wide, but it has a large area of syenite included within it. In nearly all of the country north of Grass Valley and Nevada City, extending west from a line drawn from Banner Mountain to North San Juan, to within a few miles of Smartsville, the formation is composed mainly of syenite, with parallel bands of hard metamorphic schists and slates. There are also dykes of diorite and diabase, including veins of gold-bearing quartz, some of which are being profitably worked near Nevada City.

South of Nevada City, and extending east of Grass Valley, Banner Mountain, and North San Juan, the slates predominate. The formation has a strike north-west and south-east, and a dip nearly vertical, inclining slightly to the east. The slates are very much changed in the Grass Valley district, and their true position is hard to determine. Large masses of serpentine and gabro occur among the metamorphic rocks, in places forming the walls of the auriferous veins.

The celebrated Idaho (Maryland) and Eureka mines have a footwall of serpentine and a hanging wall of diorite (greenstone). The quartz mines of Grass Valley are noted for their high-grade ores and the magnitude and permanence of the ore bodies or lodes. The serpentine belt on the east of the Grass Valley gold belt is a continuation of the same belt, which can be traced across El Dorado and Placer counties. It runs across Nevada County, and is the dividing line between the Grass Valley and Washington belts. Immediately east cf the serpentine belt is an immense belt of ferruginous steatite, varying from twenty to one hundred and fifty feet in width. This monster vein is auriferous, and is being successfully mined on the South Yuba River. The ore is, however, very low grade. East of the auriferous steatite is an auriferous formation of black slate, mica, talc, and chloride schists, with quartzite and dikes of diorite and diabase, the black slate predominating. This is commonly known as the Washington belt, and continues to the mouth of Diamond Creek. In it are numerous strong veins of auriferous quartz, some in contacts and others in fissures crossing the slates. Masses of infiltrated quartz often are found in the black slates, carrying a considerable quantity of gold. About a mile east of the Washington Mine the rocks begin to change from their slaty character and gradually become more silicious, schistose, and harder going east, to a point about half of a mile west of the Yuba Mine. East of this the formation changes gradually to protegene, alternating with narrow beds of syenite and schistose rocks. Some of the protegene belts are chlorite and others talcose. From the Eagle Bird Mine east to Fuller Lake the protegene rocks predominate. Numerous veins of auriferous quartz, from one to thirty feet thick, crop out along the hillsides, and can be traced from the tops of the ridges down the sides of the mountains and across the canyons. The belt continues northerly, crossing the Middle Yuba River into Sierra County, and, in going south into Placer County, the protegene gradually changes into a hard quartzite, alternating with schistose rocks. East of the Fuller Lakes the formation is syenite, with beds of slate and schistose rocks. This forms a belt which includes the auriferous region north of Cisco, known as the Meadow Lake District. East of Meadow Lake the same alternating slate, schistose, and syenite formations continue to the summit of the Sierras.

The principal quartz systems, from which the bulk of the gold extracted during the past two decades has been taken, are in Grass Valley and Nevada townships. In Rough and Ready is found a system of quartz veins that have been developed to a small extent. At Grass Valley, 2,470 feet above the sea level, are found two systems of veins of auriferous quartz, one running east and west and the other north and south. Both of these systems have been worked at a great profit, the product of the quartz mines of this district alone being estimated at $100,000,000.

The rich ore usually occurs in fairly well-defined bodies or pay shoots. They are, as a rule, long-drawn bodies with their maximum extension in the general direction of the dip, but with an ordinarily well-defined pitch on the plane of the vein. The pay shoots vary in length from a few feet—those generally being called pockets—up to several thousand feet. On each larger vein there are usually several pay shoots. It is common for one shoot to cease in depth or be subject to local impoverishment, but thorough exploration will usually result in finding new shoots or a continuation of the old one. There is no gradual impoverishment of the ore, and the limit of depth at

PROVIDENCE MINE, NEVADA CITY.

which pay ore is found has not been reached. High-grade ores are found in the narrow veins of the Banner Hill district, even reaching $50 to $100 a ton, with an average of $20 to $50 when worked on a large scale. In the Nevada City district the average is perhaps $15 per ton. In Grass Valley the average may be placed at $20 per ton. Of course there are many small pay shoots that show far greater values. These ores are especially characterized by their free-milling quality and the purity of the gold, the percentages of silver and of sulphurets being very small.

The ore shoot of the Eureka-Idaho-Maryland vein has been worked in the Eureka and the Idaho, and is now being worked in the Maryland. The Eureka was located in 1851, and in the course of years was worked to a depth of 1,200 feet, the mine being shut down in 1877, after vain efforts to find the continuation of the ore shoot to the west. The Eureka produced a total of $5,700,000, during several years providing 10,000 to 12,000 tons of ore per month, running from $23 to $60 a ton, with a cost of mining and milling of $10 to $15 a ton. The Idaho, adjoining it on the east, was worked but little until 1865, the ore shoot being found at 300 feet. This mine was worked until 1894 to a depth of about 2,180 feet. The total output was $11,638,000, making the combined product of the Eureka and Idaho $17,338,000. This remarkable pay shoot has been followed for nearly a mile.

The cost of quartz mining varies considerably with the thickness of veins, the hardness of the country rock, and the depth and amount of water in the mine. The generally narrow veins of Grass Valley necessitate so much deadwork that the expenses are rather high. In some mines they run as high as $6 to $7 a ton. In the large veins of Nevada City the cost is stated to range from $2.50 to $5 a ton.

Gold Hill and Massachusetts Hill, in the Grass Valley district, were the principal producers during the first two decades. The Empire has worked almost continually since 1851. The North Star was in operation until 1872, and then again from 1884 to the present time. At Nevada City the Gold Tunnel and California mines were among the earliest worked. The Providence and Nevada City mines have been worked almost continuously.

It would be impossible here to go into a description of even the most important quartz mines in any detail. They have been described at length in a number of reports, including those of Raymond, the State Mining Bureau, and the lengthy monograph of Waldemar Lindgren on "The Gold-Quartz Veins of the Grass Valley and Nevada City Mining Districts," accompanying the Sixteenth Report of the U. S. Geological Survey, 1895–6. The depth and permanency of the values of the rich ore shoots of these districts are among the main characteristics to be noted, and they may be illustrated by typical examples. The most famous producers have been the Eureka and the Idaho, already mentioned, and which are now succeeded by the Maryland, all being on one great ore shoot in the Grass Valley district.

The Eureka was located in 1851 and was worked at various intervals until 1857, but the rock failed to pay much. In 1857, it was purchased by Fricot, Rupert & Pralus, and by them worked until 1865, when it was sold for $400,000, and the Eureka Company was incorporated. In the next two years the mine yielded $1,200,000. It continued to pay dividends for several years. A hoisting works and a twenty-stamp mill were erected at a cost of $60,000. During the year ending September 31, 1871, the rock paid about $35 per ton; the mill then had thirty stamps and the yield was $567,349, of which $360,000 were paid in dividends. In 1873 the lead was lost, and in 1877 the mine was finally abandoned, having produced $4,600,000 and paid $2,134,000 in dividends.

The Idaho Mine on the same lead was located by a number of Grass Valleyans on May 9, 1863. But little work was done until 1865. On July 5th of that year a perpendicular shaft was commenced. The ledge was struck at a depth of 120 feet. A crosscut was run, but the ledge showed no gold or sulphurets. In March, 1866, work was stopped and the mine remained closed for eighteen months. In September, 1867, the company was incorporated under the name of the Idaho Quartz-Mining Company.

Work was again commenced, and the shaft continued to a depth of 300 feet. The ledge was struck in a crosscut, the first twenty loads paying $29 per load. Explorations were continued to the 600-foot level, when drifts were run both east and west, and proved the value and permanency of this great lode. The total amount paid in assessments for developing the mine was $38,549.75. In 1871, a new shaft, six by twenty feet within timbers, was commenced. This is the shaft now being used by the Maryland Company, which has succeeded the Idaho, and which is developing an extension of the pay shoot. It has a perpendicular depth of 991 feet, or 1,116 feet on the incline. At the 1,000-foot level, an incline shaft has been sunk 325 feet east of the main shaft, and this incline shaft has now reached a depth of about 2,000 feet.

The Maryland has a fine hoisting and pumping plant, and a thirty-five stamp mill with all latest improvements. Water-power is used throughout, the water being obtained from the South Yuba Water and Canal Company. The Idaho produced between $12,000,000 and $13,000,000, and paid about $6,000,000 in dividends, and the Maryland will doubtless prove its value as a bullion producer in a short time.

The North Star Mine has been one of the most productive mines in this district, and was abandoned years ago as "worked out." The mine is situated about two miles south of Grass Valley, and is 2,400 feet in depth. Twenty-three levels are turned from the shaft, extending west of the shaft a maximum distance of 900 feet, and east 1,400 feet. The ledge in the drifts will average fourteen inches in size, and the ore pays about $20 a ton. The total product of the mine, in 1893, from the 18,000 tons of ore crushed, was $340,000, and since the mine was re-opened in 1884, nearly $2,500,000 has been produced —a magnificent sum for an abandoned and worked-out mine. Previous to 1884, it is estimated the North Star, then known as the French Lead, yielded between $2,500,000 and $3,000,000. The North Star mill is conceded to be one of the finest quartz mills in the State, having forty stamps and sixteen concentrators.

The Empire has produced about $5,000,000 since the early '50s. It has reached a depth of over 2,000 feet.

One of the best mines in the Grass Valley district is the W. Y. O. D., which has recently been undergoing vigorous development. It is equipped with a magnificent plant, including a twenty-stamp mill. The Allison Ranch, Massachusetts Hill, Electric, Centennial, Wisconsin, and Badger are among the numerous mines whose story and description would be of value.

Foremost among the quartz mines of Nevada City district, both on account of production and extent of workings, is the Providence, which, for the past twenty-five years, has been worked continuously and always with very flattering results. It was located in 1858, and for a few years operated by Dingley & Co., who took out considerable bullion, but finally suspended operations after running a tunnel into the hill about 300 feet, and having also erected a ten-stamp mill. In 1870, A. Walrath, R. C. Walrath, J. V. Hunter, and others, negotiated for the purchase of the mine, the selling price being $60,000. They were allowed four months' time in which to make

SCENE IN THE FAMOUS NORTH BLOOMFIELD HYDRAULIC MINE, NEVADA COUNTY.

the first payment of $30,000, but at the end of that time did not think the mine worth one-quarter the price. They sunk a shaft 300 feet and run two tunnels south, finding ore all the way, both in the shaft and tunnels, but nothing that would warrant purchasing the property at the price asked. The shaft was sunk fifty feet deeper, and sixty feet from the shaft a very rich body of ore was struck. The first payment of $30,000 was made soon after. An extension of time was obtained for three months, at the end of which the second payment was made. From that time till the present, the mine has paid regular monthly dividends. Up to 1889, the output was over $5,000,000, and the mine is paying now better than ever. Only one-fourth of the ground has been worked, the company owning 147 acres of land, from which much of the timbers used in the mine are cut. The lode. or fissure, is of immense size, and is composed of different bodies of ore. It has been worked for over a mile, and to a depth of 1,884 feet.

Another of the great mines of the Nevada City district is the Champion. The large two-compartment shaft is down 2,100 feet, with drifts north and south from the shaft exposing the ledge, which is twenty-five feet wide in some places. Seventy stamps are dropping constantly on ore taken from the vein. The property is owned and controlled by the same company as the Wyoming and Quien Sabe, on the same vein, and the Merrifield, on a parallel vein. On the Quien Sabe are ten stamps, making a total of eighty stamps under the management of the Champion Company. In connection with these mills there is a five-ton capacity chlorination plant which reduces about one-half of the concentrates, the remainder being shipped to smelters. About 230 men are employed. The approximate monthly output is $40,000, and the expenditures $25,000. The workings of the company in the way of tunnels and drifts aggregate eight miles.

The Gold Hill, which is in the same contact as the Champion and Providence, has produced over $800,000. The Mountaineer has been operated for fifteen years or more. There are forty men employed and twenty stamps in the mill. It is a steady producer, having yielded over $1,000,000. The Reward, Pittsburg, Phœnix, Mayflower and Texas may be prominently mentioned among the other important mines of the Nevada City district.

The drift gravel mines of Nevada City district are situated along the Washington ridge, on the ancient river channels which underlie it. The ridge terminates at Nevada City, but the channel extends two miles westward and has been worked by hydraulic process. For a distance of four miles extending east from the Sugar Loaf there is a succession of mines which are being successfully worked. The channel follows closely the course of the ridge, showing that the ancient river channels followed quite closely the direction of the modern streams, which have eroded on either side and produced the ridge.

The strata of gravel sands and clays deposited by the ancient river have a depth of about 200 feet. Above this lies the volcanic capping with a depth of from 300 to 400 feet. The bedrock is a soft decomposed granite, containing hard boulders varying in size from a foot or two up to forty or fifty feet in diameter. These boulders are at times very hard, costing from $12 to $15 per foot to drift through them. The gravel stratum profitably worked by the drift process is that lying directly upon the bedrock. Its character varies from a fine, free gravel that admits of the perfect extraction of the gold by sluicing, to a hard cemented gravel full of large boulders. But in all gradations the gravel proper is composed almost entirely of quartz pebbles, fine dirt, and quartz cobbles. The gravel, when found in the channel proper, is of a high grade, yielding from $2.50 to $13 per ton. The depth of the gravel stratum extracted varies from two to five feet. The gold occurs in coarse shots, and works its way into the soft bedrock.

The mines are mostly worked by means of inclined shafts, and as there is but a small amount of water, hydraulic ejectors are used to free the mines. Large, sharp augers are used to sink holes in the bedrock, it having been discovered that much more rapid progress can thus be made than by the use of hammer and drill or pick and shovel. Powder is used in breaking the ground bored. The Nevada City channel has proved very rich in past workings, and it is said that 3,000 feet of this channel yielded $4,500,000. Gravel from the Harmony Mine at present yields from $10 to $13 per ton when crushed in the mill.

It is along this great deposit of ancient gravels that the hydraulic mines of Nevada County's past and present have mainly operated. The North Bloomfield became the most extensive hydraulic mine in the world, and there were many hydraulic operations of great magnitude supplied by costly artificial water systems twenty to sixty miles long. But the story of hydraulic mining here is now largely one of the past. A few mines are operating in a comparatively small way behind restraining dams under the Caminetti law. A number of the hydraulic mines of the past have been worked during the last few years as drift mines. There are many millions in gold in the gravels that can be worked only by the hydraulic process, and there is both hope and probability that in the future the debris problem will be solved in a way to permit a partial resumption of the hydraulic mining that once cut so great a figure in the county's production and prosperity.

SETTLING POOL ABOVE THE DAM OF THE RED DOG HYDRAULIC MINE.

The mining possibilities of Nevada County are almost without limit. Although this county is famed for the richness of its quartz veins and gravel channels, it is not lacking in other valuable minerals and products. In the western part of the county, near the Yuba boundary, a system of copper veins traverse the entire width of the county, and, at Spenceville, this vein has been worked with profit. A few miles east of Spenceville there is also a copper lode of excellent quality yet undeveloped; but the recent copper boom is bringing about an exploitation of the copper mines. Iron ore of high grade and inexhaustible quantity has been found in the vicinity of Indian

Springs. The iron belt is found to extend across the county from the Bear to the Yuba rivers, at an elevation of 1,400 feet. Manganese is also found in this region in considerable quantities. Marble of excellent quality and different varieties is found near North Bloomfield. Granite, for building purposes and for general use, can be found throughout the entire northern half of the county.

There are hundreds of mining claims that are undeveloped, and many of these will, undoubtedly, in the future become important mines. When it is understood that of the whole gold output of Nevada County the gravel mines have yielded three-fifths, and the quartz mines but two-fifths, and when the comparatively slight proportion of the quartz veins that have been even prospected below the surface is noted, the prophecy of a long and steady increase in the number and output of the quartz mines seems fully warranted. There are yet miles of gravel deposits unprospected, and the revival of the industry will turn attention to Nevada County's gold deposits, and stimulate their development to a degree unknown in the past.

Mineral Resources of Butte County.

BY
A. EKMAN.

 UTTE County derives its name from that singular elevation known as the Sutter Buttes, standing in the northwestern corner of Sutter County, near the line of Butte. This elevation, which consists of three peaks springing up from a short range of mountains, forms a conspicuous object in the centre of the great Sacramento Valley. Butte is counted among the most favored counties in California. It lies along the eastern side of the Sacramento Valley and its area is about 1,760 square miles, of which 552,960 acres are mineral, 568,640 acres timber, and 195,840 acres agricultural lands.

The entire eastern side of Butte County is formed by outrunning spurs from the western side of the Sierra Nevada Mountains, between which the different branches of the Feather River (North, Middle and South), with their many tributaries, debouch into the Sacramento Valley, after passing, for the greater part of their course, through a rich mining country, ribbed with gold-bearing quartz ledges and remains of ancient river channels. These spurs are all overgrown with an abundance of timber

A. EKMAN.

suitable for mining or any other purpose. The south and west parts of the county contain as rich farming land as can be found anywhere in the State, and the foot-hills skirting along the valley have been proved to be the finest fruit-producing lands, where all kinds of fruit and berries grow to perfection, insuring an unfailing source of supply for the miners. The railroad extends to Oroville, the county seat, and good stage roads penetrate the foot-hills and mountains. The roads of Butte County are, as a rule, very good and well taken care of.

Gold was first found in Butte County by General John Bidwell, who, in the month of March, 1848, first discovered it on the banks of the Feather River, near where the town of Biggs now stands. A week or two later, General Bidwell, in company with several other men, found gold on Butte Creek, on the west branch of Feather River, in several ravines, and on the main Feather River, at White Rock. In May of the same year, the first mining camp on Feather River was made at the mouth of Morris Ravine, three miles above Oroville. Following up the river and finding gold in many places, General Bidwell discovered a rich bar, July the fourth, the same year, and the next day "struck his camp" and took possession. This has since retained the name of Bidwell's Bar, and it proved to be one of the richest bars on the famous Feather River.

From that time until now, more than $200,000,000 have been taken out of Butte County's placer, gravel, and quartz mines. Butte County to-day offers as good mining facilities, and as good returns for the money expended in developing mines, as any county in the State of California. The many branches of Feather River will supply abundant power for mining and milling purposes, as well as a never-failing supply of water. They are, also, all gold-bearing, and river mining is flourishing to-day, and will yield much wealth in the future. Quartz and gravel mining is on the increase, and several new finds of importance have been made.

Oroville was one of the richest mining towns in the State in early days, and is to-day the center from which mining supplies are being obtained. It is a beautiful little town, with many fine churches, schools, and houses, and presents a thrifty appearance, surrounded with orange, lemon, olive, fig, and other fruit orchards on all sides, where formerly placer miners were turning over the rich gold-bearing alluvial soil. It is known that within a radius of eight miles from Oroville, not less than $82,000,000 in gold has been obtained. As a large part of the gold production of Butte County has been derived from the Feather River and its many branches, which unite about ten miles from Oroville, more capital has of late been expended in river mining. Extensive work in this direction has been done by English companies on the main Feather River, a few miles above Oroville, where the river was lifted from its bed and run in a wooden flume for a mile, while the river was mined in the summer seasons. Later on, and below the former workings, a stone-and-cement wall was built, three-fourths of a mile long, and the river was kept within its confine for many years, while the gravel on the river bottom was worked with hydraulic elevators.

A few miles below Oroville is a present scene of great mining activity. The recent gravel deposits of the Feather River, below Oroville, cover an area of several square miles. Numerous shafts and borings in these gravel beds have disclosed the fact that the false bedrock, which is a hard-pan, is from twenty to forty feet below the surface. Gold has been found in sufficient quantities to warrant building six dredgers, each capable of handling from 700 to 1,200 yards per twenty-four hours. Several new dredgers are being built on improved plans and of larger capacity. The gravel contains gold from top to bedrock, growing richer as it goes down, and the last few feet before bedrock is reached the gold grows heavier and more plentiful, sometimes reaching several dollars per cubic yard.

Mining in Butte County is not confined to gold alone. Several of the inferior metals are also produced, such as silver, copper, and lead, mainly as by-products at present, but promising to be of importance. Many other useful minerals will also be mined on a large scale, when transportation facilities are improved, such as chromic iron, of which several valuable deposits crop out. These are lodged in serpentine, and in one instance the outcropping is over twenty feet wide, and the best of it assays fifty per cent of sesquioxide of chromium. Several varieties of black and variegated marble occur, and an abundance of valuable plastic clays, hematite, asbestos, infusorial earth, kaolin, red and white mineral paints, manganese oxide, antimony and building stones. Platina occurs as a fine crystalline powder in many of the placer and gravel mines. Quite a number of valuable diamonds have been found in the tailings and "clean-ups" of the Spring Valley Hydraulic Mine; some are white, but they are generally of a light yellow color, and in artificial light seem to have more lustre, when cut and polished, than the white.

During the time hydraulic mining was permitted, some very large canals and ditches were constructed, many of these at an enormous expense. These supply water for both mining and irrigating, as well as for generating electric power. Among the largest ditches are the following:

The Spring Valley and Cherokee ditch, fifty-two miles long, costing over $500,000. The Oroville Water Company's ditches, carrying 3,000 inches, are about forty miles long, supplying Oroville with water and irrigating the Thermalito Colony; the company also generates sufficient electric power for lighting, and will in future supply 500 horse-power to various industrial operations. The Palermo Land and Water Company's ditch, sixty miles long, carries 3,000 inches of water. The

CHEROKEE HYDRAULIC MINE, CHEROKEE, BUTTE COUNTY.

South Feather Union Ditch Company's ditch, fifty miles long, is supplying water for many mines and irrigating a large area. The Butte Creek Power Company's ditch is carrying 3,000 inches of water, used for generating electric power, supplying Oroville and Chico.

Leaving Oroville, on the Forbestown road, the first mining encountered is three miles out, where two quartz veins are being developed. One at the Lady Washington Mine is a ten-inch vein of hard quartz containing considerable free gold, and another is in the Sparke Mine, three feet wide, carrying free gold and sulphurets, between walls of slate and diabase.

On the same road, about eleven miles from Oroville, is the Hurleton district. Several quartz veins are under development here. The Phœnix Mine is opened by shaft and tunnel. Quartz is crushed with a Bryan roller mill run by water-power. The Resumption Mine, which crosses the Phœnix, is being developed by tunnels. The vein is about fifteen inches wide, carrying free gold and rich sulphurets. The Mount Ida Mine, situated three miles north-east of Hurleton, has a five-stamp mill on the property and employs three men at present. Considerable prospecting is done in this district, in both gravel and quartz, and with varying results.

More mining, and on a larger scale, is done in the Forbestown district. There are numerous properties around Forbestown that have more or less development work done on them, but not many in active operation. Just at present the Gold Bank Mine and the Denver Mine are the only ones in operation. The Gold Bank is running full blast, employing seventy-five men, keeping their forty-stamp mill running steadily, and crushing 3,500 tons of ore a month. They have a hoisting works, machine shop, forty-stamp mill with sixteen Frue vanners; two canvas tables, fifty by sixty feet each; chlorination works, with a capacity of four tons daily; and a compressor, the power for which is obtained from the waste water that runs the machinery at the mine, the compressor being placed 800 vertical feet below the collar of the shaft, and about 4,000 feet away from it. The works are lighted by electricity of the incandescent system. The Gold Bank Mine has been in uninterrupted operation since November, 1887.

The Denver mine, the only other mine in active operation at present, is west of Forbestown about four miles, between Forbestown and Enterprise, and has just been started up again, after a shut-down of about a year. There is a ten-stamp mill on the property, four Frue vanners, a whiffle, and a canvas table, twenty-two by sixty feet. The Denver Mine is operated entirely through tunnels. They are employing eleven men at present.

The Beik Brothers are erecting a five-stamp mill on one of their mines, on Skin Gulch. The quartz is supposed to be rich in free gold, an unusual thing in this camp, the ledges being almost entirely sulphuret propositions.

The Carlisle Mine, on Feather River, is idle at present, but it is understood that the owners are negotiating a sale. They have a shaft and tunnels, ten-stamp mill, with four Frues, canvas table, and a six-drill compressor.

The Shakespeare Mine is now idle. At one time there was a thirty-stamp mill on it, with Frue vanners, twelve in number, a canvas table, and chlorination works.

There are numerous other properties in this district that are owned by poor men, who do the assessment work, but have no money to erect machinery and develop the mines extensively. The general characteristics of the camp are: Very hard white quartz; little free gold, most of the value being in the sulphurets, which are high-grade, as a rule averaging between $100 and $200 a ton; large ledges, for the most part flat, being about thirty-degree pitch with the horizontal.

The Enterprise mining district is situated about six miles west of Forbestown and sixteen miles from Oroville. This camp was one of the liveliest in early days. A number of rich bars were worked in the river, which yielded fabulous amounts of gold. Of late, quartz mining has been pursued with renewed activity, and the camp bids fair to equal its former importance. The Red Point Quartz Mine, situated one mile east of Enterprise, is in a 12-foot contact vein, between granite and syenite. The development work has been done by tunnel, about 200 feet long, and by a shaft. The Black Hawk Mine is an extension on the west of the Red Point and is now being developed. The Laughlin and Morrison Mine is also an extension on the same vein.

The Murray Mine, two miles south-east of Enterprise, is also in a contact vein between granite and syenite, developed by shaft and tunnel.

The Lydia Mine is one and a half miles south of Enterprise. While this vein is small, it is very rich, and is crushing quartz with a four-stamp mill, eight men being employed. The Gem Mine, one and a half miles from Enterprise, on the South Fork of Feather River, has a vein three to four feet wide, in granite, and employs eleven men.

The Slater Mines are about three miles east of Enterprise and in granite formation, and are developed by tunnels. One of the veins is now being developed and a two-stamp mill built on the property. The quartz is three to four feet wide and promises to yield a large amount of gold. Sulphurets are also very valuable in all these veins in granite. The Oro Mine is situated almost parallel with the Slater Mines, and is developed by an 800-foot tunnel. This mine was opened up twenty-five years ago by shafts. These have now caved in and the mine is at present idle. The ore yielded $9.00 per ton in free gold, and in some places a much larger amount. The sulphurets are rich, but they were not saved. About $24,000 were taken from this mine during the sinking of a 200-foot shaft and other development work on the same vein. The Paul Willot Mine is two miles south-west from Enterprise. The quartz is five feet wide where a 60-foot tunnel cuts the ledge and is now being developed by the owners. The Aveary Mine, three miles north of Enterprise, is being opened up by two men; a two-stamp mill is on this property.

The Crystal Peak Quartz Mines are situated three miles north of Enterprise, and are in granite formation. There are two veins, called

DREDGE OPERATING ON FEATHER RIVER.

No. 1 and No. 2. The No. 1 is six feet wide and developed by a tunnel now in 400 feet. No. 2 is also being opened up and is three feet wide. A ten-stamp mill, with two concentrators, is on this property. The Price river claim and the Last Chance river claims are worked in the dry seasons only. The Hoig & Dugo gravel mine is in the bed of McCabe Creek, as is also the Milligan claim. The Crystal Hill Mine is four miles north-west of Enterprise, in granite, ledge twenty inches wide, and is equipped with a four-stamp mill.

Leaving Oroville on another road, that leads to Cherokee and Yankee Hill, the first mines of importance are reached four miles from Oroville. During the past nine years, the Consolidated Gold Mines have invested a very large sum of money. Their present group of mines is known as the Banner, Clark and Coffee, Longs Bar, Amosky, and Hedges. There is a forty-stamp mill on this group. A striking feature of its construction is the use of concrete piers for the mortar blocks.

The Clark and Coffee Mine's shaft is down 500 feet, the Banner, 1,000 feet, and at present considerable development work is done on the Amosky.

About eight miles from Oroville, in a northerly direction, are quite a number of quartz veins, being slowly developed. Two miles from Hengy are a number of quartz veins, with mines in various stages of development.

The famous Spring Valley Mine, at Cherokee, twelve miles from Oroville, has been described so often in reports issued by the State Mining Bureau, that an attempt to do so here would be unnecessary. This was at one time one of the largest hydraulic mines of the State, and has produced to this date more than $13,000,000 in gold. Several smaller companies are at present working the hard gravel cement through arrastras, and are reported doing well. It is also reported that this great mine will again resume operations on a large scale, by drifting. Valuable diamonds, and a considerable quantity of platina, have been found, although no systematic efforts have been made to save anything but gold. This mine will undoubtedly produce as much or more gold than has already been taken out, if the projected drifting method is carried out.

The Yankee Hill district is six miles north-east of Cherokee, or eighteen miles from Oroville. A large amount of money has been taken from the many smaller quartz veins that crop out in this district, and the population to-day is "panning out" sufficient gold for their maintenance, and are convinced that mining in their district is hardly in its infancy yet. A number of "pockets" have been found here, some of which yielded large sums of money, one in particular yielding $48,000. Several valuable quartz properties are partially developed, and deserve a better fate than they have at present, in their complete idleness. There is a very strong vein of baryta, crossing the country for miles, which is gold and silver-bearing. This is about fifteen feet wide, and has been opened up by tunnels and small shafts. While the ore is rich enough in gold, the difficulty of milling it with a small stamp mill has caused all operations to cease at present, more on account of the high specific gravity of the baryta than any other cause.

Leaving Oroville on the Cherokee road there are a number of mines under the lava, each employing a few men. Above and around Cherokee a number of both placer and gravel mines are being prospected. Twelve miles north of Cherokee the Magalia mining district commences and extends for many miles in all directions from the town of Magalia. The famous Magalia Mine, or, as it was formerly called, the old Pershbacker Mine, is situated three miles north-west of Magalia, and employs on the average fifty men. The gravel channel is rather narrow, but immensely rich. The Diller, or Dix, Mine employs twelve men, and is worked continuously. The Pete Wood drift mine is being prospected by an English company by a tunnel 4,000 feet long, and employs twelve men. The Matheson Quartz Mine is situated five miles north-west of Magalia, and has the last three years yielded $75,000, with five men employed. The Alki drift mine, two miles north-east of Magalia, employs eight men. The Magalia Consolidated Mine, Limited, is a drift mine, situated three miles south-west of Magalia, and is prospected by long tunnels, employing twenty-five men. The Lucky John drift mine, two miles south-west of Magalia, employs eight men continuously. The Bader drift mine, one and a half miles south-west of Magalia, has been prospected for many years, and employs eight men. The West & Cummins drift mine, four miles north of Magalia, employs six men. The Howl Mine is a river mine, and employs ten men; this mine is situated twelve miles north-west of Magalia. The Durgan drift mine, two and a half miles south-east of Magalia, at present employs five men. The Butte Creek Mining Company's quartz mine is situated five miles from Magalia, and is equipped with a ten-stamp mill, concentrators and air compressors, all run by water-power; the mill and mine are lighted with electric lights of the incandescent system, and employ sixteen men.

Several promising quartz and gravel mines that are now idle in this district will shortly be re-opened. The Pomerat Quartz Mine, two miles from Magalia, contains as rich ore as can be seen anywhere, and deserves to be re-opened. The Aurora drift mines are at present idle. Quite a number of smaller quartz mines near Inskip are held by parties, and are being prospected with flattering results.

Placer County's Mining Story.

BY
IVAN H. PARKER.

SINCE the occupation of California by Americans, Placer County has borne such an important part in the history and development of the State, that it may be classed as a representative county.

The boundaries of Placer County extend from the State of Nevada and the summit of the Sierra Nevada Mountains, more than 8,000 feet above the sea-level, westward one hundred miles, to within twelve miles of Sacramento, with an elevation of less than 100 feet. In width, it extends from the Yuba and Bear rivers, on the north, to the Middle Fork of the American River, south, ranging in width from ten to thirty miles. The thirty-ninth parallel of latitude, north, passes through the county. Placer County covers 1,386 square miles, and is larger than the State of Rhode Island.

The lands in the valley near Sacramento merge into the rolling foot-hills, which gradually increase in altitude, until is reached the heavy timbered mountain meadows and snow-clad peaks of the Sierras. The scenery is grand. Here majestic snow-crowned peaks rise from deep gorges and canyons. At their base, and held in depressions of their summits, may be found beautiful lakes that rival those of far-famed Italy. On the summit borderland, between the State of Nevada and the boundaries of Placer, at an altitude of 6,280 feet, lies Lake Tahoe. Covering an area of 220 square miles, and having a known depth of 2,000 feet, it is the largest body of fresh water in the world at that altitude.

Mining has always been the leading industry of Placer County, the county deriving its name from the shallow placers, which extended from the lower plains, in the western portion of the county, almost to the summit of the Sierras. Placer County is pre-eminently rich in minerals. In its boundaries may be found the pure, sparkling water of the medicinal springs, the iron ore, the monumental marble, the granite of the quarries, the deep quartz veins, with their bonanza of golden metal, or its rich placers in the gently sloping ravines, the rugged canyons, or in the beds and banks of its rivers.

Placer County has no history prior to 1848. Gold was discovered on the South Fork of the American on January 19, 1848. The discovery of gold in Auburn Ravine, Placer County, was made by Claude Chana, who first turned the precious metal to light in the Dry Diggings on May 16, 1848, three months after the discovery in California. Samuel Seabough, in his sketches of the beginning of placer mining in California, says: "In the Dry Diggings, near Auburn, during the month of August, 1848, one man got $16,000 out of five cartloads of dirt. In the same diggings, a good many were collecting from $800 to $1,500 a day." The region soon acquired the name of "North Fork Dry Diggings," and in the summer or fall of 1849, when the settlement became more concentrated, and stores were established, was given the name it now bears — Auburn.

Following the first discoveries of gold in the ravines, river mining soon commenced to enlist the greatest attention. When first dug over, the rivers were natural ground sluices, simply beds of clean-washed gravel containing gold. Placer County, lying south of Bear River, and embracing several branches of the American, was the location of hundreds of rich river bars, where earliest mining was principally carried on, and where some of the most exciting early scenes in California have occurred. The American, having its source in Placer County, was the richest river in the State. From its bars and bed, and the deep gorges of its tributaries, flowed the golden stream that added hundreds of millions to the wealth of the world in the few years following the discovery of gold.

From 1849 to 1857, there had been made constant and large shipments of gold, secured mainly from river and bar mining. The beds of the present rivers were rich in gold, and the bars of nearly all streams contained gold. In some instances, bars, when discovered, were denuded of gravel, and the gold lay exposed upon the rough places of the bedrock. Some of the earliest miners gathered thousands in a single day from the exposed bedrock of the river bars. The river-bed was first approached by wing-dams, consisting of obstructions made of stones and brush packed with soil, extending some distance into the river and then down the current

Morning Star Drift Gravel Mine, Iowa Hill.

sufficient to drain a portion of the river-bed. Dams and races were also constructed, draining portions of the streams.

Later came the work of fluming the rivers with lumber and canvas flumes. Fluming continued during the first decade of gold mining until nearly every stream in the county had been turned from its natural channel. The only operations in fluming at present in Placer County are being made on the North Fork of the American, five miles south of Blue Canyon, where a company is engaged fluming a portion of the river this season.

The Bear River Gold Mining Company is operating a hydraulic elevator at their river mine on Bear River, two and a half miles from Colfax. The company has acquired several miles of river tailings, which cover the bed of the stream to a depth of from eighteen to twenty-four feet. The hydraulic plant has been equipped at a cost of several thousand dollars. Five hundred inches of water, conveyed in a pipe-line 3,600 feet, gives a pressure at the works of 450 feet. Col. W. S. Davis, of Auburn, has for many years successfully operated a hydraulic elevator on the bars of the Middle Fork of the American, seven miles above Auburn

Quite a number of miners have locations on the river bars, where they engage in mining during the summer. The freshets during the rainy season each winter deposit a new layer of gravel. This, to a depth of from three to six inches, is run through sluices. Some of the best returns this season were made by the owner of a claim on the North Fork of the American, seven miles below Colfax, where two men with wheel-barrows and sluice boxes took out fourteen ounces of gold in fifteen days.

Following river mining and the working of the shallow placers, came the institution of hydraulic mining in 1852, first carried on in Placer County, at Yankee Jim's. At the start, streams of twenty-five and fifty inches of water were conducted under forty or fifty feet pressure through a five-inch canvas hose with a sheet-iron nozzle. This method was improved until heavy sheet-iron pipe is now used to convey from 1,000 to 2,000 inches of water under heavy pressure, which is delivered against the gravel bank through a patent nozzle called a "monitor," or "giant."

Hydraulic mining was successfully carried on until the injunctions in the early '80s, at Dutch Flat, Gold Run, Iowa Hill, Todd's Valley, Yankee Jim's, Bath, Michigan Bluff, and other portions of the county.

The product of the hydraulic mines of the county is estimated to have reached the sum of $60,000,000.

The surface and hydraulic mining districts of the Forest Hill Divide have a record of over $30,000,000. Iowa Hill hydraulic mines produced up to 1881 a total of $20,144,570. A conservative estimate of the Gold Run and Dutch Flat hydraulic districts gives the output at $10,000,000. Following the injunction granted by Judge Sawyer, water ditches, flumes, pipe-lines, and other hydraulic mining property in Placer County, valued at several millions of dollars, fell into disuse

POLAR STAR HYDRAULIC MINE, DUTCH FLAT.

and became almost a total loss. With the passage of the Caminetti law, the United States Debris Commission has granted a number of permits to mine by the hydraulic process in Placer County, by depositing the tailings behind accepted restraining dams.

The Gold Run Gravel Company have operated two monitors during the past season at their mine at Gold Run. During the water season, which closed in July, Wm. Nicholls, Jr., of Dutch Flat, has operated his Polar Star Gravel Mine by the hydraulic process with great success. A pipe-line conveys 2,000 inches of water to the mine under a pressure of several hundred feet. The tailings are deposited behind the Liberty Hill dam in Bear River, built over twelve years ago, and which has successfully stood the test of many a winter's freshet.

Drift mining is an important branch of the mining industry of Placer County. As the rich river claims and shallow placers were exhausted, attention was turned to the deep auriferous gravels of the ancient river channels, formed at different periods, and lying high above the modern rivers. In Placer and Sierra counties, drift mining is followed more extensively than in other counties of the State.

What is known as the Forest Hill Divide is the country lying on the north side of the Middle Fork of the American River, extending from Butcher Ranch to Deadwood. The section is traversed its entire length, and crossed and recrossed, by numerous ancient channels. These channels and gravel deposits are covered by lava caps, increasing in depth from 200 or 300 feet at the mines west of Forest Hill, to over 1,000 feet in the Red Point and Hidden Treasure mines east and above Michigan Bluff.

This divide has produced over $30,000,000 in gold, which has been taken from the drift and hydraulic workings of The Dardanelles, Todd's Valley, Yankee Jim's, Mayflower, Paragon, Michigan Bluff, Weske, Red Point, Mountain Gate, Hidden Treasure, and other gravel mines. The individual and company holdings in the established channels, from Forest Hill west to Spring Garden, will afford a field for operations for the next fifty years, while mining upon the eastern portion of the Forest Hill Divide has only fairly commenced. The record of these mines, which have produced millions in bullion, only amounts to a fair prospect of the golden treasure that remains in these well-defined channels awaiting development.

The following is the list of the principal mines and districts on the Forest Hill Divide, with their returns:

Mountain Gate, White Channel	$600,000
" " Blue Channel	175,000
Hidden Treasure, White Channel	1,700,000
Weske Channel	750,000
Michigan Bluff District	5,000,000
Paragon, drift	2,300,000
" hydraulic	500,000
Mayflower, drift	1,516,250
" hydraulic	35,000
Dardanelles, drift and hydraulic	3,000,000
Forest Hill District	5,000,000
Todd's Valley, hydraulic	5,000,000
Yankee Jim's, hydraulic	5,000,000
Red Point, drift	600,000
Total	$38,176,250

The Hidden Treasure Gravel Mine, with tunnels at Sunny South and Centerville, on the upper Forest Hill Divide, is operated upon the most extensive scale of any drift mine in the United States. Electric power, in use at this mine, has proved a complete success. The plant was installed over two and a half years ago, at a cost of $30,000. The power plant is located in El Dorado Canyon, 4,000 feet from Sunny South. Water, flowing from the company's tunnels at Sunny South and Centerville, is carried in an open ditch to the reservoir above the power-house, and is dropped through a pipe with a pressure of 830 feet, generating over 200 horse-power.

The trolley system is in use, and the camp and mine are lighted by electricity. A double electric motor, weighing 6,500 pounds, with a speed of twelve miles an hour, is used for hauling gravel trains. The main mine tunnel, at Centerville, is in a distance of over 8,000 feet. The ancient channel, at the present point of working, has a width of 1,400 feet. A portion of the soft bedrock, and about six feet of the gravel, all free washing, is taken out and run through the sluices. The sluices extend down the canyon, a distance of 1,575 feet. About 2,000 cars of white gravel are mined each week. The company now has 175 men on the pay-roll. Harold T. Power is superintendent.

The Red Point Gravel Mine, twelve miles above Forest Hill, is owned and operated by a French company. Sixty men are employed, and J. A. Ferguson is superintendent. The main tunnel is in over 14,000 feet. The gravel is blue and free washing. The average width of the breast worked in the channel is sixty feet. Seven horses are used to haul the trains of gravel. The company is at present having a compressed-air plant put in, which will furnish power for use at the mine.

The bedrock at the face of the tunnel, where gravel is now being extracted, is 1,000 feet below the surface. Petrified wood is often encountered in the channel, and many huge logs of wood, resembling cedar, in a fair state of preservation, are found. The length of the period the wood has lain sealed in the channel is hard to estimate. It is found on the bedrock in a tunnel driven over two miles into the divide and 1,000 feet below the surface, which is now covered with a forest of massive cedar, spruce, and sugar pine.

The Eureka Consolidated Drift Gravel Mining Company is developing what promises to become the largest drift mine on the Coast. The mine is located two and a half miles east of the Red Point Mine

A tunnel, 2,000 feet in length, has tapped an ancient river channel running parallel with the Red Point channel, which is being developed by the company. The mine embraces 3,600 acres of land, covering six and one-half miles of channels. The company owns a water ditch, and a portion of the claim is covered with a heavy growth of timber. This company, when the mine is well opened, can easily employ over 300 men. Felix Chappellet, Jr., is superintendent.

The Paragon, Mayflower, Weske, Glen, Consolidated, Grey Eagle, Central, Bald Mountain, and other gravel mines, are located on the Forest Hill Divide.

SHOPS, TRACK, ETC., OF HIDDEN TREASURE DRIFT MINE.

The Morning Star Drift Gravel Mine is located near Iowa Hill, eight miles from Colfax, on the Iowa Hill Divide. Lieutenant-Governor J. H. Neff is superintendent. The Morning Star has become famous as a dividend producer, having paid a total of 103 dividends, ranging from $2 to $8, averaging about $4 a month per share. During the past thirty years, the Morning Star Mine has produced over $1,500,000 worth of gold bullion.

The main working tunnel is in a distance of over 4,000 feet. The tunnel has been run just beneath the channel, and the gravel is breasted and loaded into cars through chutes. The company has sixty-five men on the pay-roll. About 100 car-loads of blue, cemented gravel are mined each twenty-four hours. The ten-stamp mill, running steadily, crushes the gravel.

The Big Dipper Gravel Mine is located at Prospect Hill, near Iowa Hill, Placer County. The main tunnel is in over 4,000 feet. The mine has been successfully operated for many years and has afforded employment for a large force of men. The power plant consists of a ditch and pipe-line, which furnishes pressure for a Pelton wheel, together with a boiler for use during the dry season. Seymour Waterhouse is superintendent.

The Jupiter Consolidated Gravel Mining Company has been developing a very promising property on the divide above the Morning Star Mine. An incline, recently sunk, struck some very rich gravel on the rim of their channel. The company is now engaged crosscutting the channel, and the development of the property will proceed as rapidly as possible.

The company which owns the Occidental Gravel Mine at Grizzley Flat, near Iowa Hill, has a force of men engaged in developing the property. Wm. Cameron is superintendent. An electric plant of fifty horse-power furnishes the power. The mine is being prospected through a shaft. The Grizzley Flat channel has proved very rich in adjacent claims, and the outlook for the Occidental is very promising.

The Gold Run Gravels Company owns an extensive property at Gold Run, near the main line of the Central Pacific Railway, comprising about two miles of ancient river channel. Drifting is in progress, and the cemented gravel is crushed in a ten-stamp mill. Portions of this mine are worked by the hydraulic process, two hydraulic monitors having been operated during the past season. The company owns twenty-five miles of canals and a large amount of timber lands. Col. J. E. Doolittle is superintendent of the mine, which is owned by a London syndicate.

One of the richest strikes of recent years has been made at the Blue Lead Gravel Mine, at Dutch Flat, owned by a company of capitalists of Sacramento. S. C. Jordan is superintendent of the mine. In May, 1899, the main tunnel tapped some very rich cemented gravel on the rim of the channel. Breasting was discontinued and the owners purchased a new ten-stamp mill, which has just been put in place at the mine.

The Azalea Drift Gravel Mine, at Blue Canyon, on the main line of the Central Pacific Railway, is owned by the Blue Canyon Mining and Development Company, with J. B. Knapp as superintendent. The mine consists of 690 acres of land, well timbered, comprising more than a mile of established ancient river channel. Power for the mine is secured from the Cedar Creek Canal. Water is taken through a pipe-line with a pressure of 200 feet, to run a Pelton wheel. The tunnel being driven to tap the channel is now in 2,150 feet. The company has 40,000 shares of stock distributed among 136 share-

HYDRAULIC ELEVATOR SINKING A PIT AT MAMMOTH BAR MINE, NEAR AUBURN.

holders, nearly all of whom are railroad employés. An assessment of one cent a share per month raises a fund of $400 each month for development work. The company expects to strike the channel at about 2,500 feet.

The entire surface of Placer County, from the lower plains in the western portion, to the summit of the Sierras, is traversed by innumerable veins and ledges of gold-bearing quartz. The development of quartz ledges in Placer County has not kept pace with that of hydraulic and drift-mining properties. In a number of districts, deep mining has been carried on without a single failure. The product of the quartz mines has aggregated several millions of dollars, nearly all of which has been taken from pay shoots near the surface, which were abandoned when the water level was reached. The quartz districts offer big returns for capital, backed by good judgment and improved methods of mining.

Gold-bearing quartz was first found in Placer County, near Ophir, at an early day, and was worked to some extent by Mexicans in mortars and arrastras. All the mining done was merely by breaking the croppings and working only the richest quartz. The first quartz mill built in the county was erected in 1851, at Secret Diggings, the mine being on the Rosecranz ledge, near Ophir. In 1852, the Crœsus Hill Quartz Mining Company commenced the erection of a ten-stamp mill, driven by a steam engine of thirty-five horse-power. The site of this mill was about one mile west of Auburn. The mill was completed on January 26, 1853.

The first mill erected in the eastern portion of Placer County was built in 1855, by Messrs. Walsh and McMurtrie, on the Pennsylvania ledge, about eight miles east of Wisconsin Hill, on Shirt-Tail Canyon.

The deepest quartz-mining operations in the county are carried on by the Pioneer Gold Mining Company, of Boston, at their mine, eight miles south of Towle. The ledge is located in the mountain, on the south side of the North Fork of the American River. The mine has given out over $500,000 in gold from the slopes in the mountain side, above tunnel No. 4, which cuts the vein on the 1,000-foot level. A shaft has been sunk on the ledge in tunnel No. 4 to a depth of 250 feet, making a total depth on the vein of 1,250 feet. An electric plant was installed at this mine in January, 1898. A dam in the American River is used to drive a 353 horse-power turbine water-wheel. This wheel runs the 270 horse-power electric generator. Electric power is transmitted to the mine to run the twenty-stamp quartz mill, two compressors, and the blower. J. J. Sullivan is superintendent of the mine.

The Herman Quartz Mine, near Westville, twelve miles above Forest Hill, has been giving out splendid returns in bullion during the past few years. Dr. O. L. Barton is superintendent of the property, which is owned by San Francisco capitalists. A ten-stamp mill, running steadily, crushes an average of thirty tons of quartz each twenty-four hours. The ore is at present being taken from a shaft which connects with the 300-foot level, where the face of the ledge shows a width of ten feet. A new tunnel taps the ledge several hundred feet below the present level. The company expects soon to build a twenty-four-stamp mill at the mouth of the lower tunnel.

HOIST AT ROYAL MINE, HODSON.

Two and one-half miles below the Herman Mine is the Lady Bedford Quartz Mine. The mine has a two-stamp mill, which is equipped with a rock-breaker and self-feeder. This little mill, after being supplied with ore, pounds away unattended. It has paid the owners more than wages for the past three years, and they expect to make it earn a larger mill.

The Golden West Quartz Mine, five miles south of Blue Canyon, embraces locations on one of the largest quartz ledges in this portion of the State. The mine is the property of Reuben H. Lloyd, of San Francisco. Three tunnels have been driven in on the ledge, the longest being 275 feet. Crosscuts and present development show the width of the ledge to be over 200 feet. Numerous assays of quartz have shown values of from $5 to $100 per ton.

Considerable attention is being given to the district traversed by the immense ledge or porphyry dike, four and one-half miles southeast of Colfax. The Alameda and Annie Laurie quartz claims are the most prominent of a score of locations. The formation between the walls, for a distance of 100 feet in width, is cut by innumerable seams running diagonally across it. The ledge matter is soft, and is worked by the sluicing process to a depth of thirty-five feet from the surface. Below this depth the porphyry, or ledge matter, is as hard as average quartz ore. At the Alameda, a ten days' run in the winter, with two men and sluices, brought a return of forty-six ounces in gold. George Nissen and son, who own the Annie Laurie, secured $170 last winter from a sixteen and a half days' run. This ledge or dike varies in width from twelve to one hundred feet, and can be traced for a distance of ten miles.

Early in the '50s, before the era of quartz mining in California, Auburn Ravine, Daly's Flat, and the surface of the hill-sides adjacent to the Ophir district, brought rich returns. Much of the gold carried pieces of quartz, showing that it came from a source near at hand.

For an area of many miles, the Ophir district is thickly ribbed with quartz veins. With the exception of the Crater, Gold Blossom, Three Stars, and Hathaway, no deep mining has been done in the district. The Crater reached a depth of 800 feet, and, after yielding $750.000, the pay shoot pitched into the adjoining property. The Three Stars Mine has reached the 725-foot level, the Gold Blossom, 437 feet, and the Hathaway, 630 feet. The shafts and workings of nearly all of the remaining mines range from fifty to three hundred feet. Scores of these have given treasure from rich pay shoots near the surface, and as the cost of working increased, were abandoned. The need of capital for development has caused many promising ledges in this district to remain idle. Following is a conservative estimate of the product of a number of quartz properties in the Ophir district:

Crater	$750,000	Boulder	$20,000
Centennial	100,000	Goddard	18,000
Green	100,000	Hope	17,000
Moore	80,000	Eclipse	15,000
Doeg	77,000	Bullion	10,000
St. Patrick	75,000	Morning Star	10,000
Mina Rica	50,000	Duncan	10,000
Bellevue	50,000	Booth	10,000
California	50,000	Posterior	9,000
Good Friday	40,000	Crandall	5,000
Brush Fence	30,000		
Pete Walter	30,000	Total	$1,576,000
Conrad	20,000		

ELECTRIC POWER SAW-MILL, FOR FRAMING TIMBER.

The above figures do not include the output of the Gold Blossom, or Hathaway, both of which are examples of success in deep mining. Quartz mining in Ophir district is being given more attention at   .

present than the industry has received since the early days. The number of men now employed in the district is estimated at 200.

The Bellevue Quartz Mine, at Ophir, is an example of good returns from prospecting, as a result of pluck, well-directed development work, and a small outlay of capital. F. E. Brye, of Auburn, and P. S. Lozano, of Ophir, secured a working bond on the mine about two years ago. A shaft was commenced, and at a depth of thirty feet a pay shoot was encountered. Sinking is still in progress and the shaft is now down 170 feet. Small drifts were run east and west from the 140-foot level. From the present workings, over $29,000 in gold has been taken, the mine paid for, and many improvements made.

From the figures of the Superintendent of the United States Mint at San Francisco, on the gold yield of 1898, Placer County ranks fourth, producing during that year $1,488,022 in gold. The total output of gold since its discovery in Placer County would be hard to compute, as up to 1880 the United States Mint at San Francisco depended upon Wells, Fargo & Company's estimate, which was made by State, not county. Previous to 1880, it is now impossible to get figures, as nearly all of Wells, Fargo & Company's papers have been destroyed by fire.

The United States Mint report gives the annual product of the mines of Placer County, from 1880 to 1898 inclusive, as follows:

	Gold.	Silver.			Gold.	Silver.	
1898	$1,488,022	$5,670	$1,493,692	1887	$855,509	$555	$856,065
1897	1,524,911	6,784	1,531,725	1886	1,071,632	1,397	1,073,029
1896	1,674,844	6,690	1,681,384	1885	906,301	411	906,712
1895	1,599,634	5,272	1,604,907	1884	720,000	......	720,000
1894	1,851,216	644	1,851,878	1883	810,000	......	810,000
1893	1,351,249	616	1,351,865	1882	800,000	......	800,000
1892	1,159,079	2,119	1,161,199	1881	850,000	6,500	856,500
1891	998,494	5,291	1,004,415	1880	838,133	6,400	838,733
1890	1,003,601	1,045	1,004,646				
1889	1,245,490	1,975	1,247,665		Total for 19 years,		$21,650,415
1888	850,000	1,000	851,000				

From 1848 to 1860, the total annual gold yield of California did not fall below $41,000,000, the output one year during that decade running above $80,000,000. Hence, it would be conservative to estimate the product from 1848 to 1880 at double the above figures, or about $43,000,000.

It is estimated that there are about 200 miles of unworked auriferous ancient river gravel channels on the Forest Hill, Iowa Hill, Dutch Flat, and Railroad divides of Placer County. Basing an estimate of the amount of gold remaining, on the results obtained where portions of the deep channels have been worked, that have given from $100 to $1,000 per lineal foot, equaling a product of from $500,000 to $1,000,000 per mile, it is evident that the amount of gold already extracted is but a trifle compared with the amount of golden treasure which remains as a reward for well-directed capital.

The entire surface of Placer County, from the lower plains to the summit of the Sierra Nevada Mountains, is ribbed with quartz ledges, nearly all of which have shoots of gold-bearing ore, accompanied by

POTTERY WORKS OF GLADDING, McBEAN & CO., AT LINCOLN.

iron pyrites and galena. The Ophir quartz district is probably the most prominent. While the district has produced upwards of $2,000,-000, the product of a score of mines, the Crater, the Hathaway, and Gold Blossom are the only ones which have been worked below the 300-foot level. The Rising Sun Mine, near Colfax, and the Pioneer Mine, near Towle, are the only examples of deep mining in eastern Placer, while innumerable quartz ledges abound throughout the entire district. Water is plentiful, timber is abundant, lumber cheap, and the ledges are accessible. Considerable attention, however, has been directed toward the development of quartz properties during the past two years.   The following is a list, with the location, of the various stamp mills in Placer County:

Name of Mill.	Nearest Town.	No. of Stamps.	Name of Mill.	Nearest Town.	No. of Stamps.
American Bar	Michigan Bluff	10	Herman	Deadwood	10
Big Dipper	Iowa Hill	10	Indiana Hill	Gold Run	8
Black Oak	Weimar	5	Kidd & Johnson	Michigan Bluff	5
Boulder	Ophir	8	Lady Bedford	" "	2
Breece & Wheeler	Bath.	10	La Trinidad	Cisco	10
Buttes	Ophir	5	Malmberg	Auburn	5
Blue Lead	Dutch Flat	10	Mayflower	Forest Hill	20
Dohlonega	Emigrant Gap	8	Morning Star	Iowa Hill	10
Dardanelles	Forest Hill	5	New Mill		5
Dorer	Damascus	10	Pioneer	Damascus	10
Gold Blossom, 1	Ophir	10	Red Rock	Blue Canyon	10
" 2	"	10	Shipley		10
Golden Shaft	Dutch Flat	8	St. Patrick	Ophir	15
Gold-Run Gravels	Gold Run	10	Savage Mill	Westville	5
Gold Ring	Towle	10	Van Avery	Blue Canyon	10
Grey Eagle	Forest Hill	10	St. Lawrence	Ophir	5
Hathaway	Ophir	20			

The Drummond Mine, near Forest Hill, has two Huntington mills. The St. Lawrence, at Ophir, has one Huntington mill.

The granite quarries of Placer County are not only the most important in California, but rank with the best in the United States. Quarries are located at Lincoln, Rocklin, Loomis, and Penryn, in each instance along the line of the railroad.   The State Capitol at Sacramento, the Stockton court-house, and the famous Lick and Crocker monuments are examples of the value and beauty of Placer County granite.

An inexhaustible deposit of potter's clay occurs near Lincoln, which is utilized at the extensive pottery, tile and terra-cotta manufactory of Gladding, McBean & Company.   The works were established at Lincoln, in 1875, and are the largest west of the Mississippi river.   The company manufactures almost everything in the clay line.   The clay banks are located three-quarters of a mile from Lincoln, where there is a vein of clay thirty feet in thickness.

Chrome ore is found in Placer County.   Its discovery was made in 1876, on the Iowa Hill Divide.   Deposits were later discovered near Michigan Bluff, Clipper Gap, and Auburn.   The ore is found in irregular, disconnected masses, varying in weight from a few pounds to many hundred tons.

Asbestos deposits have been found near Iowa Hill, a number of car-loads of which were once shipped to San Francisco.   A marble quarry exists one-half mile east of Auburn.   The property is undeveloped.

El Dorado County's Mines.

FURNISHED BY THE
COUNTY MINERS' ASSOCIATION.

T was in El Dorado County that gold was discovered by James W. Marshall, in January, 1848. Authorities dispute as to whether it was January 19 or January 24. The story of the discovery and the world-wide results that immediately followed is a romantic and familiar tale, which need not be repeated in detail here.

In the era of placer mining that followed Marshall's discovery at Coloma, in 1848, and continued until the year 1856, when the richest of the surface placers had become exhausted, the old Empire County was the Mecca of the miner; towns and mining camps sprang up in a night, and for years El Dorado claimed a population of over 40,000.

With the decadence of placer mining, the large population of the first era of gold mining drifted away, leaving a reduced population that proceeded to prove the capability of the pioneer mining county to sustain successfully other industries than that which had resulted in her birth. Agricultural and horticultural pursuits were not neglected; lumbering and the rearing of cattle received due attention.

Systematic efforts were made to properly develop the deep gravel channels of the county, and, while not on a scale of the magnitude attained by some of her sister mining counties, the work of developing the mineral possibilities of the Mother Lode, that stretches across the entire breadth of the county, was not neglected. Many things conspired, until within recent years, to make the work of development slow and extremely expensive, such as the absence of transportation facilities and machinery for the successful working of low-grade ores, and the lack of cheap motive power. All these drawbacks to the successful and economical development of the great mineral wealth, that lies within the borders of the historic old county, have one by one been overcome. The completion of the Placerville and Sacramento Railroad to Placerville, in 1888, was a solution of the transportation problem, affording, as it does, cheap transportation to within a reasonable distance to all points along the great lode.

No other mining district in the West has more splendid possibilities for cheap and efficient motive power than El Dorado County. The question of power is always one of pressing importance in a mining section. In this respect, El Dorado County is well provided

THE GEORGIA SLIDE, EL DORADO COUNTY, THE SUPPOSED TERMINUS OF THE
GREAT MOTHER LODE.

for, as in a number of places water is available as a source of power, and, in the greater portion of the county, wood is still cheap and plentiful enough to render steam an economical power; but, in view of the advances which have been made in the application of electricity as a motive power, especially in mining, it is evident that it is the coming power in this county as well as elsewhere. Such being the case, this county has the solution of the power question within its borders, as its rivers and streams are capable of generating sufficient power to supply the State.

No portion of California has a finer water supply than El Dorado County. The Middle and South Forks of the American, and the various Forks of the Cosumnes, together with their tributaries and mountain lakes, the majority of which are situated favorably for storage purposes, form an unsurpassed water system, one capable of supplying, not only the needs of the county, but those of Sacramento, Contra Costa, and Alameda counties. The topography of the county is such that the water can be carried by ditches to all but a small part of it; and, in fact, a plentiful supply of water is carried to all parts of the county on the line of the Mother Lode, by the three great ditch systems on the three great ridges, or divides, within the county: the system of the California Water Company, on the North Divide; the El Dorado Canal and Deep Gravel Company's system, on the Middle; and the Diamond Ridge, or Park Canal system, on the South Divide.

The fact that El Dorado County yielded a hundred odd millions of gold, by the time that California had produced twelve times that amount, tells the story of its richness in the past, and gives a presage of its future yield, when systematic and adequate methods are used to work the bonanzas which lie deep in its quartz ledges and ancient channels. In the early days, every stream, flat, bench, and hill-side was strewn with the gold which had been released from the wearing away of the gold-bearing formations, and by the breakouts from the ancient channels which occupied so many of the ridges. The gathering of this harvest of gold revealed the existence of numerous mineral belts and districts, the most prominent of which was the famous Mother Lode, as well as a system of ancient channels containing marvelous beds of gold-bearing gravels.

Both to the east and west of the narrow belt of Mariposa slates containing the Mother Lode, lie the older, coarser, and more crystalline slates, quartzites, and limestones, known as the Calaveras formation. Wherever in these belts and districts true fissures were formed, and ore bodies deposited therein, so far as developments show, they give every evidence of being as permanently valuable as those located upon the Mother Lode.

While the shallow or placer diggings have been practically exhausted, still there are large bodies of gravel remaining unworked, partly because of the difficulty of access, and because of the cemented nature of much of the gravel in the ancient channels. However, the encouragement given to hydraulic mining by the United States Debris

Commission is having the effect of reviving that important industry, while more attention is being paid to deep gravel mining, with encouraging results.

These ancient channels have produced millions upon millions, and yet deep gravel mining, especially in El Dorado County, is only in its infancy, and offers an unequalled field for mining enterprise. As a matter of fact, the gravels worked in the ancient channels of El Dorado County have excelled those of all the other counties, save the

VIEW AT THE GEORGIA SLIDE, WHERE THE FAMOUS MOTHER LODE IS SUPPOSED
TO TERMINATE, IN EL DORADO COUNTY.

very richest in Nevada County, and, as yet, these gravels are practically untouched. When the channels of El Dorado County have been worked out on a scale commensurate with their size and value. it will be found that they will stand second to none in point of production.

There are a number of copper deposits of value in El Dorado County, notably those at Bunker Hill and Hastings Ravine, between Coloma and Pilot Hill, and the Cosumnes Copper Mine, near Fairplay.

The ores of the latter contain considerable gold. Silver is produced in combination with gold, some ores running very high in silver. While cinnabar is known to exist in several parts of the county, the Amador Quicksilver Mine, in the south-eastern part of the county, near the Amador line, is the only place where any development work has been done. The prospecting work done there showed the existence of large quantities of fair and high-grade cinnabar.

In summing up the mining resources of El Dorado County, first, it is seen that the Mother Lode within its boundaries not only shows up as well upon the surface as any equal portion of its course, and that it has rich places which will compare with any other rich places in its entire length, but that for area stoped and depth attained, it held its own with any other part at the same stage of development; second, that the other mineral belts show up well, and give promise of a great future; third, that the gravels of the ancient channels contain immense stores of wealth.

The slate quarries of El Dorado County, California, are the only ones in the State, at present known, as existing within profitable reach of railroad transportation. In fact, pure slate, possessing the requisite qualities of color, cleavage, and toughness, has not up to this date been discovered at any other place on the Pacific Coast, within the limits of the United States, the nearest available quarries, outside of El Dorado County, being those of British Columbia. It has been shown, however, most conclusively, that the slates of El Dorado County are superior in every respect to the slates of British Columbia, and are equal in quality to the best Welsh slates. This latter fact has been broadly proclaimed and proved in a very able paper, accompanied by an exhaustive analysis, presented to the Academy of Sciences, in San Francisco, by the late Melville Atwood, a distinguished member of that body, and an acknowledged authority on slates in England, as well as in America.

A portion of the El Dorado County slate belt has been exposed at intervals, along a stretch of about eight miles in length, being opened at four points, where the quality of the material warrants its production on a commercial scale. The production, however, is by no means equal to the demand, there being but one quarry now in operation, that of the Placerville Slate Company, just outside the limits of the city of Placerville. The slate belt herein referred to is known to extend from the vicinity of Kelsey, on the north, to Placerville, on the south, a distance of about eight miles, and is from 100 to 300 feet in width. On this deposit, less than one mile distant from the railroad depot, the Placerville Slate Company has uncovered a very fine body of pure slate, which is now being rapidly developed.

The quartz vein at Nashville was the first to be opened in El Dorado County (1851). The surface ore was very rich, and was worked for its free gold by the crude methods then in vogue. Early reports of the United States statistics show a yield from these old workings of $150,000. The first systematic work was begun in 1868. After

FLASH-LIGHT VIEW OF ALABASTER CAVE, NEAR ZANTGRAFF MINE, EL DORADO COUNTY, SHOWING STALACTITES OF CALCITE.

a period of three years, the mine paid for all of its development work, and plant costing $100,000, and paid $50,000 in dividends. The ore mined went from $3 to $15 per ton. Litigation closed the mine in 1871. In 1880, a patent was issued to Joshua Hendy, who, in 1882, developed the mine to the present 642-foot level. The ore taken from the last 100 feet of the shaft averaged $8.50 per ton. After opening up the lower levels, and taking out $45,500 in bullion, the financial condition of the owners caused them to suspend operations entirely. Nothing more was done with the property until 1894, when a bond was issued to a new company, who, after remodeling hoisting machinery and erecting a twenty-stamp mill, opened the mine and repaired the shaft. They took $20,000 from the ore left·between the 500 and 600 levels, part of which averaged $8 per ton. Owing to disagreements and the death of a prominent stock-holder, the bond expired, and the mine was closed without having been extended in depth.

The Union Gold Mining Company's property runs some 3,000 feet along the Mother Lode, and includes the Union, afterwards called the Springfield, and the Church mines. The history of this mine is exceedingly interesting, and there is reason to believe that its history will be repeated in many of the old mines of this county. In '50–'52, there was a mixed white and Mexican town of some 2,000 or 3,000 inhabitants upon this property, all supported by working the streams running from and through this property, and by running arrastras upon the rich float and croppings upon its surface. Prof. Silliman was the first one to examine and call attention to this property, and he induced his eastern friends and relatives to invest in and work the two properties, which were consolidated, and known as the Church-Union. This was done successfully for a year or two, when, on account of bad management, the works were shut down. A body of ore, milling from $20 to $40, was struck. Hayward and Hobart purchased the mine and worked it successfully for sixteen or seventeen years, producing several million dollars.

Hayward and Hobart changed the name of the Union to Springfield, and what they worked then comprised the southern third of the 3,000 feet of this property. It was worked to the depth of 1,600 feet, and, coming northward, the vein broke and was lost, in consequence of which they quit work. After lying idle five or six years, the present owners bought the property of Hayward and the Hobart estate, believing that the 2,000 feet between the Springfield and Church mines would prove, on development, to be as good as either.

Among the mines of the Mother Lode in this county, being developed by Scotch capital to advantage, is the Crownpoint group of five mines, situated in Diamond mining district, under the management of A. C. Morison.

The first actual development of the property began the 5th of March, 1898. Five hundred feet of shaft has been sunk, and a total of about 800 feet of drifts, tunnels, and crosscuts have been driven, together with buildings and hoisting machinery, erected for a cost of less than $25,000.   The ore is low-grade, but, on the 400-foot level, a large body of quartz, fully eight feet wide, on east vein, has just been exposed, which promises rich results.   Pay-ore, in quantity, has been exposed on the west vein also.

This property adjoins the Griffith Consolidated, also belonging to Scotch capitalists, and on the same lode as the Larkin and Marguerite mines, which are splendid representative Mother Lode bullion producers.

CYANIDE PLANT, TAYLOR MINE.

The Griffith Consolidated Mine comprises eight claims, and controls 5,200 lineal feet along the line of the Mother Lode, the rest of the claims being upon parallel veins both to the east and west. This group of claims is arranged so as to give the owners a property which is unequalled in the county, and which promises to compare favorably with any mine or mining property in the State, as it is undoubtedly located upon one of the premier mineral centers of the Mother Lode.

Until its recent sale to the Jumper Mining Company, this property was owned by a Scotch syndicate.   The main shaft of the mine was sunk to a depth of 750 feet, and many levels opened, and a great deal of high-grade ore excavated. The mine is equipped with two fine hoists and a twenty-stamp mill, all operated with electric power, from a plant a distance of four miles south of the works.

The Larkin Mine is directly north of and adjoining the Griffith Consolidated, comprising 2,600 feet of the Mother Lode, and about fifty acres of land. The development work has been in progress about three years, and includes a vertical shaft 500 feet deep. The ore-body has been continuous from the surface to the 500-foot level. The equipment of the mine includes a ten-stamp mill. The mine is on a paying basis and bids fair to become one of the permanent dividend-paying mines of El Dorado County. The mine is owned and has been developed by San Francisco business men.

One of the great enterprises of the county is controlled by the El Dorado Water and Deep Gravel Mining Company, and although it has had over $1,250,000 spent upon it already in the way of development, it is as yet only in a budding condition. The present canal system consists of a main canal forty-five miles in length with a maximum capacity of 10,000 miner's inches, and 100 miles of branches. The main canal takes water from the South Fork of the American at Cedar Rock, thirty miles east of Placerville, and in addition to the water claims located upon a large number of creeks, Silver Lake, in Amador County, Echo, Twin, Glacier, Andrain's, and Medley lakes, in El Dorado County, are used for storage purposes, to be drawn on in dry seasons. With its 350 miles of water-shed, immense storage capacity, and natural facilities, it is capable of supplying El Dorado, Sacramento, Alameda, and San Francisco counties with all of their cities, and still have a surplus. At the present time, it furnishes water for power, mining, and irrigation purposes. In addition to its water system, this company owns more good quartz and gravel properties than any one owner in the county, much of it being some of the best properties along the Mother Lode, near Placerville. In addition, the company possesses great facilities for electric power at Smith's Flat. At that point a 400-foot fall can be obtained, and that would give over 1,500 available horse-power, with 2,000 inches of water.

The Gentle Annie Mines are located one and a half miles north of Placerville, at an elevation of 2,000 feet above sea-level, upon the Mother Lode, covering a length of some 2,000 feet by 1,000 feet in width, and include four of the most important ledges or veins belonging to that system. The present development of the mines has been mainly upon the "Gentle Annie" vein, and consists of an incline shaft, 400 feet in depth, sunk upon the vein, and levels driven at right angles to the shaft. These levels aggregate some 4,000 feet in length, and most of the material extracted from them has been profitably milled in a ten-stamp mill upon the premises. The veins are

MONUMENT TO JOHN W. MARSHALL,
Erected by the State on a Hill overlooking Coloma and the Scene
of the Discovery of Gold.

from four to thirty feet in width, and carry large ore-bodies containing free gold, and an average of four per cent of sulphurets of good grade. The developments made have proved the "Gentle Annie" to be a mine of great extent and merit. A comprehensive equipment is now under contemplation for these mines.

One of the most promising mines on the Mother Lode, in El Dorado County, is the Poverty Point Mine, situated three miles north of

Placerville. The mine is at present worked by a tunnel that runs into the hill a distance of over 500 feet, tapping the ledge at a depth of over 300 feet. A fifteen-stamp mill was erected on this property last year. Two ledges have been uncovered in this mine, one on the hanging-wall twenty feet, and one on the foot-wall fifteen feet in width. No mine on the Pacific Coast offers greater facilities for the economical extracting and milling of large bodies of ore than the Poverty Point.

The Big Sandy and Gray Eagle properties possess a historical interest, because James W. Marshall owned a half interest in the first, and the whole of the second, at the time of his death. They are squarely upon the Mother Lode at Kelsey, and, when developed, will fulfill Marshall's prediction that they would become great mines. Marshall's old blacksmith shop still stands upon the Gray Eagle. These properties form an immense low-grade proposition. The double compartment shaft is down 350 feet, with various crosscuts and drifts. The mine is equipped with a ten-stamp mill.

The Gopher-Boulder property is composed of the Gopher and Boulder mines, the Dalmatia Mine, several properties of minor importance, and an electric-power plant upon Rock Creek. It is equipped with a water hoist, twenty 750-pound stamps, and two Huntington mills, giving it a capacity of forty stamps. The Dalmatia has been a famous producer, and notable for the exceptional low cost of mining and milling its ore.

The Gold Bug Mine is on Canyon Creek, near Georgetown, and is the largest hydraulic mine in operation in that portion of the county. It is owned by the Gold Bug Mining Company, of Cleveland, Ohio, who have fitted it up with a view to its operation on a large scale. Over four miles of the lower end of Canyon Creek channel is owned, and this ground was famous in the early mining days as one of the largest gold-producers of that time. It is known to have produced over $1,500,000 of gold, while much more was taken out that never was accounted for. This property has always excited a great interest among the miners who worked it in the early days, and knew how marvelously rich it was, and how imperfectly the methods of that day saved the gold. Its value lies in the gold lost in the tailings by former workings, in the virgin ground that was overlooked, and in the gold contained in the crevices and seams in the bedrock. It has received a large portion of the tailings of the Georgia Slide mines, which have been worked continuously for nearly fifty years. Gold is to be found everywhere, and if the tailings alone can be economically and rapidly handled, they will pay handsome profits.

In the early '50s and '60s over $10,000,000 in gold passed through the express office at Georgetown from the territory adjoining this property, while much more was taken out that was not taken into the account. The mine has been opened by a heavy bedrock cut 1,500 feet long, and contains a line of sluices eight feet wide on the bottom, divided into two four-foot sluices, which are paved with block and pole riffles. The hydraulic giant is supplied with water under 150-foot pressure, and uses a four-inch nozzle. The mine is operated under a permit from the California Debris Commission.

As the Georgia Slide Mine has been worked and paid dividends uninterruptedly ever since 1853, it has the honor of being the oldest continuously operated mine in the county, if not in the State. It rises 300 feet above Canyon Creek, and has a working face 200 feet in height. At this point, the peculiar character of the formation shows up to advantage, as, in its entirety, this mineral center is an immense lenticular ore-body, some three-quarters of a mile long, and 400 feet in width at its widest point. The northern end of the lens lies underneath the tailing dump of the Slide, while the southern side lies somewhere in the southern side of Oregon Canyon, which yielded $2,000,000 in the three-quarters of a mile of its course, all of which was produced by the wearing away of this ore-body. The characteristics of this ore-body indicate that it extends downward to unknown depths, like an immense chimney, and that, while there will always be free gold, a large percentage of the values will be found in the sulphurets below the limit of suface action. Sulphurets obtained from cleaning up the sluices have gone $700 per ton. All of this great mass is not ore, as there are horses, or reefs of barren material interspersed here and there through it, but it is practically a huge mass of quartz-seamed chloritic slate. There are upwards of 10,000,000 tons, above the level of Canyon Creek, which can reasonably be expected to yield a net profit, over and above all expenses, of twenty-five cents per ton.

A promising property on the east belt is being developed by Senator E. W. Chapman, at the Gold Note Mine, near Grizzly Flats. A shaft, 300 feet in depth, has been sunk upon the property, and a steam hoist and a splendid ten-stamp mill have been erected.

The Grand Victory Mine is located about two miles easterly of the main Mother Lode belt, and about six miles south-easterly from

Placerville. In the early mining days, following its discovery, in 1857, the mine was extensively worked by surface cuts and openings, from which more than 250,000 tons of ore were extracted and milled, yielding very fair returns. Commencing in 1894, the mine has been opened, by means of a shaft and tunnels, to a depth of 400 feet. A steam-power hoist, and a six-drill air-compressor, have been placed at the mine, and a thirty-stamp water-power mill, and a chlorination plant for treating the sulphurets, have been erected and operated

RESERVOIR NEAR GEORGETOWN, EL DORADO COUNTY.

almost continuously, making it one of the steady producers of the country, working a great part of the time upon ore produced by developments, by which bodies of ore, as much as sixty feet in width, have been exposed. It is one of the large, low-grade mines of the country, and so situated with regard to economy of operation, that it is the intention of the owners to put up a much larger equipment, capable of mining and reducing the ore at a maximum cost of not exceeding $1.25 per ton for all charges. When this is done, the "Grand Victory" will be one of the large bullion-producers of the State.

The Vandalia Mine is situated on the west belt, in the Pekin mining district, four and one-half miles from Shingle Springs. The mine was discovered in 1885, and the land purchased from the homesteader. The mine was worked for about three years, with a five-stamp mill, which was run on high-grade ore for over two years. The mine was then abandoned, and the machinery moved away. The present owners have developed large bodies of oxydized ore, which they are at present preparing by special machinery, and recovering the gold by the cyanide process. The ore is of such character, and the gold is in such condition, that it requires a fire assay to discover any value, and ore, showing an assay value of from $4 to $6, will not show a single color of gold in the pan. The extent of the ore-bodies, developed by the present owners, is not known yet, although a crosscut shows over 100 feet in width, and no sign of a foot-wall.

One of the most prominent mines on the west belt, and one that, by reason of the amount of development work done, the splendid improvements made, and its record as a steady bullion-producer in the past, having given more encouragement to the systematic prospecting of the mineral belt west of the Mother Lode in this part than any other mine on that belt, is the Boulder Mine, owned by Wilman Brothers. The Boulder is situated on Webber Creek, near its junction with the South Fork of the American River, about twelve miles west of Placerville. The mine is worked from tunnels driven into the side of the mountain, and by a winze from the lowest tunnel level, 600 feet in depth. The mine is equipped with a splendid twenty-stamp mill, operated by water-power.

The few mines mentioned in this short article do not comprise, by any means, the mines in active operation in this county; nor has mention been made of the many encouraging prospects being opened, many of them being entitled to be classed as mines. Its location on the Mother Lode will show that, with development, El Dorado County ought to rank second to none in the production of the yellow metal.

Amador County's Mines.

BY
W. H. STORMS.

MADOR County, though somewhat smaller than the neighboring counties on the Gold Belt, has, nevertheless, earned a well-deserved distinction for her great gold mines.

The deepest mines now operating in California are in Amador, and they are among the deepest and most profitable in the world. There are here shafts over half a mile deep, and two or three years hence will find them 1,000 to 2,000 feet deeper, as several of the most prominent mines are making preparations to sink their shafts to depths approximating 4,000 feet, and they will eventually, no doubt, go still deeper. A number of the mines now working are, in their lowest levels, far below the level of the sea.

In this county, as in those to the southward, are three recognized mineral belts. The most westerly is that in the lower foot-hills, within five or six miles of the great central plains. In this belt is found a continuation of the copper zone developed in Calaveras County, and probably identical with that occurring to the northward, in Nevada County. These ores occur in diorite or diabase, which is in a more or less altered condition (amphibolite schist), due to shearing and compression. Not infrequently serpentine is intimately associated with the occurrence of copper ores in these formations. Gold-bearing deposits also occur in this western belt, but none of these have been extensively developed as yet. A number of rich accumulations of gold pockets have been discovered in this area, but the business of pocket hunting is so uncertain in its results, that it is not followed vigorously.

The central belt is the most important in Amador County, and includes the so-called Mother Lode. To briefly, but comprehensively, define the characteristics of the Central Gold Belt would be a difficult task, for it is not a single lode, as the term Mother Lode would indicate, but a series of nearly parallel veins and lenses of quartz, with many branches and changing characteristics. These veins occur in various formations — in diabase, diorite, serpentine, black slate, amphibolite schist, etc., and, with each change of formation, is found a notable change in the character and appearance of the ore and structural features of the vein. In consideration of these many

AMADOR CITY FROM OFFICE KEYSTONE CO.

varieties, it seems more proper to refer to it as the Central Gold Belt.

The Eastern Belt is situated about ten miles east of the central belt, and, in this county, occurs largely in the granite (grano-diorite) area. In this formation occurs a class of ores which may usually be easily identified by their mineralogical features. The mines on the Eastern Belt are not extensively developed, although, in the aggregate, they constitute an important factor in the county's mineral output.

As previously stated, the field of active operation is largely along the central belt, which reaches from the Mokelumne River, on the south, to the Cosumnes, on the north. The belt at the south county line is about half a mile wide, which width it maintains to the north line, though not at all sharply defined. Numerous mines are located between the Mokelumne River and the town of Jackson, the county seat, but, until the Zeila Mine, in the southern outskirts of that city, is reached, few of the mines have as yet become famous as producers, though some of them are extensively developed, and are more than 1,000 feet in depth. The Zeila Mine has been worked for a quarter of a century or more, and, though low-grade, has always, under the careful management of Mr. W. F. Detert, been a profitable proposition. The ore zone is wide, and occurs in the amphibolite schists. These are more or less silicious, and in them occur large masses of quartz. The quartz, chloritic and talcose schists accompanying, are each auriferous, and contain disseminated auriferous pyrite. The property is equipped with hoist, forty-stamp mill, concentrating plant, and chlorination works.

A mile north of Jackson is a group of mines which have made a record, and which, from present indications, will continue to be producers for many years. The Kennedy Mine is one of this group. It was worked in the early history of mining in California, and was at that time a prominent producer. The workings extended from the surface to a depth of 750 feet. At this level, the rich ore shoot was found to practically terminate, and mining operations were suspended. For years the property remained idle, until, in 1885, the Kennedy Mining Company was re-organized and the long abandoned workings re-opened under the superintendency of Mr. F. F. Thomas, now superintendent of the Gwin Mine, in Calaveras County, which he has also successfully re-opened. As previously stated, the ore shoot was exhausted at the depth of 750 feet. Mr. Thomas continued the south shaft, and at 950 feet encountered a new shoot, which has persistently continued to the present depth of nearly 2,500 feet vertically below the collar of the south shaft. The north shaft has also been carried down, and the two shafts are connected at each level. The company is now sinking a new three-compartment, vertical shaft, 1,900 feet east of the old workings, which is calculated to encounter the vein at a vertical depth of 3,500 feet. In the meantime, the lower levels in the old workings are being continuously and successfully exploited. The direction of this mine is in the hands of Mr. J. F. Parkes. Geologically, the Kennedy vein occurs associated with the black slates of the Mariposa beds, which are intruded by tongues of diabase, which have, by compression and shearing, been locally altered to chloritic and talcose schists. The vein is a fissure and occurs in part at the contact of these formations, and in part wholly in the slates, or wholly in the schists.

The Kennedy was the first great mine in the State to be successfully re-opened after a long period of idleness. The result of this fortunate enterprise proved such an incentive to others that a large number of old mines have since been re-opened and it can be truthfully said, in almost every instance, with commercial success.

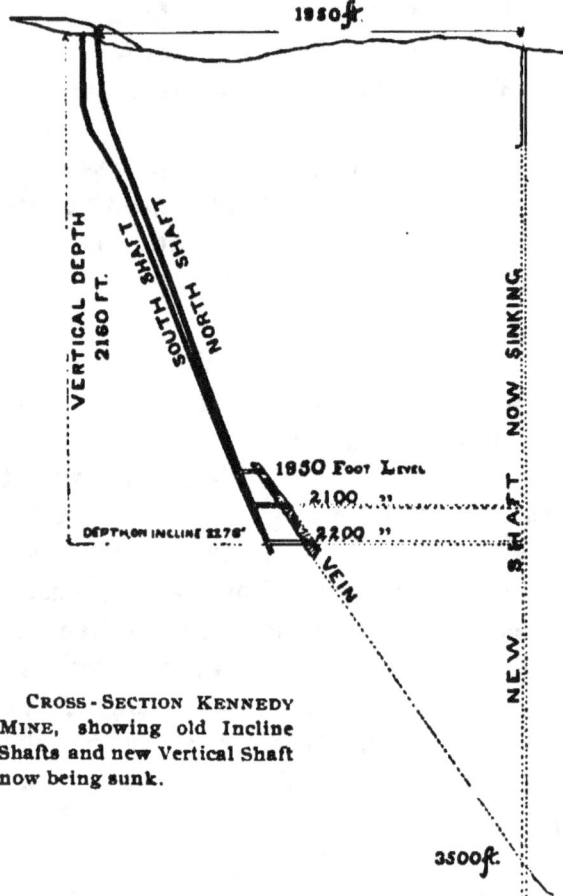

1980 ft.

VERTICAL DEPTH 2160 FT.

SOUTH SHAFT

NORTH SHAFT

NEW SHAFT NOW SINKING.

1950 Foot Level
2100 "
2200 "
DEPTH ON INCLINE 2376°

VEIN

3500 ft.

CROSS-SECTION KENNEDY MINE, showing old Incline Shafts and new Vertical Shaft now being sunk.

Immediately adjoining the Kennedy on the south, the Argonaut Mine has, within recent years, through the efforts of Mr. W. F. Detert, of Jackson, been opened from a mere prospect. It is fully equipped, and extensively developed to the depth of more than 1,500 feet. In many respects it is similar, geologically, to the Kennedy, being on the same vein. In the Argonaut is found undoubted evidence that this great vein occupies a fault plane, the displacement having occurred in the nature of an upward side thrust, the evidence indicating that the hanging wall has moved upward and southward relatively to the footwall. The Argonaut is opened through an inclined shaft of three compartments, sunk at a uniform angle of 63°. The mine is fully equipped with modern machinery, and is one of the most successful and best managed of California's gold mines. Mr. John B. Francis is superintendent of the Argonaut.

North of the Kennedy property is the Oneida Mine, which has a record made in early days. This is another of the mines re-opened within recent years by those who have unlimited confidence in the value of these veins in depth. The old workings of the Oneida were carried down to a depth of about 1,000 feet. When the new company commenced operations in January, 1896, a vertical three-compartment shaft was started several hundred feet east of the old workings in the hanging-wall diabase. It was estimated that the shaft would reach the vein at 1,750 feet, but recent development proves that the pitch of the vein below the old workings was nearer the perpendicular, and the shaft will not reach the vein at the depth estimated. It has been opened, however, on several levels by cross-cuts, and large bodies of gold ore developed. Although well equipped for hoisting large amounts of ore, the property has no mill, as yet, but when the management concludes to erect one, the old Oneida will join the ranks of dividend-payers in Amador. The Oneida is equipped with one of the most complete hoisting plants in California, being modern and up-to-date in every particular. The head-gear, designed by Mr. Hans C. Behr, of San Francisco, presents some new features in construction, and is worth more than passing notice from the visitor. It was designed with a view to distributing the material in reference to the strains, and it contains a surprisingly small amount of timber for so rigid a frame. The base is rectangular, all the vertical members sloping inward, forming a pyramidal structure, with the various upright members so disposed that the resultant strains fall within the frame and almost in line with the main compression members. The entire plant was designed with the purpose of rapidly hoisting a large tonnage from great depth with a minimum of expense.

North of the Oneida is the South Eureka, an enterprise opened through the efforts of Mr. J. F. Parkes, manager of the Kennedy. The site of the main shaft on this mine was determined by data obtained from a survey, and was started in andesite boulders which cap the ridge at this place (a remnant of an ancient river channel). The lowest workings are near the 2000-foot mark, and large bodies of ore are developed in the lower levels. Geologically, the Oneida and South Eureka are very similar to the Kennedy and Argonaut properties.

Still farther north along the lode is the Central Eureka.   This is one of the most recent "new old mines."   It was opened in a comparatively superficial manner years ago and some rich ore mined, but in 1895, Mr. W. R. Thomas, of Sutter Creek, succeeded in interesting capital in the re-opening of the old mine, then known as the Summit, and it is a pleasure to state that Mr. Thomas has a most excellent chance to add another great mine to the number already in Little Amador.

HOISTING WORKS AT THE ZEILA MINE, SUTTER CREEK.

The main shaft is down below 1,600 feet, and rich ore has been encountered on several levels.   The characteristics of the productive mines, from Jackson north to Plymouth, are well known, and the Central Eureka possesses the essential features of a bonanza mine. It is thought that the rich shoot found in the Eureka extends into the Central Eureka.

Immediately adjoining the Central Eureka on the north, and within the limits of the town of Sutter Creek, is the Amador Consolidated property, comprising the Badger and more noted Hayward's Eureka.   The workings of these mines were carried down to or about

2,400 feet — the deepest in their day on the lode. A fire occurred
on the 2,200-foot level, causing great damage to the workings, and
the property was abandoned, after having produced millions of dollars; no effort has since been made to re-open the mines, though
it is currently reported that there were large bodies of pay rock in
sight in the lower levels when the fire occurred. Considering the
history of the several properties above referred to, which, after a
period of success, followed by years of idleness, were re-opened to
meet with renewed and greater success, it is only reasonable to think
that a similar result might be anticipated in the case of the Amador
Consolidated.

One of the best managed and most successful mines in California
is the Wildman-Mahoney, at Sutter Creek. These mines had been
worked continuously for years as separate enterprises, but, about 1894,
came under the management of Mr. John Ross, Jr. The two mines
were consolidated under this management, and their workings connected. Mr. Ross made numerous important improvements in the
mills, re-arranging them, increasing their capacity materially, which
effected a higher saving of values. The diamond drill was introduced
and employed as a means of prospecting the ground beyond the workings of the mine. As a result of the discoveries made with the drill,
the company acquired adjoining properties, and is now sinking a
three-compartment, vertical shaft, thoroughly equipped with modern
devices. This shaft will be continued to a depth of at least 3,500
feet, and a large mill will be constructed to handle the greatly
increased output of the consolidated mines, when development has
progressed sufficiently to enable the new shaft to be used in supplying the new mill.

The Wildman Mine differs essentially from any of the mines previously mentioned. All of the former are found occupying a contact
of black slate and amphibolite schist, or in black slate alone, but
the Wildman occurs in diabase altered to amphibolite schist, highly
silicified, and lies between two reefs of black clay slate, practically
filling the dike. The ore-bodies are extensive, and, though comparatively low-grade, pay well, and, with largely increased capacity, the
Wildman will become one of the leading producers of the country.
Mr. John Ross, Jr., is one of the most progressive managers on the
Gold Belt.

Just beyond the Wildman property, another old mine, the Lincoln,
has recently been re-opened, and is being developed with encouraging

results, under the direction of Mr. E. C. Voorhies, of Sutter Creek. The work of rehabilitation is only fairly under way. The main shaft has passed the 1,000-foot mark, and sinking is being continued. Doubtless another year will demonstrate that the gentlemen who have undertaken this enterprise were justified in their confidence in the property.

In the Sutter Creek district are a number of important mines which are comparatively new and which have not yet earned the fame which has attended the operations of the older bonanzas. Mining is

AT THE MOUTH OF A SHAFT A HALF-MILE DEEP.

active throughout the district and new developments are of daily occurrence, and it would be, indeed, surprising if some of these new enterprises did not in time become large producers. The Baliol and Potazuba are among the newer undertakings here.

North of the Lincoln no mines are in operation until the South Spring Hill is reached. This mine and the Medan, adjoining it, are owned by one company and are extensively developed, though recently, owing to litigation, operations have been somewhat restricted.

The famous Keystone Mine adjoins the two last-mentioned properties. The mine is one of the oldest in the State, has been continuously and successfully operated for many years, and has been a great dividend payer. The Keystone ore shoot is reported to have produced about $10,000,000. Whether this statement is correct or not, the writer is unable to state, but it is a well-known fact that the mine has been a very large producer and has paid millions in dividends. The Keystone vein, like the mines at Jackson, occurs at contact of black slate and amphibolite schist. The mine is not as deep as either the Kennedy or Argonaut, and it is not unlikely that at greater depth the ore will be found continuous.

Northward from Amador City are a number of important mines which, for some reason, have been idle for several years. The most noted of these are the Bunker Hill, Mayflower, Treasure, and Gover. The latter has extensive development and very large ore bodies, but is known to be low-grade. It may yet be worked on a more extensive plan than heretofore. Mines of no better grade elsewhere are paying, but they are worked to the limit of their capacity. A large body of $3 ore cannot be profitably worked to a depth of more than 1,000 feet, with one-ton skips and a twenty-stamp mill, with a vast volume of water to handle. What mines of this class require is extensive development, first-class vertical shafts, through which hundreds of tons can be hoisted daily at minimum expense, and a large mill. Undoubtedly the Gover mill was run as carefully and economically as any mill in California when working, but operations were conducted on too small a basis for economically handling so large a proposition.

From the Gover northward to Plymouth are a number of mines, many of them promising, but as compared with the great mines of this county, they are in the prospective stage.

The early days of Amador saw two mines opened at Plymouth— the Great Empire and Pacific—which were later operated as the Plymouth Consolidated. For years these great mines poured a continuous stream of golden wealth into the coffers of the owners; but the end came, and the mines were shut down, and have never since been unwatered. The surface buildings and machinery have either been removed or fallen into decay. When closed down, these mines were but 1,600 feet deep, and there is not a mining man in Amador who thinks the great vein exhausted, for unless it proves wholly unlike the mines to the southward, which, at greater depth, have proved even more profitable than formerly, the Plymouth vein will continue far below 1,600 feet, and be found as rich and as large as above. To-day the old waste dumps are being run through Huntington mills, and are paying a profit over the expense of treatment, showing that in the flush of success good values were allowed to escape.

Northward from the Plymouth Consolidated, to the north county line, a distance of five miles, the Gold Belt extends, but is practically virgin territory. There are good prospects in that direction, but,

AT THE DUMP OF THE KENNEDY MINE.

excepting the operations at the Bay State and Philadelphia, little has been done toward opening this section.

The extensive development of the Central Gold Belt, in Amador County, has proved several important facts. The ores are all susceptible of successful treatment by the simplest metallurgical processes — amalgamation, concentration, and chlorination. The cyanide process may be applicable also, but is not extensively in use. The veins are continuous, easily mined, and extend to great depth without deterioration in values, or change in character of the ore. The greatest depth thus far attained has not developed a very high temperature, nor an abnormal flow of water. It is the opinion of the

writer that the depth to which these mines may be worked will only
be determined by the economic conditions, and it will not be a matter
of surprise to find shafts in Amador County 5,000 feet deep, and the
mines of to-day still in successful operation at that depth. The
mine managers here are making preparations to mine at 3,500 feet,
and anticipate no difficulty. Why not at 5,000 feet, or more? This
is no longer an experiment, as every member of the Institute knows,
and it will yet be found that the mechanical difficulties of deep mining
can be overcome by offsetting the shaft at stated distances of 2,500
to 3,000 feet or more, and the establishment of relay hoisting gear
at these stations, to be operated either by electricity or compressed
air. The miners of Amador have almost unbounded confidence in
the future greatness of their mines, and anticipate many years of
prosperity, which is made more than probable under the new era of
active progress.

Beside her great gold mines, Amador County has other resources.
Here, near Oleta, are quarries of handsome marble; a few miles
south-west of Plymouth is a beautiful stone (serpentine) for decorative purposes. Volcanic ashes, in great beds, afford fine building
material, and a stone as refractory as fire-brick. In fact, it is often
employed in setting furnaces.

There are also great beds of gravel containing gold — a portion
of the great system of ancient rivers that once existed in California,
and, in the beds of these ancient streams, diamonds are found with
the gold. No systematic search has ever been made for these gems,
their discovery being purely accidental, and it is not at all unlikely
that many more were washed away in the tailings than have ever
been found. The source of the diamonds has never been discovered.

In the valley, near Carbondale and Ione, are large and valuable
beds of clay and sand. The clay is used in making a great
variety of manufactured products, ranging from fine pottery to sewer
and drain pipe. Architectural terra cotta is one of the most important products.

Near Ione are several coal mines, which have been worked continuously for years. The coal is lignite, and of fair quality.

Amador is certainly one of the most picturesque of California's
counties, and affords a remarkable range of climatic condition. From
the level of the great interior plains the hills rise in ridges and
groups to the summit of the Sierras, nearly 10,000 feet above the
sea.

Mines of Calaveras County.

H. W. H. Penniman.

CALAVERAS County, which then also included Amador, Mono, and Alpine, was, like the majority of the mining counties of California, first opened to general settlement by the famous gold excitement of 1849. As rich as the richest, the tales yet told by the early pioneers bear witness to the fabulous productiveness of those wonderful surface placers.

H. W. H. PENNIMAN.

But now those golden days are long past, and the mining of the present is fast becoming less and less a matter of pick, pan, and individual luck. The Calaveras County of to-day lies a long, irregular triangle, its shortest side on the edge of the fertile plains of the San Joaquin Valley, two hundred feet above the sea, the angle opposite nestling among the great peaks of the Sierra Nevadas, at an altitude of 8,000 feet.

Separated from Amador County on the north by the Mokelumne River, and from Tuolumne on the south by the Stanislaus, both flowing in deep canyons, the drainage of the interior of the county is accomplished mainly by the Calaveras River and its tributary creeks, none of which, however, extend to the region of eternal snow; therefore, while this system, with its accompanying ditches, is wholly adequate for milling purposes, Calaveras has, until the past three years, been handicapped by insufficient water for power the year round, in many localities. This need has now been met, where it was most sorely felt, viz., along the Mother Lode, by two electric power lines.

The geological formation of the county presents many peculiarities. The predominant rocks, however, are the paleozoic sediments of the carboniferous age, giving place, in the lower part, to the sandstones and clay rocks of the Ione formation, and, in the higher altitudes, to igneous grano-diorite and hornblende rocks.

The county may be best divided into five sections for more detailed description, to wit: The lower, or Comanche-Milton, the Campo Seco-Copperopolis, the Mother Lode, the East Point, and the West Point areas, all having a general trend from north-west to south-east, except the latter, which extends eastward among the inaccessible canyons and peaks of the Sierra Nevadas.

As before stated, the country rock of the Comanche-Milton area is principally the Ione formation, overlaid in many places by auriferous Neocene and recent shore and river gravels, in which are numerous large hydraulic mines, now mostly idle, by reason of the legal restriction of hydraulic mining.

Coal, in small seams, occurs in this area, but it has never been systematically developed. Sandstones, for building purposes, are found in unlimited quantity, and, near Wallace, is a large deposit of tripolite, containing no argillaceous earth, smooth to the touch, but with wonderful abrasive properties. There are also deposits of ochre in this section.

To the east of this region lies the Campo Seco-Copperopolis area, principally celebrated as one of the richest copper belts in the State, and it is now rapidly gaining prominence for its gold production as well. The country formation is black Mariposa slate, accompanied on the east, and more or less divided, by a large area of amphibolite schist.

Large irregular dikes of igneous serpentine follow the general north-west and south-east trend of the schistosity.

At the northern end of this belt is the Campo Seco copper group of mines in active operation. The Campo Seco Mine was discovered in 1860, and an immense body of ore developed, the ledge showing twenty-eight feet in width at the 200-foot level.

The Satellite, another of the same group, was even richer than the Campo Seco, the gold and silver in the ore paying all the expenses of mining and treatment. A furnace and chemical plant were erected at a large expense, and for a time the proceeds were most satisfactory. The ores are sulphides and carbonates, and possess peculiarly corrosive properties. In 1892, operations were suspended and have only just been resumed this year (1899). A new and improved forty-ton water-jacket furnace has been erected, and the properties are again in full blast.

At Copperopolis, near the southern boundary of the county, is the Union-Keystone Copper Mine, at one time the largest copper producer in the State. It was closed down a number of years ago when the price of copper declined, and, aside from a leaching of the old ore piles, has

UTICA MINE, ANGELS CAMP.

never been in operation since. Should the present market prices of copper continue, there is every reason to hope that the property will soon be re-opened. To the west of Copperopolis, near Telegraph City, there are a number of other copper properties that in early days were heavy shippers.

The principal gold property of this belt is the Royal group at Hodson, three miles north-west of Copperopolis. There has here been discovered a series of veins following a very irregular contact of the amphibolite schist and Mariposa slates, with an igneous serpentine dike, as the probable cause of the disturbance. The greatest development has been accomplished on the Royal Mine, where four parallel lenticular veins have been opened for from 200 to 600 feet, by nine levels run from a two-compartment inclined shaft 900 feet deep, and as yet the ultimate walls have not been fully determined. There are also a number of cross veins, some of which are very wide and strong. The ore is medium grade, and carries the best gold values in the swells of the veins, some of which are forty feet in width.

Much of the ore in this belt is similar in appearance and characteristics to the ores of the Mother Lode region, but the formation is much more regular and shows less evidence of eruptive disturbance. There are a number of quarries of first-class building slate north of Telegraph City, more or less opened up, and an active demand would speedily discover many more.

Bear Mountain, an irregular ridge of diabase and porphyrite, with a general trend N 30° W, rises abruptly on the east of the Copperopolis area, attaining an altitude of over 2,800 feet at the highest point. A belt of Calaveras slates, in some places one and a half miles wide, divides the igneous formation of this range and carries some gold-bearing quartz veins, but, owing to the lack of water and to the more favored regions on either side, there has been little prospecting done.

The Mother Lode region lies east of Bear Mountain and varies in width from two to four miles. The formation is Mariposa slate, accompanied on the east, as in the Campo Seco-Copperopolis area, by a belt of amphibolite schist. In this case, however, the two formations are separated for a considerable distance by a band of Calaveras slate, and the amphibolite is also much divided by the same, especially on the eastern edge of the area, where it gradually merges into the East Belt. Serpentine dikes and limestone lenses occur in this region, many of the latter yielding a fine grade of building marble. The dual nature of this belt gives rise to two almost distinct classes of auriferous quartz ores: Those occurring in the Mariposa slates, of which the Gwin Mine furnishes the best example, and the lenticular partial replacement veins in the amphibolite schist, as illustrated by the Utica.

It is on this belt that the most extensive mining in the county is done, and to it Calaveras owes her prominence as a gold-producing county. The ore is generally of medium to low grade, but the veins are immensely wide, the ore shoots long, and the rock free milling. Steam has been the best available power, but now the electric power line of the California Exploration Company, with a capacity of 1,500 horse-power at 11,000 volts, covers the belt from the Mokelumne River to Angels, while the recently erected electric power plant of the Union Water Company, near Murphys, supplies 1000 horse-power for the Utica, Stickles, Madison, and other mills at Angels and vicinity.

The power is generated by two 500-horse-power two-phase machines, operated by tangential wheels under 570 feet head. The total length of the pipe-line is 1,750 feet. The power is transmitted eight and a half miles by the three-phase system, under 15,000 volts pressure on No. 1 B. & S. copper wire. The poles are seven by seven at top, eleven by eleven at base, and thirty feet long, selected redwood. Two cross-arms, six and eight and a half feet respectively in length, are gained to the poles. The capacity of this plant can at any time be increased, as the company has over 2,000 inches of water under the same head available at the power station. The power of the California Exploration Company is derived from the plant of the Blue Lakes Water Company, situated at Big Bar Bridge, just across the Mokelumne River, in Amador County. There three 600-horse-power Stanley two-phase generators are driven by forty-inch Doble tangential water-wheels under 1,042 feet head.

The transmission line is of thirty-five-foot redwood poles, ten by ten at the butt, and six by six at the top, set six feet in the ground.

Four No. 3 B. & S. copper wires are carried on Locke triple petticoat china insulators on two twenty-two-inch cross-arms, gained eighteen inches between centers. The line is not transposed.

At the present writing, fifty horse-power is used at the Ford Mine; twenty horse-power at the Fort Wayne Gravel Mine, San Andreas; 100 horse-power at the Thorpe Mine, at Fourth Crossing, and 200 horse-power at the Lightner, at Angels. The towns of San Andreas and Mokelumne Hill are also supplied with light. The total length of the line and connections is, approximately, twenty-five miles.

The line of the Standard Electric Company, which also takes power from the Blue Lakes plant to Stockton, is similar in construction to that just described, except that the wires are No. 1 gauge and are of aluminum. This line and the whole power plant are to be duplicated in the near future to supply San Francisco and Oakland.

It will be neither necessary nor advisable to give here a detailed description of all of the working mines of the Mother Lode belt, and, therefore, only a few of the more prominent representative properties have been selected. This list may well be headed by the Gwin

CARSON HILL, FROM ROBINSON'S FERRY, CALAVERAS COUNTY.

Ledge crosses at left of the Mountain's Apex. Line leads to Lower Tunnel, Stanislaus Mine.

Mine, situated on the northern edge of the county on Gwin Mine Gulch. This property was re-opened by the present management in 1894, it having been stripped of all good ore in sight and allowed to fill with water by the previous owners a number of years before. In three years a new shaft was sunk 1,400 feet, and levels were immediately run at 1,200, 1,300, and 1,400 feet, with others above. A forty-stamp mill was erected, and the Gwin Mine became again one of the great gold-producers of the county. Improvements have been made and development work kept up, until to-day there is not a more perfect mine and plant in the State. Forty more stamps have just been added, with space for twenty additional. The main hoist, mill number one, saw-mill, etc., are driven by 220 inches of water from the Mokelumne Hill and Campo Seco ditch under about 375 feet head. The water is then collected in a 75,000 gallon tank, and carried 4,500 feet by a fifteen-inch pipe to the bank of the Mokelumne River, where, under 415 feet pressure, a Pelton wheel, nineteen feet in diameter, runs one of the largest air compressors in the State. This machine is of the Rix horizontal type, having four eighteen by twenty-four-inch air cylinders and running at about seventy revolutions. The air is compressed to eighty-five pounds to the inch, and carried back to the mine by a six-inch steel pipe parallel to the water pipe, where it furnishes power for the hoist at the old shaft, for two pumps, one at the 1,400 and one at the 700-foot level of the main shaft, pumping about 100,000 gallons per day, and also operates three drills and a fifteen-horse-power hoist at the 1,400-foot level, besides running the new forty-stamp mill. The main shaft is now being sunk from 1,600 feet downward as rapidly as possible, the formation being pudding-stone and black slate. The fifteen-horse-power hoist mentioned above raises the broken rock to the 1,400-foot chute. The ore-body on the 1,400-foot level is 1,500 feet long as far as developed, and is continuous.

The average width of the ore milled for the first six months of 1899, by survey over the entire mine, was ten and a half feet. The average of the 1,400-foot level is wider. Everything between the walls of the fissure is milled. In the mills, each 925-pound stamp has a capacity of five tons per day, crushing to a No. 16 screen, and the tailings do not run on an average over thirty-five cents per ton. Two six-foot Union concentrators receive the pulp from each five-stamp battery. Out of approximately 10,000 tons of ore milled per month about 125 tons of concentrates are saved, running between

$80 and $90 per ton. The grading for the new mill excavation, 14,750 cubic yards (five to ten per cent earth, the balance rock in place), was done by hydraulicking, at a total cost of less than $4,000.

At San Andreas, the county seat, are the Ford and Fellowcraft mines, separated by a few hundred feet, and each having a ten-stamp mill. While not on the same vein, these properties are geologically quite similar, both being situated on the contact of the amphibolite schists and Calaveras slate, which makes a very jagged connection at this point.

The Ford is much the best developed property of the two. The main shaft is over 700 feet, with various levels. The foot-wall vein on the 700-foot level shows a width of over fifty feet of pay ore.

The Fellowcraft, now operated by the Veritas Mining Company, has been extremely rich in pockets. It is being thoroughly opened up, and has every indication of a paying property. The entire plant is operated by steam.

The Fort Wayne Gravel Mine, on the outskirts of San Andreas, is an entirely different type of property. An ancient river-bed, 100 feet below the present surface, and heavily capped by andesite tufa, is being thoroughly opened up by bedrock drifts, and large quantities of high-grade gravel have been found. A steam hoist is used to bring the waste and pay gravel to the surface, where the latter is dumped into a bin, from which it is fed to a Cox Pan, operated by electricity, and having a capacity of 100 tons per day.

The Thorpe Mine, near Fourth Crossing, is geologically situated on a band of Calaveras slate, enclosed in the amphibolite schists, and on the western contact of these two. While the ground is well mineralized over a wide area, the best ore has been found on an east and west vein, eight to forty feet in width, cutting the slates at nearly 45°, the country rock having the regular north-west and south-east trend. A modern plant has been erected upon this property, and the ore-bodies are being systematically opened and prospected. The main shaft (inclined) has three compartments, and is now being sunk from the 700-foot level downward. The mill consists of thirty 850-pound stamps, dropping seven to eight inches 100 times per minute. The Thorpe is another property that, like the Gwin, was a producer in early days. The ledge was discovered in the '50s, when the surface ground was sluiced away in placering, and the ore worked in an arrastra. Later, shortly after 1870, a ten-stamp mill was erected, but the high cost of mining and milling in those days left very little net returns, and the mine was closed down, to remain idle until 1894, when the present developments were commenced.

The Demorest Mine is situated about one and three-fourths miles south-west of the Thorpe property, on the contact between the amphibolite schists and Mariposa slate. The vein is strong, varying in width from four to twenty feet, and is somewhat of the character of the Gwin. The mine has been fairly well opened to the 400-foot level by an inclined shaft, with a number of crosscuts and drifts.

THORPE MINE, FOURTH CROSSING.

The ore is a bluish white quartz, rather ribbony, and carries three to four per cent sulphurets. A five-stamp mill on the property serves to thoroughly test the ore-body. The entire plant is operated by steam.

The Lightner Mine, at Angels, is unique in being the first producing property in the county to be operated entirely by electric power. Situated immediately north of the Utica, the ore-body is of precisely the same character, and, as at that mine, the total absence of a waste dump is a peculiarly noticeable feature. The main shaft is vertical, 400 feet in depth, and has three compartments. The gallows frame is 104 feet high, and contains the ore bins and rock breaker. The mill has forty 850-pound stamps, making 105 drops per minute. Five tons per stamp per day are crushed through No. 1 tin screens. The sulphurets are saved by fifteen four-foot and six six-foot Frue concentrators, giving clean concentrates and low tailings. Five two-phase Westinghouse induction motors furnish the power for various purposes.

The Utica, Stickles, Madison, Gold Cliff, and a number of other less well-known mines, under the same management, with four mills aggregating 180 stamps, and a large chlorination works, form one of the largest mining plants in the world. The town of Angels owes its size and present prosperous condition almost wholly to this enterprise. The company also controls the Union Water Company, and, therefore, practically owns the most complete water system in the county, as well as the electric power plant already described.

The ore-body of the Utica and Stickles mines may be more properly described as a zone than a vein. The stopes are from ten to, in some places more than, one hundred feet wide, in altered diabase or amphibolite schist, and infiltrated massive quartz occurs in lenticular veins and bunches throughout. The enclosing schists have been badly crushed, and are gold-bearing as well as the quartz, though in the wider parts of the zone there are some streaks of barren rock. The ground is difficult to hold, and a great forest of immense round timbers is used monthly. The Gold Cliff and its south extension, the Madison, lie south-west of the Utica on another, but quite similar, zone. At the Gold Cliff, the oxidized, altered diabase, with laminated quartz throughout, has been mined for many years on the surface, leaving a great gaping cut in the hill 150 feet deep, 200 feet wide at the top, and over 500 feet long. A shaft has now been sunk to connect with the Madison works. In the Madison, the pay ore is found on the west side of an immense massive quartz vein, sixty feet wide in places, carrying little value. The pay streak is from six to twenty feet in width, and, like that of the Gold Cliff, of amphibolite and talcose schists, with thin layers of quartz and calcite, carrying about four per cent of sulphurets. These small stringers of quartz gather in places to small veins and bunches, which seems to increase the gold values of the ore in such spots. The Utica has sixty stamps, the Stickles sixty, the Madison forty, and the Gold Cliff twenty; this latter mill is, however, not used at present. The remaining three are operated either by water or electricity, the Utica and Stickles by tangential wheels under 420-foot pressure, and the Madison by a Leffel double turbine under 38-foot, using the waste water from the other two.

In an able paper, read before the American Institute of Mining Engineers, at the Buffalo meeting in 1898, the superintendent of all of these mills described at length their operation and treatment of the ores, since which time there has been no important change, except the introduction of electric power. The Blake crushers and ore bins are in the gallows frames. Each twenty stamps has also a 700-ton

SMALL TEN-TON EXPERIMENTAL CYANIDE PLANT AT THE MELONES MINE.

ore bin at the head of the mill. The stamps weigh 835 pounds each, drop seven inches, 105 times per minute, and crush about five tons per day through No. 1 tin screens. The mortars are narrow, lined with ten-inch discharge, and fitted in front with triple splash boards. Fifty-four four-foot Frue, sixteen four-foot Union, and sixteen three and five-sixths-foot Tulloch concentrators save the sulphurets, which are treated by the regular chlorination process in the company's plant, which also does custom work for the vicinity.

There are a number of old properties south of Angels that have been large producers in times past, and some of which are now being re-opened. There is also an increasing number of new ventures, mostly, however, in the prospective state.

By far the most important of these new enterprises is the Melones group, at Robinson's Ferry, on the Stanislaus River. This is in a measure the re-opening of old and tried properties, but on a scale so much larger that the old workings are comparatively but prospect holes. The group comprises the Reserve, Last Chance, Melones, Enterprise, Mineral Mountain, Keystone, and Stanislaus mines, is about a mile in length, and extends from near the north bank of the Stanislaus River to a little beyond the summit of Carson Hill, which rises abruptly over 1,000 feet.

Two tunnels, 1,600 and 3,000 feet in length, have been driven under the hill with numerous crosscuts and drifts, and a third is now being run, the ultimate length of which will be about 5,000 feet. These tunnels and connections have opened up, literally, millions of tons of medium to low-grade ore, the values being determined by actual mill test with a ten-ton Kincaid mill, one Frue concentrator, and canvas tables.

The geological formation is typical of the amphibolite section of the Mother Lode, and most closely resembles that found at the Gold Cliff and Madison. The famous pockets and placers of Carson Hill are too well known to require an extended review here. A 120-stamp mill is under construction, to be operated by Knight turbines, under fifty-four feet head. Six thousand inches of water in the Stanislaus River is available for the greater part of the year, and never less than 3,000. A flume has been constructed from a point two and a quarter miles above the mill. The 5,000-foot tunnel will deliver the ore at the head of the mill at a small expense, and, with free water-power and immense bodies of mill-tested ore blocked out, the future of this property certainly seems an assured success.

The East Belt, so-called, is a broad area of Calaveras slates, lying east of the Mother Lode, and containing throughout its width numerous quartz-filled fissures, igneous dikes and limestone lenses. The general contour of the country is more abrupt than on the Mother Lode, giving greater facility for tunneling. The veins are, as a rule, smaller, the ore shoots shorter, and the ore medium to high grade. The trend of the slates and fissures varies from north and south through north-west and south-east to east and west, and the drainage, flowing south-east, cuts squarely across the formation in many instances. The ancient river channels, capped with lava, seem to have had their source near the eastern edge of this area, and, while wherever they were easy of access, or had been washed down by recent erosion, considerable gold has always been found, the portions harder to drain have received comparatively little attention, although including the Mother Lode region where, with the exception of the Cataract, these East Belt channels join an intricate system from Mokelumne Hill, near North Branch. There are over fifty miles of these old river-beds in the county.

LIGHTNER MINE AND MILL, ANGELS CAMP.

The typical quartz mine of this region is the Sheep Ranch, which is now being re-opened, after a number of years of idleness. The ledge averaged but eighteen inches in width, but was so uniformly rich in free gold that, for a number of years, regular monthly shipments of specimen rock were made to a manufacturing jeweler in San Francisco. The property was worked to a depth of 1,250 feet, with but few crosscuts, the ore-body being horizontally continuous for 1,000 feet. The mine was closed down on account of the flow of water and increasing depth, which necessitated the installation of a new hoisting and pumping plant. The new management immediately erected a substantial gallows frame of twelve by twelve and fourteen by fourteen inch timber, twenty by twenty-two feet square and sixty-four feet in height, at the collar of the old shaft, and retimbering was commenced. A tunnel on the Pioche Mine, an adjoining claim, was continued, tapping the shaft at nearly 300 feet, and thus providing an outlet for the surface drainage. The old stopes are to be re-opened, and the medium-grade ore, thousands of tons of which were left in place, will be broken down. The country will be crosscut at a low level for parallel veins, and, in a word, the whole property will be developed according to modern ideas. A mill of twenty 900-pound stamps is under construction.

East of Murphys, near the Stanislaus River, are the Taylor and Collyerville mines, now being successfully worked by the cyanide process. These properties lie on a contact of limestone and quartzite phthanite, and have yielded, since their discovery, many fine gold specimens. On the northern edge of the East Belt, perhaps the most typical property is the Petticoat, at Railroad Flat. Though not yet a steady producer, the ore-body is being well opened up, and mill tests, made at various times in a near-by mill, have been wholly satisfactory. A modern mill is to be erected in the near future. At the time the Sheep Ranch was first in operation, there were many paying mines in the East Belt, but for one reason or another, they were closed down, until all of the present enterprises are either entirely new or a re-opening of old properties. For companies of limited capital, this section undoubtedly offers superior inducements to the Mother Lode, but, being quite difficult of access, and sparsely settled, it has not received the attention of late from bona fide mining men that its resources deserve.

In many places the limestone lenses are altered to building marble. Above Murphys is an immense deposit of black hematite, and near by another, equally large, of chromic iron, while gypsum and carbonate of lime are also of frequent occurrence.

The West Point area may be said to be roughly divided from the East Belt by the South Fork of the Mokelumne River. The country formation is diorite granite, changing toward the east to a hornblende rock, and there including a rough, deeply soiled region, practically uninhabited save by wood choppers and shakemakers, and known besides only to itinerant cattle and sheep men and a few hardy campers.

The western part of this area around West Point has ever since its discovery been one of the richest sections of the county, both in pockets and milling ore. The pay rock is entirely different from anything else found in the lower altitudes, much of it being extremely base with galena, antimony, arsenic, bismuth, etc. Shipments of sorted ore are being constantly made to the San Francisco smelters by private parties, and quite a number of paying properties are in operation. A number of years ago there was more activity than at present, but for some time past there has been a steady improvement, which bids fair to continue indefinitely.

The Paragon group, situated on the Licking Fork, three miles south-east of West Point, has been well opened by a number of tunnels, and with a ten-stamp water-power mill makes regular monthly bullion shipments. The plant is lighted by electricity from a small private dynamo, and has telephone connection about the works. The Keltz Mine, on the North Fork of the Mokelumne River, four miles north-east of West Point, has been a producer of shipping ore, at intervals, for the past five years. It is on a simple fissure in granite, and has been developed by tunnels. The ore is high grade, heavily sulphureted, and during two years after purchase by the present owners paid for itself ten times over.

Only two properties have been mentioned to illustrate the two plans which have both been successfully followed in this section: either working a small mill upon high-grade rock, or sorting and shipping the ore as it comes from the mine. The veins, of course, are subject to the usual uncertainties of eruptive granite fissures, but the high grade of the ore, and its frequent occurrence, as proved by the prosperity of a community dependent almost wholly thereon, as at West Point, in a great measure offsets the inherent disadvantages of the geological formation.

In the foregoing sketch of the mining industry of Calaveras County, supplemented by a few general notes upon the geology, characteristics, etc., it has not been the object to paint the situation in the glowing colors of the promotor's prospectus. There are many so-called mines in Calaveras, as in all mining countries on earth, that never were of any value, and never will be, but it certainly is a fact, that this county also has hundreds of partially developed mines that have indications which would fully warrant an expenditure upon them sufficient to prove their value. Enough illustrations have been given, with sufficient detail, to enable a person acquainted with the business to form a clear conception of what is being accomplished upon the leading properties. All estimates and statements made in this article are the result of careful investigation, and may be considered absolutely reliable.

Mining in Tuolumne County.

BY

BOARD OF SUPERVISORS

AND

MINERS' ASSOCIATION OF TUOLUMNE COUNTY.

UOLUMNE County is situated on the western slope of the Sierra Nevada Mountains, in about the geographical center of California. It is bounded on the north-west by Calaveras, on the north by Alpine, on the north-east by Mono, on the south by Mariposa, and on the south-east by Stanislaus.

It has an area of 2,232 square miles. There is still open for settlement 367,845 acres of vacant land. During the last ten years the county has enjoyed an era of prosperity which has brought it into second place among the gold-producing counties of California. In 1890 it had a population of but 6,000; to-day there is a population of over 20,000, with an increasing activity in mining which predicts a most prosperous future.

A few years ago there was not a stamp dropping on the Mother Lode, but the opening of the rich shoots in the Rawhide infused new life into the Mother Lode, and stamps were soon working at the Dutch, App, Santa Ysabel, Jumper, and Eagle-Shawmut. The success of these mining ventures revived confidence in the Mother Lode and brought courage to capital, so long needed for its proper development. This has resulted in a steady increase in the number of stamps, and also in the amount of development work under way.

The East Belt, too, has received an impetus from the success of the Black Oak and Providence, two new producers among the many old producers of nearly a generation ago. All along the East Belt, from Big Oak Flat and Groveland to far above Arrastraville and Cherokee, new ground is being explored and old re-opened. The outcome of all this activity will be a large addition to the many producing mines, which have made Tuolumne so famous. It was ore from its mines that won for Tuolumne the gold medal at the Mid-winter Fair, in 1894, and the silk banner awarded as the first prize at the Miners' Jubilee Fair, in 1898, to the county in California making the best mineral display. These have been the only exhibitions in this State where prizes were competed for by the mining counties.

On the Mother Lode, the Rawhide is now being worked at a depth of 1,700 feet; on the Pocket Belt, the Bonanza, at 1,700 feet; and in the East Belt, the Deadhorse, at 1,500 feet, which shows the permanent character of the veins.

The local conditions in Tuolumne County for economical mining are exceptional, even for the most favored mining sections in California. A railroad system traverses the county, crossing both the Mother Lode and the East Belt to Carters, where it connects with a narrow-gauge road into the great timber belt further east. Another branch road is under construction along the Mother Lode to the north county line.

The climatic conditions of the Sierra foot-hills permit continuous operations of the mines at all seasons of the year. The three forks of the Tuolumne, the Stanislaus, and the lakes in the mountains furnish an abundance of water for power for operating the mines, many mines having their private water systems. Others are supplied by the Tuolumne Water Company, which has about 250 miles of ditches and five large reservoirs in the county. An electric power plant furnishes power to the Rawhide, App, Jumper, and other mines, besides supplying lights to the towns of the county. The new plant is equipped to furnish 1,500 horse-power, with a ditch and pipe capacity of 3,000 horse-power. All of this power is furnished without in the least diminishing the water supply of the county. It is just that much additional power. There is also an abundance of fuel for steam-power plants.

Another new industry is being established in the county which will benefit the miner. An immense lumber plant will soon be in operation, capable of turning out 200,000 feet per day. They are now building a narrow-gauge road from Carters into the great timber belt, from which the logs will be brought to the mills at Carters. With this addition to the saw-mills already working, timber is sure to be supplied at rates more satisfactory than in most mining sections. With cheap power, cheap transportation, cheap timber, and all the

favorable conditions of climate and location, and with the wealth of
mineral ground, the mining industry will certainly thrive in "Old
Tuolumne."

But there are other industries which must not be overlooked.
Some of the best agricultural land in the State is found in this county.
The finest fruits are raised here. No better marble can be found in
the State than in the extensive marble quarries of this county; and
there are also other building stones which can be quarried with profit.

HYDRAULIC MINING IN TUOLUMNE COUNTY.

Altogether, the mining industry in Tuolumne County is in a very
prosperous and flourishing condition, with an assured future.

The eastern part of the county belongs to the granite region. To
the west is the slate region, the country rock being argillaceous,
silicious, or talcose. The slate region might be divided into two belts,
viz., the clay slate and the talcose slate, the latter west from and
parallel to the former. The whole surface may be regarded as being
divided into three great belts parallel to each other, and whose course
is north and south; these are the granite, the clay slate, and the
talcose slate. The limestone belt, hundreds of feet in thickness, is
nearly coincident with the central or clay slate. Belts of serpentine
cross north and south through the western slate formation, and, for

a long distance, one of them runs parallel with and near the west wall of the Mother Lode, which course is north, about 35° west, through the county.

The Mother Lode has been defined as a mineralized zone of varying width, sometimes reaching over a mile, and extending continuously through the counties of Mariposa, Tuolumne, Calaveras, Amador, El Dorado, and, possibly, further north. Through a portion of this distance it is characterized by the presence of one main fissure of remarkable proportions, while in other places, where the conditions were different, it consists of a number of parallel fissures, filled by quartz veins. The lode is associated with a narrow and almost continuous belt of black slate, called the Mariposa beds, while the veins occur either in the slate or on the contact between it and the diabase dikes, known among miners as "greenstone."

The Mother Lode enters Tuolumne County from the south, at Moccasin Creek, and leaves it at Robinson's Ferry, following, in a general way, the north-west and south-east trend of the mountains, the veins conforming, as a usual thing, to the course of the stratification of the enclosing rocks, with a dip of 40° to 80°.

From Moccasin Creek north there is scarcely any quartz to be seen along the course of the vein for several miles toward the Tuolumne River. However, it is highly probable that the lode is continuous, for the walls are regular and well defined, and, wherever the rock is well exposed, the foot-wall slates are broken and twisted in every direction, and present an appearance similar to that where the fissure crosses Moccasin Creek. The hanging-wall, nearly to the Tuolumne River, is a granite dike, which terminates in a prominent mass about half a mile from the river. The lode crosses the river at an altitude of 700 feet. On the north bank it is fifty feet wide. The hanging-wall is felsite. The fissure is filled chiefly with mariposite and magnesite. At a distance of a quarter of a mile eastward, there is a deposit of limestone about 200 feet thick. At the Mary Ellen Mine, the second north of the river, the vein is shown to be from two to six feet wide, with a gouge. North of the Mary Ellen, for some distance, the vein does not show much on the surface; the fissure is probably continuous, though serpentine and granite occur in its course.

Toward Jacksonville, diabase first appears. It is sometimes fresh, but is more often jointed and decomposed. It is succeeded by serpentine, which lies directly in the course of the vein. The serpentine is split into two arms, at its northern extremity, by a knob-like mass of granite. The Willieta Mine has been worked in this granite. The vein on the surface of the Clio Mine is four and a half feet thick. The hanging-wall is slate. The foot-wall is a decomposed, jointed rock, once crystalline. The decomposed rock on the foot-wall is 500 or 600 feet wide, and extends along the river nearly to Jacksonville.

Northward to the Republican only a little quartz appears, though there is a continuous fissure, with a thick gouge. At the Orcutt, the gouge is sometimes six feet thick. The quartz on the Webster, the next claim north, is from eight to ten feet thick, and is quite massive, though on the hanging-wall it is mingled with broken slate. The slates are very much broken, and crossed by many clay seams. Small faults are numerous. The seams on the hanging-wall often contain small, rich pockets.

A little east of the Mother Lode, in this section, there appears, what is called by the miners, the east vein. It varies in distance between fifty and two hundred feet from the lode. The vein is generally nearly barren where the country rock is filled with clay seams. At the juncture of one seam with another, or with a stringer of quartz, pockets are usually found; they are almost invariably in the hanging-wall.

At the Eagle-Shawmut, the vein assumes immense proportions on the north side of Blue Gulch, the fissure being filled with thirty feet of quartz, which outcrops in two tall, tower-like masses fifty feet high. At this mine a rich pay shoot was found in the zone of slate east of this vein, which carries a high percentage of auriferous pyrite. Seven large and rich pay shoots have been found in this slate zone, during the last few years, in other mines. This fact has led miners to more carefully explore the hanging-wall country. In mines where the quartz is so low grade as to be almost worthless, there are to be found pay shoots in the slate zone.

Sullivan Creek crosses the vein a mile south of Stent, and here there is a good exposure of both hanging and foot-wall rock.

At the Mazeppa, Jumper, and Golden Rule mines, the pay shoots are found in the zone of slate. The following description of the Golden Rule may be taken as characteristic of the Mother Lode in Tuolumne County. The main drift runs north and south from the tunnel, and is nearly 1,000 feet long on a zone of chlorite schist, containing gold and auriferous pyrite, and, along the foot-wall side of the zone, narrow veins of quartz and calcite, also rich in gold; associated with this streak is a dike which is often found to contain a reticulated mass of veinlets of quartz and calcite, also rich in gold. A crosscut has been run 200 feet west, which exposes two other gold-bearing zones or veins, separated by twenty feet of diorite. The hanging-wall country rock, for 1,000 feet or more east of the easterly vein, is composed entirely of amphibolite schist, with small veins of quartz and dikes, and has been sluiced off years ago. The west of the foot-wall country consists of diorite, separated from the lode by a strip of serpentine over 100 feet wide. The succession of rocks, as exposed by a crosscut through the serpentine, is a large mass of greatly altered ankerite with mariposite, separated from the serpentine by a heavy gouge. A dike of diorite separates the ankerite mass from a second similar zone. Lying east of the ankerite zone is a succession of diorite and diabase rocks, partly altered into amphibolite schist, which, in the zones above described, are gold-bearing. On the surface a large quartz cropping is seen on the hanging-wall side of the ankerite area, which has not yet been found in any of the underground workings. At this place may be seen an example of the magnitude which the Mother Lode sometimes assumes. Quartz

A FLUME OF THE TUOLUMNE WATER CO.

Mountain is about 600 feet wide, nearly half a mile long, and 250 feet high. It is composed wholly of quartz and vein matter, the latter consisting of mariposite and dolomitic material. In the middle, and forming the summit, is a great body of massive quartz.

North of the App the lode splits into two veins, fairly well defined. At the Dutch they are about seventy feet apart. Numerous small quartz seams intersect the slate between the two veins; the lode almost pinches out where Wood's Creek crosses it, the west vein alone appearing. For more than a mile north of Wood's Creek the lode is very prominent, forming the central portion of Whisky Hill. At the northern end of the Trio it is 300 feet wide. At the Alabama it contracts to sixty feet. The lode is here crossed and covered by Table Mountain. The Rawhide Mine lies north of Table Mountain. The fissure is 150 wide and contains considerable bodies of quartz. Here the serpentine crosses the course of the vein.

East of the Rawhide the country rock is chlorite schist. On the west the serpentine has a width of 2,800 feet and is quite massive. Three-quarters of a mile north of the Rawhide, the serpentine turns abruptly to the west. The serpentine, which is half a mile wide at

IN THE TIMBER REGION OF TUOLUMNE COUNTY.

this point, gradually tapers to a point and disappears, about a mile and a half away. The Alameda is a little north of the point where the serpentine turns. The walls are slate, and the ledge is about three feet thick.

The following is the output of a number of Mother Lode mines: Trio, $250,000; Mooney, $400,000; Knox and Boyle, $550,000; Dutch, $320,000; Little Gem, $200,000; Crystalline, $75,000; Alabama, $150,000; Sweeney, $85,000; New Era, $150,000; Hitchcock, $30,000; Cardinelle, $400,000; Paterson, $340,000; Valparaiso

Group, $320,000; Golden Rule (Tuttletown), $350,000; Golden Gate, $1,400,000; Heslep, $1,000,000; Orcutt, $100,000; Neuebaumer, $30,000; Clark, $60,000; Clio, $60,000; Stanley, $35,000; Tarantula, $25,000; Norwegian Group, $130,000; Ames Mine, $40,000.

The output of the following mines, which are worked on the largest scale, and credited with being the largest producers in the county, could not be obtained: Rawhide, Jumper, Eagle-Shawmut, Santa Ysabel, and Golden Rule.

Following is the number of stamps on the Mother Lode in this county: Rawhide, 40; Jumper, 60; App, 20; Trio, 10; Golden Rule, 10; Santa Ysabel, 20; Dutch, 20; Little Gem, 10; Crystalline, 15; Alabama, 40; Golden Gate, 20; Álameda, 20; Bown, 20; Street, 6; Norwegian, 4; Arbona, 10; Maryatt, 20; Eagle-Shawmut, 40; Paterson, 20; Pina Blanco, 5; Mammoth, 10; Stanley, 10. Total, 330.

The East Belt, in Tuolumne County, includes all mines in the granite region. Ten miles east from, and nearly parallel to, the Mother Lode is the Eureka or main fissure belt. Farther east is a series of veins that might be designated as the contact belt of veins. Still further up the mountain flank there are veins in granite, and, in the high Sierras, the Tioga mineral district. The East Belt, in this county, produced millions by the old method of milling. The high-grade sulphurets were allowed to go to waste, and only the free gold was looked for. Some of the mines were immensely rich, the Soulsby alone producing $5,500,000 in free gold. When it was no longer profitable to work the ore for free gold, the mines were closed. For years the early producers of the East Belt remained closed, but one after another of them is being re-opened, with most satisfactory results. With modern milling methods, this vast mineral district will soon equal, if not far exceed, the Mother Lode in its output of gold. For the prospector and miner of limited means there is no better field. In these rugged mountains there is still left open a large amount of good mineral ground, which has been prospected but a very little.

The veins of the East Belt are found in slates, mica schist, quartzite, and other metamorphic rocks, but more often occur in grano-diorite. The veins are irregular in their strike and dip, and are found at all angles with a wide variation. The ores of these veins have a striking similarity, and can be easily distinguished from ores found in different formations.

They carry a high percentage of rich sulphurets, running from $50 to $1,000 per ton. Galena is often an index of value. Some of the zinc ores are also high grade. In some localities, tellurides of gold are found. The veins are narrow, as a general thing, but frequently make great kidneys. In these kidneys the best pay has been found in many mines. Sometimes the vein is frozen to both walls, sometimes to one wall, and again free from both. The veins often pinch to mere seams, which are hard to trace at times, but which re-open again if followed.

MARBLE QUARRY IN TUOLUMNE COUNTY.

Following is the number of stamps on the East Belt: Star, 10; Riverside, 10; Confidence, 20; Providence, 10; Keltz, 15; Eagle, 2; Nervey, 5; Tibbits, 5; Hazel Dell, 4; Green, 10; Rising Sun, 5; Eastern, 10; Dreisham, 6; Laura-North Star, 10; Carlotta, 4; Louisiana, 10; Eureka, 20; Grizzly, 20; Black Oak, 30; Soulsby, 15; Belleview, 10; Hibbing, 3; Gold Hunter, 10; Buchanan, 20; Ferguson, 2; New Albany, 10; Basin Slope, 5; Mt. Zion, 10; Mt. Jefferson, 10; Kanaka, 10; Longfellow, 10; Joe Hooker, 5; Seminole, 10; Lady Washington, 10; Graham & Condon, 5; Big Oak, 10; Mexican, 2; Cullers, 5. Total, 368.

Following is a conservative estimate of the East Belt output: Soulsby, $5,500,000; Confidence, $3,250,000; Draper, $2,500,000; Gilson, $1,250,000; Grizzly, $1,500,000; Excelsior, $450,000; Buchanan, $600,000; Hunter, $300,000; Keltz, $300,000; Riverside, $180,000; Green, $200,000; Spring Gulch, $250,000; Consuela, $50,000; Rhode Island and Philadelphia, $100,000; Belleview, $200,000; Lamphier, $90,000; Starr King, $100,000; Sonnett, $55,000; Shanghai, $25,000; Basin Slope, $100,000; Carlotta, $40,000; Wyoming, $30,000; Stockton, $25,000; Louisiana, $95,000; Virginia, $40,000; Providence, $100,000; New Albany, $200,000; Grant, 10,000; Star, $225,000; Mt. Vernon, $55,000; Mississippi, $500,000; Mt. Zion, $50,000; Mt. Jefferson, $300,000; Kanaka, $45,000; Longfellow, $500,000; Nonpareil, $60,000; Belcher, $55,000; Nichols, $10,000.

When it is remembered that this splendid showing was made long before the modern methods were introduced, it can readily be seen how Tuolumne's output will be increased when some more of those old producers are equipped with improved and modern machinery. Active development work is now being done on the Rhode Island and Philadelphia (the Goldwin group), Grizzly, Draper, Kanaka, Excelsior, Green, Carlotta, Mt. Jefferson, Longfellow, Laura-North Star, Buchanan, Uncle Sam, Lost Fox, Star, and many others.

Tuolumne County is famous for its large number of rich pocket mines. Bald Mountain and the vicinity of the Bonanza Mine have produced more rich pockets than any other section in the world. Jackass Hill and Whisky Hill are also noted for their pocket mines. The richest pocket mine in the world is undoubtedly the Bonanza Mine, situated within the city limits of Sonora. More than $2,000,000 have been taken from this mine, which is still being worked. Many other pocket mines are being worked, and they add considerably to the county's output. Many millions have already been produced from these mines. The pocket mining district is in the metamorphic slate near to the limestone belt, between the Mother Lode and the East Belt.

Following is the estimated output in round figures of some of the best known pocket mines: Bonanza, $2,000,000; Jackass Hill, $400,000; Colby, $200,000; Ford Lead, $190,000; Garrett Claim, $175,000; Austrian Lead, $160,000; Sell Lead, $160,000; Simonich Claim, $165,000; Arnold Claim, $110,000; Peterson Claim, $100,000; Magruder $100,000; Fox Claim, $100,000; Bald Mountain Claim, $100,000; Sugarman Claim, $350,000; Carpenter, $80,000; Morris

Claim, $75,000; Tanzy, $65,000; Tainter & Grew, $50,000; Thebian Twist, $50,000; Golden Era, $50,000; Brown Claim, $50,000; Mandich Claim, $50,000; Pedro, $50,000; Bliscock & Putnam, $50,000; Birney,. $60,000; Saratoga, $20,000; Fairview, $50,000; Tulloch Lead, $15,000; Crystal Hill, $10,000.

There are two ancient river channels in California, buried under lava, which form what are known as Table Mountains, one known as the Tuolumne Table Mountain, in Tuolumne County, and the other known as the Butte Table Mountain, in Butte County. The Tuolumne

IN THE GRAZING REGION OF TUOLUMNE COUNTY.

Table Mountain extends across the country like a gigantic black wall, composed of basalt, with nearly perpendicular sides, and with a top that is bare and almost level, and presents one of the most marked features of the county. It was formed in a primeval age by a flow of lava, probably poured from a valcano near Silver Mountain, in Alpine County, and ran south-westerly, in nearly the same course as the present Stanislaus River, which has cut through it in several places. It filled the bed of an ancient auriferous river, with a fork in the basaltic stream, about fourteen miles above Columbia.

After the lava had cooled and hardened, the softer rock on either side was, in the course of succeeding ages, washed away to the depth of from 500 to 800 feet on the western side, and from 200 to 500 feet on the eastern side, leaving not only the lava, but, in many places, the gravel of the old river-bed above the general level of the country. In sinking from the level top of the mountain, a layer of basalt, about 140 feet thick, had to be penetrated; then came a layer of volcanic sand, about 100 feet thick; then about fifty feet of pipe clay, and then the auriferous ground of the old channel, some twenty-five feet thick.

Professor Whitney estimates the amount of denudation which has taken place since the flow of lava took its present position, at not less than 3,000 or 4,000 feet of vertical height. The excessive hardness of the basaltic capping of the mountain has protected it from any appreciable amount of denudation and erosion.

The discovery of the auriferous character of the bed of this ancient river was made accidentally, by some placer miners working in the vicinity of Shaw's Flat, in 1854, at a point near the rim-rock of the channel, where the lava capping had been denuded.

From the several claims located at this point, sixty or seventy pounds of gold per day were taken. Forty thousand feet of tunnels have been run in this mountain, at a cost of not less than $800,000. At the present time, but little work is being done on the channel, although it still affords abundant opportunities to capital.

A well-defined copper belt is found in the west side of the Mother Lode, in the metamorphic slates. It has the general trend of the auriferous slate belt. From its favorable location and conditions, it offers a good field to those interested in copper.

The Tioga district is situated on the summit of the Sierras, on the border of Tuolumne and Mono counties, near Bennetville. The mines are in a slate formation. The principal claims are the Great Sierra, Sheepherder, and Golden Crown. Many locations have been made on the lode, which can be traced for many miles. The ore is said to be high grade, the value lying principally in five or six per cent of sulphurets. There is an abundance of timber, and an excellent chance to obtain water-power, but the inaccessibility of the mines, high rates of freight, and the cold climate of the high Sierras, have so far operated against this district.

It is an undisputed fact that the placer ground of Tuolumne was the richest ever found. While the rivers, creeks, and gulches were rich in comparison to similar ground elsewhere, the limestone belt far exceeded any other locality in the yield of gold. Nowhere have there been found so many large nuggets within such a limited area. A few of the most notable finds are as follows: In 1848, a seventy-five-pound nugget was found in Wood's Creek; a fifty-pound nugget, worth $8,500, was found half a mile east of Columbia; a $5,000 nugget was found at Holden's Garden, in 1851; a $7,000 nugget was found at Gold Hill, in 1859; a thirty-five-pound nugget was found in the streets of Sonora, in 1850; a mass of quartz and gold was found at Holden's Garden, which sold for $30,000; a seventy-two-pound nugget was found at Columbia, in 1854; a nugget worth $2,750, was found at Martinez, in 1853; a twenty-three-pound nugget was found

IN THE LIMESTONE BELT OF TUOLUMNE COUNTY.

at Pine Gulch, in 1854; a sixty-six-pound nugget was found near Columbia; a $2,000 nugget was found at American Camp; a 450-ounce nugget, worth $6,750, was found at Columbia; a twenty-three-pound nugget was found at Wood's Diggings; a twenty-eight-pound nugget was found at Sonora, in 1850; a thirty-three and a half pound nugget was found in Columbia, in 1854.

The above list will give some idea of the richness of the ground in coarse gold. The following list is an estimate of what each locality has produced. These figures give only the actual amount sent from the districts by express companies. They do not represent anywhere near the amount yielded, because millions were taken from the county of which there is no record.

Columbia and Springfield, $55,000,000; Goldsprings and Nigger Gulch, $7,500,000; Yankee Hill and Knickerbocker Flat, $3,500,000; Mormon Creek, $2,225,000; Pine Log, Experimental Gulch, and Italian Bar, $3,500,000; Sonora, $11,000,000; Saw-Mill Flat, $2,500,- 000; Horse-Shoe Bend, $300,000; Brown's Flat, $4,500,000; Shaw's Flat, $6,000,000; Kincaid Flat, $3,000,000; Campo Seco, $5,500,000; Poverty Hill and Chili Camp, $4,000,000; Jamestown, $3,500,000; Algerine, $2,500,000; Sullivan's Creek, $3,000,000; Chinese Camp, $2,500,000; Montezuma and Picayune Gulch, $1,500,000; Don Pedro's Bar, Jacksonville, and Stevens' Bar, $9,000,000; Groveland, Deer Flat, and Big-Oak Flat, $25,000,000; Table Mountain Humbug, $500,- 000; Moccasin Creek, $150,000. There were many other rich localities from which millions were produced, but there is no existing record concerning them. The following were some of the most prominent: Six-Mile Bar, Willow Bar, Green-Spring Run, O'Byrne's Ferry, Central Ferry, Peoria Flat, Moorehouse Bar, Confidence and Davis Flats, Blanket Creek, Turnback Creek, Cherokee, and Rough-and-Ready.

The limestone belt, on which were found the early placers noted for their immense yield from 1850 to 1855, runs through all the southern mining counties, and can be traced continuously for one hundred miles. Its course is north-west and south-east, and in width varies from half a mile to four miles. At Sonora it is narrow, while at Shaw's Flat and Columbia it is several miles in width. Throughout its entire length it was noted for the richness of the placer deposits.

The limestone bedrock has been deeply worn by the action of the swiftly running water carrying boulders and debris, which have cut and carved it in the most singular and fantastic shapes to a depth of many feet. In many places, remarkable underground caverns of unknown extent are found, excavated by percolating waters charged with carbonic acid. There is a vast cavern near Columbia known as the "Crystal Palace Cave," and another near Cave City, El Dorado County, which has never been thoroughly explored.

East of Columbia there was a hole in the limestone 250 feet deep, at the bottom of which was found a running stream, four feet wide and twelve feet deep. Another hole led into a stream which made its outlet near Jamestown, eight miles below. At Gold Spring and Springfield are remarkable springs supposed to be outlets of subterranean streams flowing through limestone fissures. Several distinct strata of limestone exist, and it is said that one or more fossil encrinites have been discovered, but there is no indisputable evidence to show the place of this deposit in geological history. It is undoubtedly of deep-sea, marine origin, and shows distinct stratification and alternate layers of gray, blue, and white.

As to the comparative age of the different deposits, it may be

said that the limestone underlies Table Mountain and the slate ante-
dates the limestone. From the existence of granite dikes, forced up
through the slate and lying in contact therewith, it is known that the
slates are older than the granite. First, in age, therefore, comes the slate, then the granite, then limestone, and finally the Table Mountain deposits.

BANNER WON BY TUOLUMNE COUNTY FOR BEST MINING
EXHIBIT, MINING FAIR, SAN FRANCISCO, 1898.

The richest portion of the limestone belt has been found on the east side of Table Mountain, and it is very probable that the placers owed much of their wealth to the scattering and distribution by water of portions of the Table Mountain channel not protected by lava.

It is hard to estimate the millions of dollars produced, but the express companies' records show that $55,-000,000 was obtained between Columbia and Springfield. It is safe to say that $100,000,000 would not be far from the true amount. The ground has been mostly worked to water-level, but there is an immense amount of auriferous gravel below water-level, filling the deep pot-holes and honey-combed limestone. Considerable prospecting is being done in several locali-ties with a view of draining the flats and mining the ground on a large scale. It is fair to presume that the placers of Tuolumne, once the richest in the world, will soon add a substantial sum to the annual output of the county.

Mariposa County's Mineral Wealth.

BY
W. H. STORMS.

IN the early history of mining in California, Mariposa was one of the most prosperous of the counties in which mining was the chief industry. The placers were of the most superficial character, generally speaking, and were quickly exhausted, but the facility with which the gold was obtained offset, in a great degree, the lack of permanency. Some of the old-time names have gone into history, and will live for many years to come, though the towns themselves have already mostly disappeared. Carson, the early seat of county government, has long since fallen into decay, nothing remaining to mark the spot but a few crumbling walls of stone and adobe. Mormon Bar, on Mariposa Creek, is only a memory. The rich placers of Hornitos were long since exhausted, and Bear Valley and Princeton, once flourishing towns, are quiet little hamlets, where a few people live and work the small mines of the neighboring hills.

The banks of the Merced River were at one time lined with the camps of as rough-and-ready a set of pioneers as ever made history in any region, but all that now remains, to bring to the recollection of the passer-by the days of old, are unsightly piles of tailings on the bars, and here and there a toppling chimney or a falling wall. The principal town to-day is Mariposa, where are located the county buildings. The once good-sized town, notorious in early history, is now a quiet village of a few hundred inhabitants. In the southern part of the county, mining is almost at a stand-still, but in the section north of the Merced River there is more activity.

In the early days, quartz-mining in this county was actively and successfully carried on. South of the Merced, the Princeton, Mariposa, Pine Tree and Josephine, Oso, Mount Ophir, and other mines, produced millions of dollars in gold, but the Fremont land grant was "floated" over the most valuable of the mineral area, covering more than 40,000 acres of as good mineral land as could be found in the State. Long continued and expensive litigation followed, and though some mining has since been done on the grant (now the property of the Mariposa Mining and Commercial Company), little of importance has been accomplished in recent years.

A large area of the region north of the Merced was consolidated

THE MAY ROCK, NEAR BEAR VALLEY.
The most remarkable Outcrop on the Mother Lode, being 82 feet high,
125 feet long and 20 feet thick.

into what is known as the Cook Estate, and this, also, after several years of prosperity, was deserted and for years remained unproductive. This property, covering 20,000 acres, or thereabouts, has been taken over by the Merced Gold Mining Company, of Boston, Mass., and development has been in progress, and extensive improvements made during the past four years.

It should not be thought that, because the mining industry of Mariposa County has been stagnant for several years, it is without resources, for such is not the case. No section of California to-day affords better opportunity for legitimate mining enterprise, conducted on a broad scale, for, without doubt, there are mines lying idle to-day in this county which, in the hands of experienced and skillful management, could be made to pay handsome profits on large investments. It must be understood that the opportunities for large returns on small investments are few indeed. The region requires the best equipment and modern methods, as well as a thorough treatment of its ores. When these are applied, the great ore-bodies will become profitable. Take, for instance, the Princeton vein— worked to a depth of about 600 feet, and producing $4,000,000 in gold. No one familiar with the great depth and continuity of the veins of California believes the Princeton exhausted. There are similar mines in the State, elsewhere, which have been worked successfully to three and four times the depth of the Princeton. The Pine Tree and Josephine mines have immense ore-bodies, containing a grade of ore which, by modern methods, should yield a profit.

The great mine at Hite's Cove, with a record of $2,250,000, is only worked down 900 feet below the croppings. This mine is on the East Lode, in the same belt as numerous other mines which, in years gone by, were successfully operated under adverse conditions by crude methods, and finally closed down. The aggregate production of these mines reaches millions of dollars. Although this eastern district has been idle for years, new life is being instilled into it, and it may be looked to as one of the portions of the State which will furnish some pleasant surprises.

The mines of this East Lode, in Mariposa County, are mostly in black slate, and the fissures are usually accompanied by a dike rock. The mines have been worked to various depths, few of them exceeding 600 feet.

Geologically, Mariposa has some very interesting features. It has at least four distinct mineral zones, or belts. The most westerly lies along the low foot-hills bordering the San Joaquin plains, and contains, principally, ores of copper and gold.

Several miles east of this belt is the so-called Mother Lode, which has become famous throughout the world. This belt terminates abruptly in the southern part of this county, near the old town of Bridgeport, in a great mass of white quartz 100 feet in width. It is not auriferous, or only slightly so at this place. From this point it extends northward through Mariposa into Tuolumne, Calaveras, and

Amador counties, and into El Dorado, but in the latter county it becomes scattered, and its identity as a "lode" is lost. There has been considerable controversy about the termination of the Mother Lode in Mariposa County, but there is no evidence to indicate that it goes further south than the massive outcrop above referred to.

The clay slates, which with diabase, diorite, and serpentine form the principal rocks along the Mother Lode in Mariposa, are abruptly terminated in this county by the intrusion of a large mass of grano-diorite from the southward. This rock has been thrust in with some violence, and along the entire line of contact metamorphism of the older rocks is pronounced. It has flexed the Mariposa beds and accompanying greenstones to the south-westward, and the south-easterly trend of the Mother Lode is found sweeping around to the south and south-westward, conforming to the altered strike of the inclosing rocks. This might be considered as indicating that the veins of the Gold Belt are older than the intrusion of the grano-diorite; there is much doubt as to this.

The Mother Lode in Mariposa County develops some striking characteristics. These features are persistent through a distance of not less than fifty miles from its southern terminus, at Bridgeport, to Carson Creek, on the north side of Carson Hill, in Calaveras County. The vein, or more properly lode, resembles a huge dike more than anything else. It is not absolutely continuous, but is nearly so, one break occurring just north of the Merced River, and another in Tuolumne County, just north of the Mariposa County line. The lode varies in width from a few feet to a broad zone over 300 feet wide. The material of the lode consists of carbonate of lime and magnesia (dolomite), and in part also of carbonate of iron (ankerite). Throughout this mass is scattered more or less abundantly the beautiful green, scaly mineral, named by Prof. Silliman "mariposite."

The peculiar characteristics of this great dolomitic vein, which is such a prominent feature of the southern portion of the so-called Mother Lode, are no better shown anywhere along its length, perhaps, than at Coulterville and vicinity. Just below the village, on Maxwell Creek, the great vein crosses that stream, which has cut a gap through it about 100 feet wide. On the south side of the creek is located the Louisa Mine, and immediately north of it is the Margaret, on the opposite side of the creek. Where the lode crosses the creek it has a width of 300 feet, and consists of an immense mass of ankerite, through which is disseminated the green mariposite. Large, lens-shaped masses of quartz outcrop boldly from the ankerite, being somewhat harder than the latter. The lenses are irregularly distributed, but occur mostly along the hanging wall and near the center of the vein. The large veins or lenses of quartz are separated by equally large or larger zones of the dolomitic mineral, which is interlaced in every direction by quartz veinlets and veins of varying size, making the entire mass a mineral zone or lode proper. One of the quartz lenses is nearly twenty feet in width, outcrops to a height of twenty-five feet, and is 300 feet long. A shaft sunk on the foot-wall side of

QUARTZ CROPPINGS, LOUISA MINE, COULTERVILLE.

it to a depth of sixty feet showed it to be thinning out; but there is no doubt that it would be found replaced in depth by another lens of similar character. South of this large cropping a small vein branches out into the hanging wall, which is diabase, striking northward and increasing in width until it disappears underneath Maxwell Creek. On the opposite side of the stream a large vein appears, which apparently is the northward continuation of the one referred to.

Through the center of the lode is a quartz vein ten to twenty feet in width, and still west of it another but smaller vein. The entire western portion of the ankerite mass, constituting about one-third of the whole width of the zone, is a perfect network of small quartz veins, stringers, and bunches of quartz.

A prominent feature of interest is the union of two of the largest quartz lenses near the center of the lode by a third large vein, which crosses the intervening ankerite diagonally. The two large veins are about 120 feet apart. Beginning near Maxwell Creek, on one of the large outcrops referred to, a careful examination discovers a seam in the quartz, along which gold may be seen almost every foot of the way. This gold seam can be followed some distance in a southerly direction to where the diagonal branch above referred to leaves it. This latter also shows gold along a similar seam, leading to the other large quartz lens parallel with the first, and here, again, is a gold seam which may be followed if care be taken. Outside of the gold thus occurring, none was observed elsewhere by the writer, though it was said that prospects could be obtained by crushing certain portions of the rock and carefully panning it. Southward from the section above described, the several large veins converge toward a central point on the ridge, which rises higher and higher, terminating in an immense mass of quartz at its apex. Southward from this point the quartz is less prominent, and the ankerite constitutes the major portion of the lode until another occurrence of quartz lenses is reached, which, in each case, whether following the lode north or south, is much the same.

Gold sometimes occurs in the ankerite and mariposite, when seamed with quartz. A brecciated, crushed condition of this rock seems to favor the gold, or, at least, rock of this character contains more gold than that which is massive and solid. The entire mass is low grade, and ankerite wholly free from quartz is practically free from gold in this mine.

The hanging-wall country at these mines is massive diabase, and the foot-wall a diabase dike, separating the ankerite lode from the black clay slates of the Mariposa beds. These latter are a mile wide, and are intruded in numerous places by dikes of diabase and diorite.

The next prominent outcrop on the lode, north-westerly from Coulterville, is at Pinon Blanco, about three miles distant. This hill rises prominently, its larger axis lying with the lode, and being defined by conspicuous outcrops of quartz, similar to those at the Louisa. The country rock at Pinon Blanco is somewhat different from that at the Louisa. At the former, a coarse gabbro is prominent, some crystals of pyroxene in semi-decomposed rock masses being two by three inches in width and length. Serpentine is also in contact with the lode on the foot-wall side. The ankerite mass is more than 300 feet wide. Portions seem to be almost pure dolomite, the iron constituent being absent. Some of this rock contains mariposite in abundance, and in other parts it is missing. The quartz, at a certain point on the eminence known as South Pinon Blanco, contains small crystals of tetrahedrite, and this rock is rich in gold.

RUINS OF THE OLD MARIPOSA MILL.

The oxidation of the gray copper has produced a large number of small blue specks of azurite. This characteristic has been noticed at several mines on the Central Gold Belt, notably at the Alameda Mine, the Rawhide Mine, and Rawhide No. 2, in Tuolumne County.

The East Lode, as previously remarked, is east of the Mother Lode about twenty miles. The veins are mostly in black clay slates of the Calaveras formation (presumably carboniferous). Some of the veins occur in grano-diorite, but the greater number are in slaty rocks. North of the forks of the Merced River the mines are numerous, but, generally speaking, are not in operation. The county is rugged, consisting of high ridges separated by deep canyons, with steep, almost precipitous slopes. However, this region is being investigated, and we may look for a revival of the mining industry in that field with satisfactory results.

The most important mine thus far opened on the East Lode of the gold belt in Mariposa County is the Hite Mine, at Hite's Cove, on the south fork of the Merced River. The mountains are very steep and rise from 2,000 to 4,000 feet above the river. The hills are densely clad with chaparral and greasewood, and the higher ridges support a growth of yellow pine and a few oaks. There is but one road into the region, that from the direction of Mariposa, made at great expense. The rock formations exposed in the vicinity are those characteristic of the Calaveras formation, consisting of clay slates, mica schist, quartzite, crystalline limestone, and more or less silicious beds of slaty structure. These rocks are greatly folded, shattered and crushed, and indurated by pressure and intruded by dikes of diorite, granite, and felsitic rocks.

The Hite vein strikes N. 70° W. and dips north-easterly at about 75° in the clay slates. It maintains a remarkably straight course for a long distance, but is accompanied by numerous branches, mostly into the hanging-wall side. The slates near the vein are foliated, metallic, and lustrous, usually soft, and often crushed to a gouge-like condition. Rock of this character is very heavy and necessitates frequent repairs to timbering. Whenever this rock presents a smooth, perfect slaty cleavage, it is always at a distance from the fissure. The latter condition gradually changes to that of the foliated mass above described upon nearing the crevice, whether it be ore-bearing or not. One remarkable phenomenon connected with the Hite vein is the splitting off of a large vein which makes out into the foot-wall side in a sweeping curve, and returning to the main fissure again in a distance of about 600 feet from the point of departure. In depth it also re-unites with the main vein at a depth of about 600 feet. The enclosed mass of slate has the form of a large plano-convex lens, having a diameter of 600 feet and a thickness at the center of 45 to 50 feet. The segment of slate is enclosed between two veins of quartz, each of which is gold-bearing, and, as the results of operations showed, was rich in gold. The surface croppings of milk-white quartz, a few inches to twelve feet in width, were studded will dull-yellow gold, and from the day of its discovery the owner was practically a rich man, for, though almost penniless at the time, he soon had pounded out in a hand mortar enough gold from the rich quartz to supply every necessity and later to build a mill. The vein has since produced nearly $2,500,000.

The fissures have been extensively developed to a depth of 900 feet below the croppings through crosscut tunnels, the lower of which reaches the vein at a depth of 600 feet, and is nearly 1,400 feet long, driven through hard slate of the foot-wall country. A large dike of felsitic rock accompanies the vein, lying in the hanging

CROPPINGS OF THE MARIPOSA VEIN.

wall, and is usually in contact with the vein. In the lower workings, the vein varies from four to twenty-five feet in width, the latter obtaining at the union of the hanging-wall vein with the branch split off in the foot.

The quartz commonly includes more or less slaty material, a feature common to veins in slate rock everywhere. A banded appearance of the quartz, probably induced, in part at least, by the original slaty structure of the rock, which has been replaced by silica, is very noticeable and persistent throughout the mine. This banded or ribbon structure may possibly have been intensified by pressure. The physical appearance of the quartz is not always a safe index of its value, for two pieces of rock may be alike to all appearances, and still differ greatly in value. The mine has not had any development worthy of mention in the hanging-wall side, but there are abundant indications that good shoots of quartz may be found there. It is a mine well suited to prospecting from the old workings by means of the diamond drill.

The mine has been idle for some time, and is presumed to be worked out.* The fissure is still strong, with all the physical conditions which characterized it at the higher levels, and it is only reasonable to anticipate a repetition of these ore shoots at greater depth. Should they not occur in the same fissure, they may be found to have lapped, as is frequently the case.

Another important mine in this county, which also occurs in black slate, is the Princeton, on the Mariposa estate, at Princeton. It was worked in the '60s, and is reported to have produced more than $4,000,000 to a depth of 600 feet. Little at this time is known of the old mine workings other than that the gold was derived from a single large shoot, presumed to be worked out. The last operations conducted at that mine were not creditable to the management, as the workings and shafts were filled up with waste rock in the process of gutting the mine, and there is no probability that it will ever be re-opened through the old openings. A dike of light gray felsitic rock also accompanies the Princeton vein.†

The Sierra Belt, in Mariposa County, is not well known. There is, however, a mineral belt running along near the crest of the Sierra Nevadas, in this county, which extends northerly and southerly many miles. These veins and ore deposits are large, considered base, and have been given little attention. The fact is they are simply sulphide ores—sulphides of iron, lead, zinc and copper chiefly, containing gold and silver—and some of these ore deposits, situated elsewhere, would be operated. A lack of transportation facilities in this region has retarded their investigation and development, but undoubtedly in the near future they must attract favorable notice. The belt lies at an altitude of 8,000 to 10,000 feet, and the seasons are short. Water-power in the neighboring canyons is abundant, and a magnificent growth of pine is found up to 9,000 and even 10,000 feet.

* Since writing the above, a corporation has been organized to re-open and further develop the Hite Mine.

† The Princeton, Pine Tree and Josephine Mines are now being re-opened by the Mariposa Commercial and Mining Company.

Northern California's Resources.

BY

M. F. DITTMAR.

HE base character of Northern California's ores has, to a great degree, retarded the development of the latent mineral wealth of this important district in the past. Not alone has development been retarded, but, in spite of the output recorded year after year, investors have been led to believe that the district was prominent only for its uncertainties.

The early placers were very rich and added largely to the world's supply of gold during the '50s. At the present time the great deposits of gravel, especially in Trinity and Siskiyou counties, offer opportunities in hydraulic mining unequalled anywhere else on the globe. The ledges that fed these great deposits of auriferous gravel, hidden away in the mountain fastnesses, are still in the primacy of development. In spite of this, however, two of the three counties comprising the main area of the mineral belt of Northern California (Shasta and Trinity) are in excess of the $1,000,000 mark for 1898, and the remaining county (Siskiyou) ranks first of all the counties that fall below this mark.

When speaking of the Northern California mineral belt, that section of country lying north of the lava-covered country about Lassen Butte is referred to. This peak of the Sierra Nevadas is in the southeastern portion of Shasta County and has an elevation of 10,595 feet. Lassen Butte is an extinct volcano, and in past ages buried the auriferous formations beneath its lava, breaking the link which would otherwise unite the gold fields of Northern and Central California.

The entire eastern portion of Shasta County is covered with a lava flow, and until the head of Cow Creek, twenty-five miles north-east of Redding, is reached, little indication of mineral-bearing formation is encountered. From this point, the mineral zone extends for miles to the north-west and to the south-west, including the greater portion of Shasta, Trinity, and Siskiyou counties.

In Lassen County, the gold-bearing formation crops out occasionally through the lava. This is particularly true of the Hayden Hill district, where several very good properties have been developed, in spite of the remoteness of the locality from lines of transportion.

The character of mining in Northern California is very diversified, and, in all classes of mining, the work is reaching a magnitude not exceeded elsewhere in the State, and in the matter of hydraulic mining, not equalled in the world.

During the past few years development in copper has worked wonders, and the mining world is beginning to realize that in Northern California, lying adjacent to the railroad, with large rivers traversing the immediate district, with nature's power at ready command, and with great forests of yellow and sugar pine available for all purposes, a belt of sulphide ore has been discovered, rich in copper, gold and silver, and encountered in immense bodies.

The great plant of the Mountain Copper Company, located at Keswick, six miles north of Redding, was the first to reduce these values on a large scale. This mine and plant are described in detail in another article in this book.

This great mine is but one of a large number that will be developed on the belt, swinging away like a huge crescent, with Iron Mountain, fifteen miles north-west of Redding, at one end, and the Afterthought Mine, twenty-five miles north-east, with its thousands of tons of ore available, at the opposite end. At present, the properties showing best under development are the Balakalala and the Stowell group, a few miles north of Iron Mountain; the Hearne group, several miles further north, on Squaw Creek, recently bonded, and now being developed by Boston capital; and, still further to the north, the Mammoth group of mines, also under development by Boston capital. Near the last-mentioned property, Lewishon Brothers are developing ground showing very favorable surface indications. These mines, on an air-line, are about eight miles from Iron Mountain; many other promising locations, besides those mentioned, cover the intervening territory.

In the vicinity of the confluence of the Pitt and Sacramento rivers, the surface indications extend eastward, showing strong on

both sides of Pitt River, and crossing the divide to the headwaters of Stillwater Creek. Here a group of mines was recently purchased by the Northern California Investment Company. A German syndicate, represented by Georg Bayha, is developing this group, known as the Black Diamond.

Copper City, a few miles further to the north-east of the last group mentioned, represents the greatest amount of development work done on the east side of the Sacramento River. The Bully Hill property has been developed during the past few years, until the quantity and value of ore in sight fully warrant the erection of a large smelting plant. The property was developed by Mr. James Salee, one of the former owners of the Iron Mountain Mine, and transferred by him to Captain De Lamar, for a sum approximating $200,000. Development work is being prosecuted with all possible energy, in order to ascertain, as soon as practicable, how large a reduction plant the mine will warrant. This property promises to take rank, in the near future, with the big producing mines of the West.

About eight miles south-east from Copper City is the Furnaceville district, which includes the Afterthought and Donkey mines, which show large bodies of sulphide ore. Capitalists are investigating the merits of these properties, with a view of placing them in line of production with the rest of Shasta County's great copper belt.

Lying adjacent to the copper belt, and extending from Cow Creek, on the east, to Clear Creek, on the west, is a district rich in gold-bearing quartz. Some mines throughout this district have been operated on quite an extensive scale, but, as a rule, the ordinary methods of crushing and concentrating resulted in a heavy percentage of loss, and this fact, to a great degree, precluded the development of our gold mines. The introduction of smelters has, to a great degree, overcome the disadvantages under which the quartz miners were laboring. Silica being required for flux in the smelting of sulphide ores, a ready market for the product of the quartz mines is assured.

The impetus given to this class of mining is already manifest in the greatly increased gold product, which seems destined to lead that of any other county in the State. A notable development, resulting from shipping ore to Keswick, is the Mount Shasta Mine, in the Clear Creek district, eight miles west of Redding. A shaft, 300 feet deep, with drifts on the 120 and 200-foot levels, shows a good ledge of very

VIEW OF THE SMELTING PLANT OF THE MOUNTAIN COPPER CO., AT KESWICK.

high-grade ore, running from $50 to $400 a ton in value. From this mine, about twenty-five tons of ore are shipped daily to Keswick. It has returned the purchase price to the owners, paid for the equipment, and now pays a regular dividend, yet this mine lay undeveloped for years, solely because the owners saw nothing to encourage its development and operation.

Leaving the immediate vicinity of the base-ore belt, we find the gold formation extending to the extreme south-western part of Shasta County where, in the Harrison Gulch district, the southern limit of the Northern California Gold Belt can be defined. One of the best gold producers of Northern California, yielding from $30,000 to $40,000 per month, is the property of the Midas Gold Mining Company, located in this district.

North-west of Redding, about twenty miles, is the French Gulch and Deadwood districts, extending from the Gladstone Mine to the famous Brown Bear, at Deadwood, which has to its credit a product of $6,000,000. The ledges of this district occur mainly on slate and porphyry contacts, and include such mines as the Washington, the Niagara, the New Brunswick, and others of lesser moment, but of great promise. The gold taken from the surface ores throughout this section was, to a great extent, free; but, as depth is attained, the values are held by bases, and, in most instances, it is necessary to ship ore to smelters in order to obtain the best results.

Extending from the base-ore belt to the north are the important Squaw Creek and Dog Creek districts. In the former is located the well-known Uncle Sam Mine, which was abandoned after a yield of over $1,000,000. This property has been resuscitated, however, by new owners, and thirty stamps have again been placed in the old mill with thousands of tons of ore available. In the Dog Creek district, some twenty miles farther to the north, are located numerous promising ledges of high-grade ore that will certainly develop into rich producers in the immediate future. In the north-western part of the county a mineralized district extends into Trinity, and development work is rapidly bringing it to the front.

From where the gold belt extends into Trinity County, between French Gulch and Deadwood, to a point forty miles to the north, there is very little to encourage the prospector or miner. In the district, however, comprising northern Trinity and south-western Siskiyou, there is much for the pioneer in mining ventures to base high hopes upon.

On both sides of the Salmon River Mountains, a range dividing Trinity from Siskiyou County, development work is disclosing some of the richest quartz in the State. In the Coffee Creek district, of Trinity County, is a granite belt, traversed by numerous ledges of rich telluride ores. The district is in its infancy, very little prospecting having been done on the head of Coffee Creek until within the past few years.

IN THE OLD HAYES RED HILL MINE ON THE TRINITY RIVER,
Now belonging to the Compagnie Française Placers Hydrauliques.

A few miles below the mouth of Coffee Creek, some heavy hydraulic operations are in progress, particularly at Trinity Center, where giants are operated on a seventy-foot gravel bank, washing out about an acre of ground a month. This belt of auriferous gravel extends for some distance to the south and west of Trinity Center, and has reached the height of its development at Junction City, near Weaverville. The La Grange Mine, at this point, is credited with being the largest hydraulic plant in the world. The company owns over 3,000 acres of gravel land, and, where operations are concentrated at present, the bank is 300 feet high. Last year the big ditch was completed, and deserves more than passing description. This great artificial water-course conveys 9,000 inches of water from a source that assures a constant flow, and, in construction, presented many difficulties, which have been successfully overcome.

Starting from the confluence of Deer Creek and Stuart's Fork, eight miles of flume conveys the water along the north side of the main stream to the commencement of the pipe line. This line, 4,852 feet in length, conducts the water across Stuart's Fork, and is the largest inverted siphon in the world.

PROSPECTING A GRAVEL CHANNEL BY BORING, SISKIYOU COUNTY.

At its lowest point, a bridge spans the stream, and, at a pressure of 1,041 feet, the large thirty-inch pipe of one-half-inch steel withstands the pressure of 452 pounds to the square inch. Ascending on the opposite side of the stream, the water is carried a mile and a half through a flume. From this point a tunnel, 8,396 feet long, conveys the water through a mountain to another flume which, four miles in length, carries it to the Rush Creek dam, at the head of Rush Creek or Chaumont Quitry ditch. From the reservoir it is piped to the mine.

The disposal of the debris, which has been the cause of much loss and annoyance in the central part of the State, is a problem of easy solution on the Trinity and Klamath rivers, in Trinity and Siskiyou counties. These streams, with numerous tributaries, pour their waters direct into the Pacific, and the mountain gorges, traversed by them, support no agriculture demanding protection, as is the case in the rich alluvial valleys of the Sacramento and San Joaquin.

Siskiyou County is experiencing more rapid development of permanent mining resources than ever before, and comprises all character of operation in gold properties, as well as important exploitation in copper. The district adjoining the Salmon River Range on the north is almost identical with the district lying in Trinity County on the south side of the range. The higher altitudes are developing rich quartz properties, and the lower reaches offer opportunity in extensive hydraulic mining, which is taken advantage of only to a limited degree. Among the prominent producers of this section can be mentioned the Black Bear, with a production of $2,500,000, and the Gold Run, which has achieved a prominence in the past two years entitling the mine to being classed with the richest quartz properties of Northern California.

Development work has placed a number of other properties of this section among the producers, and the county is sure to exceed the $1,000,000 mark for the year. Great deposits of auriferous gravel cover the western portion of the county, along the course of the Klamath River, and extend toward the ocean. A number of large properties are being successfully operated, as well as many smaller ones. The river-bed is also being successfully mined by dredging and wing-damming.

In the vicinity of Yreka, the county seat of Siskiyou, is a district to a certain extent isolated from the rest of the county's mineral belt, and a number of rich quartz properties are here operating, notably the Schroder Mine, recently sold.

The gold yield of Siskiyou County is placed in excess of $50,000,000 since the days of '49, and these figures, which are based upon reliable sources, show the merit of the section. The entire district of Northern California is what may be said to be easily accessible. Good mountain roads reach almost every district.

While this article has treated mainly of the precious metals and copper, Northern California presents many other forms of mineral wealth. Cinnabar is being successfully and extensively mined at Cinnabar, Trinity County, and the output of the Altoona Quicksilver Mine is large. Discoveries of quicksilver have recently been made in Shasta County, which may result in prominent development in the future. Building and ornamental stones abound, and are being successfully quarried at various points. Talc has been mined and shipped from Shasta County. Coal measures exist, and large bodies that promise to be of future value have been discovered in northern Shasta and southern Siskiyou counties, as well as in the more remote portion of Trinity County. Iron abounds, and a large deposit on Pitt River has attracted attention for some time. Mineral springs are numerous, and the industry of bottling waters from Northern California springs has grown to considerable importance.

GEORGE L. CARR'S ARRASTRA, NORTHERN TRINITY COUNTY.

The possibilities of gold-dredging in the Sacramento, the Trinity, and the Klamath rivers, with their tributaries, deserve prominent mention. Dredging in a small way has been in progress on the Sacramento and the Klamath for a number of years, and, while successful, the want of a dredger that could handle a large amount of gravel, and thoroughly clean the bedrock, was felt. The machines of the Postlethwaite type seem to fill this want, and the large dredger launched on the Trinity River last year solved the problem. The near future will see this class of mining successfully and extensively conducted on California streams, and particularly will the rich river-beds of Northern California be exploited by means of the dredger. No part of the West offers opportunities for the investment of capital on a scale so extensive as obtain in Northern California, a fact that is being recognized by some of the heaviest mining investors of the world.

Plumas and Sierra Counties.

LUMAS and Sierra counties have, since 1849, been among the leading gold-producing counties of the State. The industry has much declined in late years, but the future promises a large increase of mining activity, for which the extensive undeveloped resources offer opportunity and invitation.

Because of their similarity of position, topography, geology, and resources, they may be considered together as constituting one important gold-producing region. They lie in the heart of the Sierras, and at high altitudes. Sierra is a small county, sixty miles long and thirty wide, with an area of 830 square miles, bounding Plumas on the south and Nevada on the north, reaching from the State line across the mountain range to a line on the western slope about 3,000 feet above the sea, the lowest point in the county having an altitude of 2,000 feet. Plumas County similarly includes the crest of the Sierra Nevada Range, reaching seventy-five miles·west of the State line, and having an area of about 2,000 square miles. These counties have an extremely rugged and picturesque surface, due to much erosion, high elevations alternating with deep canyons, and very little of the surface lying less than 4,500 feet above the sea. Sierra County is mainly made up of mountains and canyons, the only valley worthy of the name being Sierra Valley. Plumas County has a trifle larger proportion of agricultural land, there being several small and fertile valleys, including Genessee, Big Meadows, Indian, American, and Butt. There are several prominent peaks in both counties. Sierra Buttes, in Sierra County, altitude 8,950 feet, is one of the landmarks of that part of the State. In Plumas, Pilot Peak, on the southern border, has an altitude of 7,605 feet, and, on the northern border, Lassen Peak rises 10,570 feet.

All this region is splendidly wooded and watered, no other section of the State rivalling it in these features. Heavy forests clothe the

surface below the timber line. Sierra County is drained by the North
and Middle Forks of the Yuba River, and by the Feather. Plumas is
drained entirely by the Feather. These main streams have a network of forks and tributaries, which reach through deep canyons
upward into the higher altitudes, where nestle a great number of
lakes constituting natural reservoirs.

While here and there little valleys afford opportunities for farming, fruit growing, and stock raising, and while the forests provide a
vast lumber supply, mining is, and from the beginning has been, the

CLEANING BEDROCK AT THE SPANISH RANCH HYDRAULIC MINE, PLUMAS COUNTY.

chief industry and the foundation of prosperity. It will long continue the chief industry, and the outlook is bright for a large increase
in mining activity, as attention and energy are directed to the extensive undeveloped mineral resources.

It was soon after the discovery of gold that the prospectors of
1849 pushed their way into this high region, following the streams
from the lower courses first attacked. There was a stampede to this
region in 1850, and in 1851 miners swarmed along all the streams,
which were rich in surface gold, and yielded vast quantities of the
precious metal during those early years. No part of the early mining
history of the State has more of romance and picturesqueness than
that which belongs to those early diggings. The chief mining centers
then, as now, were coincident with the principal streams.

The rich surface placers early declined, and the miners' attention was turned to the vast deposits of auriferous gravel in the ancient and generally buried river channels which were soon discovered. These ancient channels are a chief and characteristic feature of the gold resources of both these counties. In both Plumas and Sierra counties are the upper courses of these ancient channels, which have been so extensively worked and so productive in Nevada and other neighboring counties. In the western part of Sierra County several large gravel channels cross north and south, and have produced vast quantities of gold. One extensive channel runs for thirty miles along the divide separating Slate and Canyon creeks, reaching into both counties along their common border. It was largely the erosion of these ancient deposits that enriched the recent streams. Deep deposits of lava overlie much of these old gravel deposits, as they do much of the territory which carries gold-bearing quartz veins.

ON SCALES RIDGE, IN THE SIERRA COUNTY
MINING REGION.

The presence and richness of these ancient deposits invited early exploitation by the processes of hydraulic and drift mining, and so it was that hydraulic and drift mining began in the '50s, and early developed a large scale of operation. Plumas and Sierra counties were long distinctively hydraulic and drift mining regions, and they shared with Nevada County the greatness and glory of hydraulic mining when it was at the height of its development. Some of the largest hydraulic mines in the State were in these counties. In 1881, there were 1,000 miles of mining ditches in Plumas County. In that year there were twelve quartz mills in the county.

The stoppage of hydraulic mining by the courts in the early '80s was an especially severe blow to both these counties, and from that time the mining industry experienced a decline. After the cessation of hydraulic mining, more attention was given to working these deposits as drift mines, and some mines were notable successes. The Bald Mountain Drift Mine, in Sierra County, was located in 1864, and was worked through many tunnels reaching in from the side from 150 to 500 feet. About 250 of these tunnels were driven. The mine was about worked out during the '70s, and in ten years had yielded $2,000,000. The Bald Mountain Extension was then developed, and in 1881 the channel was struck in a tunnel 4,000 feet long. This has since been one of the most important drift mines of this region. It is owned by the Bald Mountain Drift Mining Company, of Downieville, and is equipped with an electric-power plant for hoisting, pumping, and lighting. The tunnel is now 6,500 feet long, and the width of the breast in the channel is 200 feet.

One of the notable mining propositions of Plumas County is that of the Alturas Mining Company, which owns about four miles of the bed of Slate Creek, along the southern border, and which is controlled by C. W. Hendel. This property yielded $1,000,000 in the early workings. It extends from the head of the torrential course of the stream upward, and has caught vast quantities of unworked cemented gravel and tailings from the drift mines above it, in the twenty square miles of ancient deposits developed. It is estimated that there is $10,000,000 in gold in this stretch of creek-bed awaiting capital, which shall drive a tunnel at the lower end sufficient to provide fall for sluicing, and supply other features of a needed plant on a proper scale.

The great auriferous slate belt traverses these counties, carrying many gold-bearing veins. Across the western part of Sierra County, and on through Plumas, the trend of the slates and greenstone is north-west and south-east, and the dip 45° to 60°. In both counties these formations are extensively overlaid by deep lava deposits, but, in many cases, the auriferous slates have been brought to light by erosion. Portions of Plumas County are traversed by heavy low-grade quartz veins, showing bold outcrops and occasional shoots of pay ore. One of these prominent quartz veins runs several miles east from the mouth of Rush Creek. The principal quartz-mining districts of Plumas County are the Greenville, Dixie, and Jamieson Creek, in the vicinity of Indian, Mohawk, and Genessee valleys.

The oldest and most famous quartz mine in Plumas County is the Plumas-Eureka, owned by the Sierra Buttes Mining Company, of London, which also operates the Sierra Buttes Mine, in Sierra County.

The ledge of the Plumas-Eureka was discovered in 1851, on Gold Mountain, and a large association of miners tried working it in arrastras. It was then the Eureka, and, before long, $100,000 was paid for mining and milling operations that were failures. The Sierra Buttes Mining Company bought the Eureka and some adjoining claims, in 1871, and proceeded on an extensive scale. It has been a profitable producer ever since, and is now operated with a fifteen-stamp mill. The Green Mountain is another old and prominent quartz

THE HEAD OF THE DIVIDE BETWEEN SLATE AND CANYON CREEKS, WITH THE SIERRA BUTTES IN THE DISTANCE.

mine still producing. It is worked through a tunnel 6,000 feet long, with 9,000 feet of drifts, and has a thirty-stamp mill. The Jamieson, with twenty stamps, the Crescent, with thirty stamps, and the Diadem and Drury-Pacific are among the other prominent quartz mines of Plumas County.

In Sierra County, the main mining district is in the vicinity of Sierra City, in which district there are many quartz veins. Sierra County has a number of valuable quartz mines, and some that have yielded handsomely. The most notable quartz mine in this county is the Sierra Buttes, belonging to the Sierra Buttes Gold Mining Company, of London. The Sierra Buttes group has paid dividends from the start, and produced $15,000,000. It was bought for $1,000,000, in 1870, and, in 1882, was operating ninety-six stamps, in three mills, and had yielded $7,000,000.

Space will not permit a detailed description of mining properties of note, or even a naming of many. The present extent of the industry may, perhaps, be best illustrated briefly by a statement of

SHOWING THE INTERESTING METHOD OF OPERATING THE CONTINUOUS-LINK CARRIER AT THE ALTURAS MINE WITH AN IMPULSE WHEEL, THE USED WATER SUPPLYING THE PIPE.

the number of mines and mills listed in the Register of Mines, now being compiled by the State Mining Bureau, and soon to be published. This list does not include prospects on which considerable development work has not been done.

For Plumas County the showing is: Drift mines, 84; placer, 31; hydraulic, 52; quartz, 73; miscellaneous, 21 — total, 261; mineral springs, 10; stamp mills, 25; arrastras and patent mills, 7.

Sierra County: Drift mines, 34; placer, 3; hydraulic, 29; quartz,

48; miscellaneous, 3—total, 97; mineral springs, 1; stamp mills, 17; arrastras, 1.

In 1898, Sierra County produced $399,063 in gold, and Plumas, $369,609. In Sierra, $98,279 came from placers, $48,610 from hydraulic mines, $194,230 from quartz mines, and $58,463 from drift mines. In Plumas, the portions were: Placer, $126,140; hydraulic, $79,797; quartz, $141,415; drift, $22,257.

It will be seen that gold mining comprises practically all there is of the mineral industry in these counties at the present time. This does not show that there are no other mineral resources of value in these counties. There are others, but lack of transportation facilities and other causes have heretofore discouraged their development, as such causes have retarded the otherwise natural increase of gold production. Plumas has rich and extensive copper deposits, and now many copper propositions are being exploited. In the near future, Plumas will be a producer of copper, and, undoubtedly, on a large scale. Both coal and copper exist in quantities in the Genessee Valley, and in the Mohawk Valley is a fine deposit of kaolin. The Sierra Iron Mine, at the head of Gold Valley, shows 600,000 tons of magnetite, with all conditions favorable for production if transportation is supplied. A few months ago, the first discovery of corundum in California was made in Plumas County. Marble is among the other economic minerals of Plumas County. Sierra County, likewise, has varied mineral wealth.

These counties are among those which offer exceptional inducements to the prospector and to the seeker of mining investments, and they are among the counties with an assurance of a prosperous mining future. There are vast quantities of auriferous gravel in the many miles of ancient channel, not only unworked, but unprospected, and destined to yield millions. Under the Caminetti law, there is a beginning of a hydraulic mining revival that will increase in the future and add to the general prosperity. There are hundreds of prospects undeveloped for want of capital, which is now turning to this region. There is a perfect mountain climate, affording health and enjoyment, incomparable forests, an inexhaustible and well distributed water supply for power or other purposes, and there is a great deal of unexplored and virgin ground from which mineral wealth will come.

Yuba County.

AMONG the gold-producing counties of the State with a splendid past, a quiet present, and a quite promising future in regard to its mineral industry, is Yuba County. The contrast between the past and the present is very marked. In the old days of the rich surface placers, and, later, in the heyday of hydraulic mining, the statisticians annually placed millions of dollars to Yuba's credit. In 1898, its mineral product consisted wholly of gold and amounted to $166,865, of which $141,144 was produced by "placer" mining, $4,453 by hydraulic mining, $4,268 by drift mining, and $17,000 by quartz mining.

The 617 square miles of Yuba County are comprised in an irregularly bounded and elongated area. The western portion lies in the heart of the rich Sacramento Valley, and possesses a climate, soil, and water supply favorable to the agricultural and horticultural industries, which are now Yuba's chief glory. The eastern portion climbs well into the Sierra Nevada foot-hill region, and is first rolling and then rugged, and well wooded and watered. This portion of the county lies just north of the great mining region of Nevada County, which it separates from Butte County, and here the geological formations of diorite, slate, serpentine, etc., are favorable to quartz mining. There are many quartz veins, with frequent shoots of pay ore.

Through the county runs the Yuba River, coming from the rich auriferous regions of Nevada and Sierra counties, and in early days its course through the county included some of the richest placers discovered, and many millions were taken out. The ancient channels, which proved so rich in the counties above to the east, also extend into Yuba County, and in the past they saw some of the biggest hydraulic and drift mining operations in the State.

Quartz mining has never prospered as it might have done, and as it will do. It began early, saw many failures, and in 1870 the quartz mining product was only $10,680. The principal quartz mining district is near Brownsville, where a number of mines, closed in the '70s, are being re-opened. The old Pennsylvania Mine produced several million dollars between 1861 and 1870, and within two years has been re-opened, displaying a new ore shoot in the 800-foot level. There are a number of good and promising quartz mines. The lower Yuba offers exceptional opportunities for gold dredging, and this form of the industry will flourish in the future.

Thirty-Inch Cast-Iron Pipe Pressure Line.
Yuba Electric Power Company's Plant, near Yuba.

Alameda County.

YING on the shore of San Francisco Bay, far from the gold regions of the State, is Alameda County, whose important mineral interests add materially to the wealth and variety of its products and industries. The two most important features of this county's mineral industry are coal and salt. It has the largest and the only important coal mine in California, and the extensive salt works on the bay shore, which produce thousands of tons of salt yearly by the evaporation of sea water, constitute the chief feature of the State's salt industry. In 1898, Alameda County led the counties of the State in the production of salt ($155,812), in coal ($176,250), and in manganese.

In the southern part of the county are several manganese mines, the product of which varies from year to year. There are extensive deposits of chromic iron, the production of which varies greatly with the market and the foreign supply. There are valuable sandstone quarries near Niles. Natural gas has been discovered near Pleasanton. The production of rubble rock is large, and unworked quicksilver deposits occur. Valuable clay beds support the operation of quite extensive potteries, and, in the interior of the county, in the region of Mount Diablo, there are numerous oil indications.

The Tesla Coal Mine, at Corral Hollow, in the eastern part of the county, constitutes the dominant feature of the county's mineral industry since the recent development and operation of the mine on a large scale. The valuable coal veins were early discovered, and lay undeveloped until, in quite recent years, the San Francisco and San Joaquin Coal Company acquired an area about one and a half by five and a half miles in extent, and invested a great amount of capital in developing the mines and establishing a plant on a large scale, and on the basis of the highest possible efficiency. Several hundred men are employed. The surface plant includes huge bunkers, a railroad, thirty-six miles long, to the head of navigation at Stockton, the model proprietary town of Tesla, with bunk houses, eighty cottages, water and electric systems, library, hospital, etc. There are seven valuable seams of coal developed by long tunnels and drifts. An electric generating plant at the mine, and a plant for the manufacture of fuel briquettes, at Stockton, are being elaborated. The operation of the mine and plant is an interesting study.

View of the Tesla Coal Mine, at Tesla, Alameda County.

Fresno and Madera Counties.

RESNO and Madera counties lie together, a very little south of the center of the State, and enjoy common features of topography and resources. Fresno County has, until very recent years, included Madera County, which has been carved out of its north-eastern part for political reasons. These counties have not in the past been of high rank among the mineral producing counties of the State, but there are many interesting features to this big mining field, and these counties are among those of this State with a mining future far greater than their past. Fresno County has, during 1899, leaped into high rank in the production of oil, through the development of the Coalinga oil field, and it is possible that the near future may see it the leading county of the State in its yield of petroleum.

Fresno and Madera counties have for their eastern boundaries the summit of the Sierra Range, at an average altitude of nearly 14,000 feet. Coming westward, there is the usual Sierra Nevada region of rugged heights and deep canyons, heavily timbered. Then there is the extensive foot-hill region, reaching to the plain of the great San Joaquin Valley, and drained and abundantly watered by the Kings and San Joaquin rivers and their tributaries. This fertile valley floor, sixty miles wide, is here one of the gardens of the State, and in it Fresno County proudly displays the great raisin district of America and her extensive vineyards and orchards. Madera, but one-third the size of Fresno County, finds its eastern boundary at the center of the valley. Fresno County crosses the valley and includes the arid foot-hill region of the eastern slope of the Coast Range. The resources of the counties are as varied and interesting as the topography, climate, and geological features involved in this geographical range.

The mineral features of this region are many and important, but have long been neglected, and are but now receiving active attention. In crossing the foot-hill region, these counties include the great auriferous belt of the Sierra Nevada Range, and also, lower down, its copper belt, which reaches from Fresno County away north to Nevada County and beyond.

In Mariposa County, just north of Madera, the "auriferous slate belt" of the Mother Lode region practically terminates, and there is found the southern end of that fairly continuous and well-defined mineralized fissure called the Mother Lode. This auriferous slate belt, however, splits up into arms, which extend southward into granite and other formations in Madera and Fresno counties. Each includes a series of mineralized veins. Other formations add other features to

TUNNEL ENTRANCE, FRESNO COUNTY, CAL.

the gold resources of these counties. These resources have not been much developed, or even much prospected. It may be said that this big section of the western slope of the Sierra Nevada Range is awaiting both developing and prospecting, and that the future will see many valuable gold mines worked where nothing has yet been done. Grub Gulch is one of the important districts of Madera County. There are here a number of small, high-grade veins which warrant development. Here the pay shoots are mainly in mica schist. One of the largest gold-bearing quartz veins in the State runs from Grub Gulch south into Fresno County, but few mines on it have received important development, the chief ones being the Mammoth, Starlight-Riverside, and Savanna. In the south-eastern part of Madera County are the Hildreth, Fresno, and Potter Ridge districts, Jackass Mountains, and the Minarets. There are other gold districts of note and promise in both counties.

Years ago, gold-mining operations were more extensive than now, especially in what is now Madera County. The Abbey is credited with a production of over $1,000,000. Through faulty judgment and other causes, some failures resulted, and this aided the general reac-

A TYPICAL UNDERGROUND FLASH-LIGHT, MADERA CO.

tion that attended mining for several years. The record for 1898 of $94,884 in Madera, and $27,557 in Fresno County, shows no increase of actual production, but there has been, especially in the last year or two, a marked increase of activity in prospecting and developing in several districts, and a number of new properties are coming to the front, several old ones are being re-opened, and capital is taking hold of the many promising prospects.

There is much activity in the copper belt in these counties, and several mines of importance are reaching the point of production. This belt runs through the low hills skirting the plain. In Fresno

County, the Copper King Mine has been considerably developed during the past two years. The vein is several feet wide, and assays have shown eleven per cent copper, $11 in gold, and five to eight ounces of silver. With a large amount of unworked ore on the dump, the mine was recently sold to an English syndicate, which plans a 100-ton smelter. The same syndicate has bought the Heiskell Copper Mine, on the northern boundary, which was opened fifteen years ago, but never operated. Several copper properties are being exploited in Madera County. This year, the Ne Plus Ultra brought $125,000, and a smelter is now being erected at Madera.

Petroleum has suddenly become, during 1899, the chief mineral product of Fresno County. The oil-bearing formations of the San Joaquin Valley, described elsewhere in this work by able authorities, extend across Fresno County, in its eastern part, along the Coast Range foot-hills. Oil indications are many, and, since 1863, efforts have been frequently made to develop productive wells. The first success came in 1898, in the Coalinga field, where remarkable flowing wells, yielding the lightest and most valuable fuel oil produced in California, were developed by going to sufficient depth. In the summer of 1899, the few productive wells, in an area of half a mile wide by one mile long, were reported to be yielding 60,000 barrels per month. An oil excitement has followed, with a multiplicity of companies and numerous prospecting operations, which are likely to extend the producing field at Oil City and to develop other important fields.

There is a variety of other mineral products not at all or but little utilized as yet. Madera has a famous and inexhaustible granite quarry at Raymond, producing an unsurpassed quality of gray stone. In the Minaret Mountains, near the Mono County line, is one of the largest and richest deposits of iron ore in the United States. The ore is hematite and magnetite, running in grade from sixty-four to sixty-six per cent. One vein exposes a mass 300 feet wide, 1,500 feet high, and two miles long. Lack of transportation, fuel, etc., makes it of only future economic value. Undeveloped silver mines are near. In its western part, Fresno County has extensive beds of gypsum. In that region and near Coalinga, two coal mines were extensively developed years ago. but abandoned on account of commercial conditions. Elsewhere are large deposits of graphite. Large ledges of magnesite, much chromite, limestone, bismuth, antimony, and hot springs are among the extensive mineral features of these counties which await the future.

Southern California.

OUTHERN California is a vast mineral region, with a geographical, geological, and climatic identity of its own. It is one of the great mineral regions of the United States, and, despite the fact that it is one of the oldest parts of the West historically, and notwithstanding that here gold was first discovered on the Pacific Coast, and that here the mining industry of California had its beginning, it is still in the primacy of its mineral development. It has a remarkable variety and distribution of mineral wealth, and is destined to enormously increase its mineral output and its mining population within a very few years.

The term "Southern California" generally applies to the counties to the south of the Tehachapi Mountains, which unite the Sierra Nevada and Coast ranges, enclosing the upper end of the great central valley of California, and topographically dividing the State into regions of distinct characteristics. A belt of varying width along the coast to the west of the mountains, which continue the Coast Range southward, is the remarkably salubrious, fertile, rich, and populous region which has chiefly given Southern California its fame. In this region are the oil and asphaltum fields, but, with the exception of these products, the mineral resources of Southern California chiefly lie eastward, distributed over a vividly contrasted region of rugged desert of nearly three times the area of the "Italy of America," by the coast, and larger than many States of the Union.

This region includes the greater part of San Diego, Riverside, San Bernardino, and Los Angeles counties. On the coast are Santa Barbara, Ventura, and Orange counties. Kern County is divided by the Sierra Nevada and Tehachapi mountains, and less than a quarter of it lies on the south and eastern sides of the mountains, but the most important mining district of Kern County, that of Randsburg, lies in what is termed Southern California, to which its output is credited, and this county is considered here with the others. Inyo County lies wholly to the east of the Sierra Nevada Mountains and north of Kern, but its likeness to the rest of this big, arid mineral county, and its close relationship through routes of travel, give it classification as a part of Southern California.

The importance of the mining industry as a whole, in Southern California, is shown by the record of its mineral production in 1898, when the total value of all substances mined was $6,420,672. Of this the gold product was $2,301,233, silver, $113,730; total of the precious metals, $2,414,963. The leading product was petroleum,

YELLOW ASTER MINE AND MILL, RANDSBURG.

the value of which was $2,376,420. Borax, to the value of $1,153,000, was produced, and the asphaltum product is given at $482,175. This is a large increase over 1897, and the result of a rapid annual increase during recent years. With the active development of the mineral resources now going on, and promising to long continue, there is every prospect that the future will see annual records greater than the present by several millions.

The area in which the greatly diversified mineral wealth chiefly exists is the desert region to the east of the coast mountains. It is an area of rugged mountains and bare and forbidding hills and sandy plains, divided by a series of mountainous elevations into the

Mohave and Colorado deserts. Mines are scattered all over this area, which includes a large number of recognized mining districts. The geological formations are entirely different from those in the mineral regions in the central and northern parts of the State, and continuous mineral belts of one general trend to any considerable distance are lacking. There is a wide range of the geological formations from Paleozoic to Tertiary, and a great variety of rocks of igneous origin. In this big storehouse of mineral wealth are found gold, silver, borax, iron, copper, tin, lead, sulphur, salt, soda, baryta, gypsum, asbestos, onyx, marble, antimony and a variety of other mineral products.

The climatic conditions present the most formidable difficulties to mining and prospecting. Of wood there is practically none. Streams and springs are few and far apart, as are the wells that have been developed here and there, and the intense heat of the desert makes water as precious as it is scarce. Transportation is difficult and costly. These conditions have retarded the mineral development, as active and extensive as it has been in recent years. Lines of transportation are the great need of the important mining districts scattered over these many thousands of square miles, and it is the supplying of this need over railroads and highways that will do most to cheapen production and stimulate development.

California mining may be said to have had its beginning with the discovery of gold in 1812, at the Spanish Mission of San Fernando, in Los Angeles County, according to some authorities. The early padres gathered some placer gold with the aid of Indian labor during those early years, and in 1828 some gold was shipped from San Diego. But the padres suppressed knowledge of the existence of gold as much as possible, and their production was small, and evidently ceased for many years. In 1841, gold placers were discovered in San Feliciana Canyon, about forty miles north-west of Los Angeles, and in a few weeks several hundred of the inhabitants of the coast region were in the gold fields. The production was not notable, and elicited no notice from the outside world. It was six years from this discovery to that made by James W. Marshall, far to the north, which, because of its instant and world-wide results, is properly taken as the actual beginning of California's golden era.

In the early '50s, the spread of gold hunters brought about a more extensive working of the placers in Los Angeles County, and many thousands were taken out, sometimes by winnowing the gravel in blankets. Very soon after the California rush from all over the world arrived, prospectors began pushing out into those parched desert solitudes beyond the populated coast belt, and they have been heroically traversing it during all the half century that has elapsed since then. There is not in the history of gold mining a more pictur-esque and romantic chapter than that provided by the thousands of lone and unknown searchers for gold and silver who have brought to light the mineral wealth of this remarkable region, and there is no chapter portraying more human persistency, endurance, suffering, and disaster in the search for gold. The rest of the American story of desert prospecting may be found in Arizona and New Mexico, but the tale of this feature of American mining life deals chiefly with the Colorado and Mohave deserts, and the terrible Death Valley country fur-ther north, in Inyo County, where, when the water gave out, or the way was lost, or the sandstorm came, hundreds have died deaths of agony and added white bones to the scant debris of the desert. The price of human life is still being paid in this way for the rapidly growing knowledge of the riches that Nature has hidden in these warded soli-tudes. In the conditions presented to the mining industry, this region and some of the mining regions further north offer one of the marked contrasts of which California is prolific.

Yet this is but half the truth. Men, well and safely equipped, now penetrate to any part of this region, and, while the heat of sum-mer is extreme, the air is so dry that, with a good water supply, effort is easily maintained, and labor is as efficient here as anywhere. The great need of this region is transportation. This is partly sup-plied by the Southern Pacific Railroad, which crosses the Colorado Desert; by the Santa Fe, which crosses the Mohave Desert, and by the Randsburg Railroad, which reaches twenty-six miles from Kramer into the richest and most productive district yet developed, and which, to this extent, is a great aid to several mining districts beyond Randsburg, to the north and east of the terminus. The Carson and Colorado Railroad reaches the edge of this region by its route through Nevada. In the borax industry, traction road engines have partly removed this difficulty. The lack of cheap water and fuel for power is being in part supplied by gasoline engines, and, in the future, electric power transmission lines, from the Sierra Nevada region, will further supply this want, as development creates a demand warrant-ing such investments.

The entire mineralized desert region thus spoken of in a brief and general way, is traversed by many low ranges of hills and mountains. It includes two of the lowest depressions in the land area of the world—the basin of Death Valley, in Inyo County, and that of the dry Salton "Lake," in San Diego County, each being approximately 300 feet below the level of the sea. Overlooking Death Valley, the hottest place on the American continent, is Mount Whitney, less than fifty miles away, one of the highest points on the continent, and clothed in glaciers and eternal snows.

BATTERY ROOM, YELLOW ASTER MILL, RANDSBURG.

This great area is irregularly ribbed with quartz veins occurring in various formations, presenting interesting geological studies, and is dotted with deposits of a variety of metals and economic minerals. Many of the gold-bearing ore bodies are of high grade, and the ores are more often refractory than free milling. Many are adapted to the cyanide process, and it is in this part of the State that the most cyanide plants have been installed up to this time, many being now in operation.

Speaking still of this whole desert region in a general way, it may be said that while valuable mines were developed and operated many years ago, active and general development began only three or four years ago. The discovery of the rich Randsburg district, in 1895, turned new attention to this mining country, and, supplemented by the general revival of mining, greatly stimulated prospecting, investment and development, from the Mexican border north, and from the Colorado River westward. In the past three years there have probably been more mines and prospects sold than during the whole mining history of this portion of the State. In this field, the capital that has recently gone into mining has come mainly from Southern California itself, and from the East. Eastern capital has been drawn to this field with especial success. As a result, a great many properties are being developed, and will soon be added to the list of producers. There are a number of profitable placers, many but recently discovered, but, owing to the lack of water, they have been almost, without exception, worked on a small scale by the "dry washing" process, which consists in putting the gravel through a machine somewhat resembling a fanning mill, the dirt being blown away and the gold caught by dry riffles.

In gold production, Kern County is pre-eminent among those enumerated as comprising Southern California. In 1898, its precious metal product was $1,024,473, nearly as much as all the rest of these counties, and its total mineral product was $1,129,573. This pre-eminence was due, mainly, to the development of the Randsburg district, lying in the Mohave Desert, in the south-east corner of the county, which reaches across the main mountain range, and rests in the arid region described.

This small and rich district is nearly fifty miles north-east of Mohave, at the eastern side of Kern County, close to the line of San Bernardino County. Although lying close to the old-established routes of travel to the borax fields, and the rich silver mines of Inyo County, no mineral discoveries were made until 1895, when prospectors, working at dry-washing, found croppings of rich ore. A rush followed, hundreds of claims were located, and, in 1896, Randsburg had a population of 3,000. The district is in a low range of mountains, an uplift of diorite, flanked by heavy beds of metamorphic rocks, and abounding in porphyry dikes. Quartz veins run in all directions through the diorite, and show frequent and often large shoots of rich ore. The early development operations were conducted by miners without capital, who, despite the cost of supplies, were often able to develop their claims by long shipments of rich ore. It was early said that the district was not a "deep" one, but this has been thoroughly disproved by a few mines which have been opened to depths of several hundred feet, and which have, in some cases, displayed larger and richer ore-bodies at this depth than above. The early lack of milling facilities was soon supplied by the erection of custom mills, at points within a few miles, where wells existed, and now several stamp mills reduce the ores at comparatively low cost.

The most important mine in the Randsburg district is the Yellow Aster, which includes a large group of claims, of which four are now producing through four main tunnels. One tunnel has cut twenty-

A DRY WASHER ON THE DESERT, NEAR RANDSBURG, KERN COUNTY.

six veins, the main one varying from two to forty feet in width, and avaraging $40 a ton. Large bodies of ore exceed $100 a ton in value. The depth now reached is 800 feet, and there is no sign of values ceasing or main veins pinching out. In 1898, this mine produced about $500,000, the ore being shipped to Barstow for reduction, all ore running less than $30 a ton being left on the dump. It is estimated that the ore now in sight amounts to $5,000,000. Early in 1899, a thirty-stamp mill was put into operation, and the mine's product for 1899 will show a large increase. Steam-power is used, and water is piped six miles from wells.

Other important developed mines are the Wedge, Little Butte, Kinyon and others, which have demonstrated the greatness of this district. Many promising properties remain undeveloped. The Randsburg Railroad has given accessibility and cheapening of operations to this and adjoining districts. At Barstow, on the line of the Santa Fe Railway, a fifty-stamp custom mill was put into operation in 1898, and is being a great aid to the development of the district, ores being shipped over the Randsburg Railway.

The Johannesburg district, adjoining the Rand, and the Stringer and Rademacher districts to the north, are among the other important

QUARTZ LEDGE AT THE FACE OF A DRIFT.

districts in this part of Kern County. The last two are undeveloped, but give rich promise when capital, power, water and transportation are supplied.

Kern County has great mineral riches in the San Joaquin Valley side of the mountains. There, along the slope of the Sierra Nevadas, the conditions and resources resemble those of Fresno County to the north. Gold was early found, and in the past a number of important mines have been developed. The county is here also enjoying a mining revival. It has extensive antimony mines awaiting capital, and is now the chief producer of antimony. There are great deposits of sulphur and a variety of other minerals. Important among the county's resources are oil and asphaltum, the fields lying about the head of the San Joaquin Valley.

San Bernardino County has 20,000 square miles, and more diversified mineral wealth than any other county in the State. Gold, silver, copper, borax, tin, lead, iron,.salt, soda, gypsum, antimony, sulphur, asbestos, marble and onyx are among its minerals. In its northern part are immense and inexhaustible deposits of borax, and borax was the chief item in its record of $1,644,152 in 1898. A vast deposit of high-grade iron ore, far from transportation, is among its resources which await the future. There are several important and promising gold districts, notably the Dale district, in the southern part, which, in spite of present high cost of operation, produces over $6,000 a month. Dale City is a milling center, to which ores are hauled from a number of mines within a few miles. There is a large number of other gold-producing districts, in which much activity is now displayed, and which show many valuable mines and promising prospects, but which are yet badly handicapped by the adverse conditions common to the region.

Riverside County, stretching across the desert to the south of San Bernardino County, is similar to the latter in mining conditions and mineral resources. There are a number of districts which will take leading positions in the future. Much recent prospecting has resulted in a number of good discoveries, and, although the gold product for 1898 was only $190,572, out of a total mineral product of $247,022, great possibilities exist, and the present development of a number of properties will give an early increase. Valuable coal beds exist near Elsinore, and pottery clay, marble, and other building stones are among the present products.

San Diego County, lying across the southern end of the State, has two notable features of the industry: One is the Golden Cross Mine and Mills, with 140 stamps, and the other is the salt deposits in the depressed basin of the Colorado Desert. The salt mines, which include beds deposited by evaporation in what was once an arm of the sea, produce large quantities of a fine quality of salt.

The Golden Cross group of mines is an interesting example of the successful working of large bodies of low-grade ore. The mill is yet the largest in the State. The mine is in the Cargo Muchacho district, in the eastern part of the county, about twenty miles north-west of Yuma. The occurrence of the gold is remarkable in that it does not come from veins, but mainly from a country rock, intruded by dikes of granite. Operations began in 1893. There is some high-grade ore, but most of the ore worked averages from $2 to $5 a ton. The prop-

erty has been successfully operated, but partnership disagreements brought about a receivership, which will soon end. Water is brought eighteen miles from the Colorado River. There are, throughout this region, many other extensive bodies of low-grade ore which offer similarly safe and profitable propositions to capital. This county has a number of important mines and districts, which yielded $673,996 in 1898, and they are attracting the lively attention of capital.

Inyo County has a great past record in silver and lead production, as well as in gold. In past years it has seen the production of many millions. In the days of silver mining it was the chief silver-producing county of the State, and has produced about two-thirds of the total silver product of the State. Since the fall of the price of silver, production has greatly declined.

The county is the third largest in the State, and lies wholly east of the crest of the Sierra Nevada Range. It is very arid, and it has always been and will be chiefly a mining county. It contains the famous Death Valley, and a large surrounding region difficult to traverse. The Inyo,

THE COLD-WATER ANTICLINE ON MT. CAYETANA, FROM SESPE CANYON.

Panamint, and Argus are the principal mountain ranges composing the rugged heights, which alternate with alkali plains. The county occupies a very isolated position, and has been much neglected, but it is filled with mineral riches to which a much revived attention is now being given in spite of the transportation difficulties. The total product of the county in gold and silver has been about $12,000,000, and, in 1898, it produced $137,107 in gold and $73,503 in silver. A number of properties are now being developed by mining capitalists, chiefly in the Panamint and Argus mountains. For many years, the product of borax in Death Valley has been large. It contains all the varied minerals found in the desert to the south.

Los Angeles County contains a number of small gold mines, but its chief mineral product is petroleum. Three important oil fields have been developed in this county, one in the city of Los Angeles, and the Puente and Whittier fields. Oil was discovered in the city of Los Angeles in 1893. In Los Angeles, considerably over 1,000 wells have been driven, and, in 1898, the product of this field was 1,100,000 barrels. A new extension of this field is now being developed. The Puente and Whittier fields are seeing much activity, and an increase in product. Its oil fields gave Los Angeles County a total mineral product, valued at $1,732,357, in 1898.

While the remaining counties have a number of features as mineral producers, petroleum and asphalt are the only products of present importance. The oil and asphalt resources of California are so extensively treated in accompanying articles, that little need be said here of this feature of the industry. The oil product of these counties is rapidly increasing, and the year 1899 has developed a boom in the oil fields that is causing more prospecting for oil than ever before. Santa Barbara has boundless asphalt resources that are being rapidly developed. Santa Barbara's chief oil field is about Summerland, on the ocean shore, where the oil-bearing strata are followed for hundreds of feet out under the sea. Ventura is one of the chief oil-producing counties, and the development of a valuable field at Fullerton, Orange County, has lately brought that county to the front.

This sketch of Southern California's mineral resources but feebly conveys a comprehension of the great extent of this mineral region, and of its mineral resources, mining industry, and future. It is estimated that $26,000,000 is invested in milling and water plants, and other features of mine development and operation. There are several thousand mines and prospects scattered all over this big territory, with comparatively few developed to a demonstration of their values.

The California Miners' Association.

BY

PRESIDENT J. H. NEFF.

THE gold-mining industry has probably never in any other large mining region received such effective and sustained assistance from a voluntary organization as that which the California Miners' Association has, during eight years, given to the industry in this State.

The Association has never been a technical one, concerning itself with the science and practices of mining; nor a commercial and financial one, concerning itself with mining properties and investments; nor has it had anything to do with labor problems. Most other mining organizations of the world have been formed for these and similar purposes. This Association has, from its beginning, existed mainly for the purpose of securing legislation in behalf of the whole mining industry, and otherwise promoting and protecting the legal rights and privileges of miners. In many other matters of general importance, it has represented the whole mining interest and expressed its policy, and it was the first association organized in the United States, in the general interest of the mining industry.

Representing the concentrated influence of a great part of the State's population interested in mining, the Association has made that influence strongly felt in the Halls of Legislation. Through being composed of men representative of the best ability and character in the California mining field, adopting conservative measures and conciliatory methods, and working with energy to secure its ends, it has maintained the confidence and respect of all people and interests, and been able to do much for the industry. It has maintained its vitality for eight years, and is to-day stronger in membership and influence than ever before.

Besides its legislative work, the Association has been a factor in bringing about the present revival of interest in mining in California,

J. H. Neff, President California Miners' Association.

and in creating the renewal of a wide-spread public sentiment favorable to it. This has, perhaps, been its most valuable service. This change has, of course, been due largely to the world's general turning to the gold fields in recent years, a movement in which California has naturally shared; but the existence and work of the California Miners' Association has unquestionably contributed largely to this new and satisfactory condition. Since 1892 there can be observed a disappearance of prejudices against gold mining as a field for investment, and a growing public appreciation of the fact that the mining interests are as worthy of the fostering care of the State as the interests of agriculture and horticulture. There has been shown in quite recent years a greater readiness to legislate in the interest of the miners, and to concede appropriations asked in behalf of the industry, and the Press of the State has almost uniformly shown a larger and more kindly interest in this field.

In a general statement of what the Association has accomplished, a prominent place should be given to the present national movement in behalf of an Executive Department of Mines and Mining in the Federal Government. Very early in its career, it formulated what has appeared to be the first generally noticed demand for this recognition of the American mining industry, and since then it has, by committee work, printed arguments and Congressional efforts, played a leading part in its advocacy at Washington, at various sessions of the Trans-Mississippi Congress, the general mining conventions at Denver, and in the Press.

The success of the Association has been partly due to the representative basis upon which it is organized. It is a federation of seventeen local miners' associations, of which sixteen are county organizations, while one, recently organized, represents the combined interests of Southern California. The counties in which these affiliating associations exist are: Shasta, Trinity, Siskiyou, Butte, Yuba, Plumas, Nevada, Sierra, Placer, El Dorado, Amador, Tuolumne, Calaveras, Santa Clara, Alameda, and San Francisco. The total membership is about 9,000. The members of these associations are also members of the California Miners' Association. The first thirteen of these counties comprise the regions of greatest mining population and activity in the northern part of the State, and along the Mother Lode. Santa Clara is the leading producer of quicksilver.

The support of the counties of San Francisco and Alameda is due to the fact that they contain a large and increasing number of men who have money invested in mining properties, or whose business interests are bound up with the industry and depend on its general prosperity. This is especially the case in San Francisco, the center of the manufacturing, commercial and financial operations of the Pacific Coast mining field, a city founded by the mining industry and one whose prosperity is largely dependent on it. In San Francisco, the Association finds its strongest financial and moral support, as heartily and freely as is such support given by the leading mining counties.

The Southern California Miners' Association was organized in April, 1899, and is part of the State organization which has thus received new strength. The present active and efficient Secretary of the Association, Mr. E. H. Benjamin, expects success in organizing associations in other counties, in which efforts many of the members of the Executive Committee are interested.

The membership of the county organizations is not limited to any class of miners, or to mining men at all. Mine owners, mine superintendents and working miners, who are benefited by the general prosperity of the industry, make up a majority of the membership in the mining counties. Besides these, business and professional men, and men of any class who are concerned in the prosperity of localities which are dependent mainly on mining, give the Association the support of their memberships and dues. These county associations, through their conventions and executive committees, are aids to the purely local as well as the general interests of the industry.

The San Francisco Association is made up of leading mine owners, manufacturers, business and professional men, and others whose interests are, directly or indirectly, mainly in the mining field. The Association thus represents a large proportion of the mining population of a mining State, and includes the most energetic and intelligent members of it. The Association in no way concerns itself with individual or business interests, but gives unity and energy to the elements of the State's population described.

The only financial condition of membership in the mining counties is the payment of $1 a year dues. Of this, seventy-five per cent goes to the Treasury of the State Association. In the San Francisco Association, the dues range from $5 a year for individual membership to $25 and $50 for mining companies and business concerns. By this plan, the State Association receives several thousand dollars a year for its expenses, and this regular income is supplemented by

W. C. RALSTON,
VICE-PRESIDENT CALIFORNIA MINERS' ASSOCIATION.

voluntary contributions to large amounts, always successfully solicited by finance committees when special needs arise.

The county organizations hold annual conventions, at which are elected delegates to the annual State Convention of the Association. The basis of representation provides a State Convention of several hundred members, and these conventions have always been fully attended and composed of intelligent and enterprising men. The State Convention usually lasts three days. At these conventions, the chief problems concerning the miners of the State, which are within the scope of the Association, are freely discussed; reports of committees on the work of the Executive Committee during the year are acted upon; the policy of the Association on various matters, to be followed by the Executive Committee, is declared, and special features add variety and interest to the proceedings. These conventions are animated by a loyalty to the mining industry, and serve to stimulate such loyalty and interest, and give a sense of strength and solidarity to the mining people of the State.

The Annual Convention expresses the wants and policies of the miners concerning legislation, etc. The direct work of the Association during the year is accomplished by the Executive Committee, composed of two members from each county, chosen by the county delegations, and of a membership at large, appointed by the President. Special committees, on special lines of work, are appointed from the Executive Committee. The Executive Committee has large discretion, and, at the several meetings held each year, takes such action, in the name of the Association, as circumstances require. The officers and Executive Committee for 1899 are as follows: Jacob H. Neff, President; William C. Ralston, Vice-President; S. J. Hendy, Treasurer; Edward H. Benjamin, Secretary.

Executive Committee: At Large—W. W. Montague, San Francisco; A. Caminetti, Amador County; J. M. Walling, Nevada County; Thomas Mein, San Francisco; Professor S. B. Christy, Berkeley; John M. Wright, San Francisco; Harold T. Power, Placer County; Dan T. Cole, San Francisco; Lewis T. Wright, Shasta County; Joseph Sloss, John Birmingham, C. W. Cross, George Stone, Julian Sonntag, W. S. Keyes, E. A. Belcher, Andrew Carrigan, Edward Coleman, Curtis H. Lindley, Charles G. Yale, Lewis F. Byington, San Francisco.

Alameda County—Frank A. Leach, Felix Chappelet; Amador County—J. F. Parks, E. C. Voorhies; Butte County— A. Ekman, W. P. Hammon; Calaveras County — W. L. Honnold, David McClure;

El Dorado County—A. H. Ten Broeck, H. E. Picket; Nevada County—B. S. Rector, James McBride; Placer County—T. J. Nichols, Amos Stevens; Santa Clara County—Charles C. Derby, R. R. Bulmore; San Francisco—J. F. Halloran, S. Mooney; Sierra County—Frank R. Wehe, J. O. Jones; Shasta County—M. E. Dittmar, J. M. Gleaves; Siskiyou County—H. H. Hunter, Andrew G. Myers; Tuolumne County—J. G. Hopper, W. G. Long; Trinity County—John McMurray, P. Paulsen; Yuba County—Joseph Durfee, James O'Brien.

To briefly and further sum up the work of the Association, it may be recalled that it conceived and successfully labored for the "Caminetti Law" of 1893, providing for the partial resumption of hydraulic mining in California under Federal control. It has secured appropriations of $250,000 each from the Federal and State governments for restraining dams to confine mining debris in the upper courses of the navigable streams, and, along with these partial results, it has created a public sentiment favorable to the resumption of hydraulic mining under restrictions that will protect fully all other interests, and has secured governmental interest in and recognition of hydraulic mining that promises to result in the much larger measure of success hoped for. It has been largely, if not mainly, instrumental in securing a reversal of an oppressive policy of the Land Department toward the mining interests in the matter of public mineral lands, and, after years of struggle, it has apparently reached the eve of success in its efforts for Congressional legislation, saving to the miners vast areas of mineral lands in this State, which lands have been steadily absorbed by corporations. It has, through the efforts of its Committees on Legislation, secured the passage by the Legislature of a number of acts in the interest of mining. It has given its co-operation to such mining enterprises as Exposition Mining Exhibits, and the Jubilee and Mining Fair of 1898, celebrating the fiftieth anniversary of Marshall's discovery. It has heartily joined in the work of the Trans-Mississippi Congresses and the International Mining Congresses at Denver, and its representatives have taken leading parts in those representative gatherings.

The California Miners' Association had its origin in a condition of the gold-mining industry peculiar to California. The earliest motive of its formation, eight years ago, was a determination to make a concerted effort to do something to revive the hydraulic mining industry, after it had been prostrated for years. The story of hydraulic mining is an important chapter in the industrial history of the State, but, as it is ably outlined elsewhere in this book, the story need not be told here in detail, except as it directly relates to the history of the Association. The story begins with the birth of hydraulic mining in Nevada County in the '50s; it includes its steady development through the years until the decade of the '80s began, when over four hundred mines in the Sierra Nevada region were producing several million dollars annually, and represented the investment of $100,000,000, and it includes the long and bitter struggle with the agricultural interests of the valley.

During the decade of the '60s, the debris from the hydraulic operations in the mountains, gradually borne to the lower courses of the streams, attracted serious attention, and, as the deposits began to work very serious injury in shoaling the navigable portions of the rivers, and by being deposited on low-lying agricultural lands near them, the issue became a live one, and drew increasingly formidable opposition to hydraulic mining. In the '70s, much study was given to the problem, the solution of which was thus early declared by engineers and others to be in dams, which would confine the material coming from hydraulic mines, supplemented by natural erosion, to the upper courses of the streams.

During the decade last mentioned, hydraulic mining increased, as did the opposition to it, and, by 1880–82, it had reached the zenith of its glory, while the opposition to it attained its greatest power in the Anti-Debris Association of the Sacramento Valley, which was liberally supported for several years by county appropriations and private subscriptions. The miners recognized the fact that injury was being done, but pointed out the overwhelming importance of the industry, and asked that the injury be prevented without sacrificing enormous property interests and the chief source of the prosperity of populous mining regions. Between 1878 and 1882, the State Engineer and a Board of Federal Engineers made detailed investigations and reports. It may be mentioned, as illustrating the importance of the issue, and of the first cause of this Association, that the Federal engineers estimated that, by spending $500,000, there could be $335,000,000 extracted by hydraulic mining in the region considered, without damage to the valley interests. Years before this, Raymond, in "Mineral Resources of the United States," had estimated that, in California as a whole, from $1,000,000,000 to $1,500,000,000 could be produced by hydraulic mining. The engineers mentioned estimated the damage to agricultural lands at $3,294,035. These

EDWARD H. BENJAMIN,
SECRETARY CALIFORNIA MINERS' ASSOCIATION.

engineers were among those who recommended the construction of restraining barriers as a sufficient protection to the valleys.

When the culmination of the struggle against hydraulic mining came with the celebrated "Sawyer Decision," January, 1883, in which the North Bloomfield Company was restrained from depositing tailings in a tributary of a navigable stream, hydraulic mining was the dominant feature of gold mining in California, and its future held possibilities of further expansion, were it not for the debris problem which stood in its way, the solution of which the miners were unable to secure.

The result of the Sawyer decision, through the many other suits filed, was to close at once practically every hydraulic mine in Nevada, Plumas, Sierra, and adjoining counties, and in the Mother Lode counties to the south. They remained closed for years, and the industry practically remains prostrate yet. The effect on the mining counties may be illustrated by the statement that, in the ten years from 1880 to 1890, the assessed valuation of property in Nevada County shrank $10,000,000, population dwindled, and the rate of taxation doubled. Many mining towns were practically deserted, highways showed little life, and, as interest in mining of every form was at an ebb, there was a long season of complete depression.

During the long period, while the mining regions were suffering from this enormous sacrifice of property interests, and from this general inactivity in the mining field, together with a general popular prejudice against gold mining, accompanying the ascendency of the agricultural and horticultural interests, the miners realized that hydraulic mining could never be resumed except under restrictions that would fully protect other interests. They bowed, with few exceptions, to the decisions of the courts without question, and asked and hoped only that a just and legal way be opened for a partial return of the prosperity they had lost. Soon after the Sawyer decision, suggestions that the mining and agricultural interests should agree on a plan that would be of common benefit, by the improvement of the lower portions of the rivers and the restraining of further deposits above, were made and for some years they were put forth at intervals. But the hostility engendered by the past struggle remained strong and uncompromising; every such suggestion was vigorously cried down by press and people of the Sacramento Valley, and as nothing could be accomplished for the mining interests, in the face of opposition from a large section of the State, the outlook remained hopeless to the miners, and compromise was impracticable. During these years there was much discussion of the situation. The Legislature again recommended the problem to Congress; Governor Perkins and then Governor Markham recommended in messages that efforts be made to do something for hydraulic mining, and, in 1888, Congress provided, by an act introduced by Congressman Marion Biggs, for a second investigation by a Board of Federal Engineers, which board was known as the Biggs Commission.

In also another direction, the industry was suffering from disfavor quite in harmony with the rest of its troubles. In its construction of the laws governing the disposition of public lands and the acquirements of mining rights, the Interior Department, through the Land Commissioner, had adopted a policy hostile to the miners, whose interests and legal rights were sacrificed in the interest of railroad corporations and other claimants to agricultural lands. It was the existing departmental ruling that public land should be presumed to be agricultural and not mineral, unless, not only an actual but a profitable production of mineral had been made in the tract in question, and the burden of proof was placed on the protesting mineral claimant to show that it was chiefly valuable as mineral land. A great portion of the public domain in the undeveloped mineral regions of the Western States, which was in intent and fact reserved to the mining industry by Congress, was thus rapidly passing by patent to aided railroad companies, and away from the prospectors and miners. This worked an especial wrong in northern and eastern California, where lie long stretches of the rich lava-buried beds of ancient rivers, peculiar to this State, awaiting prospecting operations that must come slowly and expensively. Yet, under this ruling, a section of public land in a known drift-mining region, and in general line with a known buried river channel, would be classed as non-mineral unless gold had been actually and profitably mined from it.

This was the situation in the fall of 1891. There had been no concerted efforts by the hydraulic mining interests to secure the aid and legal protection they wanted, nor to back up individuals who were interesting themselves in the problem, and there were no special indications that the old hostile force was less strong and uncompromising than it had been in the past. But it was soon to be shown that time had worked softening changes on both sides af the controversy, and that the time had come when it was possible for both sides to counsel together for their common benefit.

In November, 1891, a few enterprising men of Auburn, Placer County, conceived the plan of getting the miners of that county to form an organization which should expand into a State Association, to do whatever might be possible to revive hydraulic mining.   A few leading men met in a preliminary conference at Auburn, November 18th, and the sentiments of those present, and of those who wrote in response to a circular call, appeared to warrant proceeding with the plan. Resolutions were adopted reciting the grievances of the miners, declaring that the time was ripe for those interested in the industry to organize for defense, and calling a convention at Auburn, November 28th, to make the concentrated voices of the miners heard.

The writer had the honor to be the chairman of that first, largely attended convention, as he had been of the preliminary conference, and he well remembers what an earnest but conservative body it was. The preamble and resolutions, adopted after strong but harmonious discussion, explain well the spirit and purposes attending the birth of the California Miners' Association.   The preamble declared that the apparent conflict between the mining and agricultural interests, in reference to hydraulic mining, had resulted in the confiscation of property and the annihilation of a great industry; that the conflict might be adjusted by a wise and conservative harmonizing of the interests affected, a practicable solution of the problem having been shown by Government engineers, and it set forth the hostile policy of the Interior Department regarding mineral lands.

It was resolved that a State Convention be called for the purpose of securing, through our Congressional Representatives, appropriations for the construction of the restraining barriers, long recommended by Federal engineers, together with enactments legalizing the hydraulic mining industry, which then had the protection of no statute law. Two other purposes were declared: To secure such modification of the mining laws as would prevent the Interior Department from practically nullifying the intent of Congress in enacting them; "and to secure sufficient appropriations to thoroughly dredge and otherwise improve the Sacramento, San Joaquin, and Feather rivers," and it was declared that the general prosperity demanded a speedy settlement of the pending issue.

These three purposes animated the call for the State Convention that resulted.   It was to make a strong, definite, and concerted move to secure the only means by which hydraulic mining could be partly resumed; to secure, at thé same time, the improvement of the navigable portions of the streams involved, and to ask the valley interests to join in a harmonious effort for the common good, and, third, to secure changes in the mining and land laws, and their interpretation—a matter in which other States were also interested.

The Auburn Convention had resolved itself into the Placer County Miners' Association. A strong address to the people of the State accompanied the call for the State Convention, and it won a quick and unexpectedly cordial response.

Other counties held conventions to elect delegates, and many were appointed on invitation of the call. There were thirty-five counties represented in the large body that formed the first State Convention in San Francisco, January 20th, 1892. Representatives of the valley interests had been invited, and several leading ones came, yet doubtful of the policy they should pursue. A new sense of strength and hope and energy filled the convention. The policy, declared in a conciliatory spirit, quickly won the approval and co-operation of the valley representatives, and from that day the opposition of the valley interests to the miners ceased to be general and formidable.

The convention adopted strong resolutions in line with those adopted at Auburn, and declared a fourth purpose in providing for a committee "to formulate and promote amendments to the mining statutes of the United States." There was adopted a memorial to Congress, and a committee was directed sent to Washington to promote the desired legislation, the main purpose being to secure appropriations for the rivers in the joint interests of the miners and farmers. It was resolved to form a permanent association "for the purpose of promoting the mining industry in California," and it was recommended that associations be formed in the mining counties.

Such was the situation out of which the California Miners' Association sprang, and such was the manner of its birth. Hydraulic mining furnished the chief of its springs of life, but when the representative miners of the State got together with energy and common purposes, they quickly broadened the scope of their organization and it has steadily broadened ever since.

The subsequent story of the Association will be sketched more briefly. The first result of the successful convention was a quick and complete revolution in public sentiment throughout the State toward the hydraulic miners, and toward the industry in general. The miners had shown strength, enterprise and a spirit of self-help; they had declared a policy that was plainly for the common good and that won ready admiration and support; and their fellow citizens of the valley withheld the blows that had been readily given. Leading newspapers quickly gave prominence and approval to the miners' cause. Boards of Supervisors and commercial bodies throughout the State adopted resolutions supporting the miners' programme. This was especially true of San Francisco, the Chamber of Commerce and Board of Trade early indorsing the movement At the very start, the Association acquired a prestige and influence that it has never lost.

The Executive Committee, created by the plan of organization that has been adhered to ever since, at once set about carrying out the will of the first convention. Committees on Finance and Legislation were among those appointed. The latter proceeded to form a Congressional bill, embodying the recommendations of the Federal engineers, and the Caminetti law enacted a year later was upon the lines then agreed upon. The Finance Committee found spontaneous support. The Placer County Association quickly grew to a membership of 600, and the Placer County Supervisors appropriated $1,000, of which $300 went to the River and Harbor Improvement Association of the valley. Nevada County Supervisors appropriated $1,000, and the County Miners' Association raised the same amount. Other counties gave strong support. The San Francisco Board of Trade voted $1,000 to the California Miners' Association, and other contributions were liberal. With this encouragement, a committee, consisting of Niles Searles, J. K. Luttrell, Robert McMurray and J. B. Hobson, was sent to Washington, where two bills relating to hydraulic mining had been introduced by Congressman A. Caminetti. The labors of the committee and of the delegation, and the substantial harmony with the representatives of the River and Harbor Convention, did not achieve success that session.

When the second convention of the California Miners' Association met in San Francisco, November 15, 1892, there was the same enthusiasm and determination, and the outlook was fairly bright for the success of the hydraulic mining measures, which were made temporarily the chief business of the Association. There had $14,322 come into the treasury, of which San Francisco gave $7,405, and of which $10,631 paid the expenses of the workers at Washington, who gave their services gratuitously. The story of the long struggle for the Caminetti law, and for subsequent legislation, need not be recounted, but it is one of energetic and persistent action, long sustained, and aided by men who cannot be given credit here. It is sufficient to say, that the second convention provided for carrying on the campaign which resulted in the passage, in March, 1893, of what is known as the Caminetti Act, which is described elsewhere, and which was passed without carrying appropriations, the amounts of which, asked for the rivers below and the dams above, were $1,670,000. The act as passed was a compromise measure. Since then, the Association has concerned itself with amendments and with securing appropriations to make it fairly effective as an aid to the industry. Meantime, an appropriation of $250,000 for the construction of restraining dams was secured from the State Legislature, contingent on the appropriation of a like amount by Congress. Such an appropriation was secured in 1895, but the whole amount remains unexpended, the Federal engineers composing the California Debris Commission being still engaged in surveys of sites for the dams.

The second convention adopted a memorial to Congress on the mineral lands question, and the third likewise took action, but the matter was not energetically taken up until 1894, when it was given into the hands of a committee, of which A. H. Ricketts was chairman. This committee did valiant work. Mr. Ricketts argued the issue with President Cleveland, the Secretary of the Interior, and the Commissioner of the General Land Office, and, during 1895, protests against the patenting of 6,000,000 acres of land to the railroad were filed in the name of a member of the Association. In 1895, a bill providing for the classification of public lands in this State, of which 7,000,000 acres were estimated to be mineral, and to be properly the heritage of the prospector and miner, were introduced in Congress. During a long and persistent campaign for this measure, an appeal to the people of the State was issued, maps and detailed investigation were made under Mr. Ricketts' direction, appeals and briefs were filed at Washington, and much labor here and in Washington was expended in efforts to secure for California legislation similar to that classifying the public lands of Montana and Idaho. But the protests were dismissed, corporate influences remained powerful, and, so far, the bill, pressed at each session of Congress since 1894, remains pending, and the patenting of mineral land to the railroad goes on.

Each year the annual convention, and then the Executive Committee, have had in hand actively these prominent issues, as they are likely to have for some time to come. An initial appropriation has been secured in the interest of hydraulic mining, and the Government is committed to its aid. When the appropriation is expended, further appropriations will be needed, and they will be labored for. The prospects for the mineral lands bill seemed to promise success at the coming session, but this campaign will be resolutely kept up to the end, and to this issue the mining revival has given greater importance throughout the State.

Several conventions have discussed the Federal mining law of 1872, which, in itself and through constructions of the Court, is incomplete and in need of various amendments, and have called for the reconstruction and codification of the mining laws. This matter is to be one of the most prominent ones before the convention of 1899. One of the ends steadily pursued by the Association has been the creation of a Department of Mines in the Federal Government, as already mentioned, and this will also remain one of the purposes of the Association.

During these years, the persistent labors of the Association have been performed by members working unselfishly, loyally and without financial recompense for the good of the mining industry. It would be impossible to give due credit here to all, and no effort has been made to that end. They are well known to those for whom they have labored. In ways not here described, the Association has, year by year, represented the whole industry, and extended the fields of its interest and efforts. The San Francisco meeting of the American Institute of Mining Engineers, and the account of California's mineral industry in which this sketch appears, are among the evidences of the scope of its interests and usefulness. The Association has recently seen the industrial soil, in which it planted a revived public interest, well watered by streams of the new life and enterprise in the mining world; and, in the greater future which we have faith is at hand for the California miner, the Association cannot fail to exhibit a still keener life, and a further expansion of its membership, work and power.

The State Mining Bureau.

OR nearly twenty years the State Mining Bureau, of California, has been a great aid to the mining industry in this State, and its benefits have been widely shared outside of the State. In theory, it is an institution in the working of which the practical shall predominate over the purely scientific. The work of the Bureau has conformed quite well to this theory, and among the similar institutions which the State of California has established to foster particular industries, it has probably maintained the greatest degree of popularity and consequent readiness to give it liberal public maintenance.

It was established by the Legislature in 1880. It was the result of an energetic agitation, conducted by Joseph Wasson, then representing Inyo and Mono counties in the Assembly, in behalf of a practical institution to gather and disseminate information about the mineral resources of the State, and to otherwise aid the mining industry in a practical way.

Henry G. Hanks, a man of high scientific attainments and much practical experience in the mining field, became the first State Mineralogist, and held the office for six years. From a small beginning, the work and quarters of the Bureau steadily enlarged.

During these years, several men of ability have, as State Mineralogists, successfully carried on the work on the general lines first laid down. Some years ago the institution was put under the direction of five trustees, who have a very general supervision of the executive work of the State Mineralogist. For several years the Bureau has received, as it does now, $25,000 a year from the State, exclusive of provision for publications. Though designed chiefly as a practical aid to the industry, it has never performed paternal functions to the extent displayed by similar governmental institutions in some of the British Colonies and elsewhere.

Its publications constitute one service known and appreciated the world over. Ten annual reports were published from 1880 to 1890, and since 1890, three biennial reports have gone forth, the last being

Museum of the State Mining Bureau, Ferry Building, San Francisco.

that for 1895-96.   These reports give a mass of detailed descriptive information about the mineral resources of every county in the State. Expert men have each year been kept exploring the mining field, and many special investigations have been made, and special papers prepared by some of the most eminent mining engineers, geologists and metallurgists of the State.   Besides these reports, some of which are quite voluminous, a number of special bulletins have been published. Some of them have attracted wide attention, notably "Methods of Mine Timbering," by W. H. Storms, 1894; "Gas and Petroleum Yielding Formations of the Central Valley of California," by W. L. Watts, 1894; "California Gold-Milling Practices," by E. B. Preston, 1895: "Mine Drainage, Pumps, etc.," by Hans C. Behr, 1896; "The Oil-Yielding Formations of Los Angeles, Ventura, and Santa Barbara Counties," by W. L. Watts, 1897.   The work of the Bureau includes complete statistics of the mineral products of the State, and these have been published in special bulletins.   All of the reports, but those of 1891–92 and 1895–96, are out of print and command high prices in the second-hand book market.   Most of the bulletins are exhausted.

A very important function of the Bureau has long been the giving of practical information to mining men.   In accordance with existing law, it is required to furnish free qualitative analyses of all mineral specimens submitted from this State, and there is a well-equipped laboratory for this purpose.   During eight months of 1899, over 600 such analyses were made.   About 300 letters of inquiry are monthly answered, and hundreds of personal inquiries are gladly and carefully replied to.

During 1899, the Bureau was installed in large and permanent quarters in the Ferry Building, San Francisco.   Its valuable library includes 6,000 volumes, and its extensive and interesting museum has a growing collection of 12,000 specimens, including every mineral known to occur in California.   Much active field work will be pursued during the coming two years, and the publications soon to be issued include a Register of Mines, giving by counties, in tabulated form, descriptions of every mining property now worthy of mention in the State, and accompanied by elaborate topographical and geographical mining maps.   There will soon be issued bulletins on the oil resources of the State, with maps, the geology of the Randsburg district, etc.   Publications are free, for postage, to residents of the State, and charged for to others.

California as a Field for Mining Capital.

BY

W. C. RALSTON.

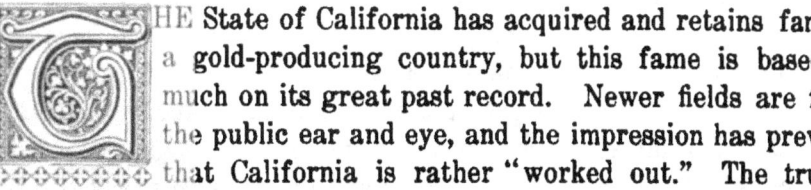HE State of California has acquired and retains fame as a gold-producing country, but this fame is based too much on its great past record. Newer fields are filling the public ear and eye, and the impression has prevailed that California is rather "worked out." The truth is that not only is California not worked out, but, when the extent of its unworked resources is considered, the declaration that has been 'made by many intelligent mining men, that "the surface has only been scratched," may be truthfully repeated.

This State has, since 1849, produced over $1,300,000,000 in gold, a record which no other mining region can show. In one of its early years its product was over $80,000,000, and to-day it is about $16,000,000. It is yet producing more than the Klondike. The comparative falling off of its product in recent years, coupled with the late general depression in the industry to which this was greatly due, has been largely the cause of, the impression spoken of. But this falling off loses its apparent significance when it is remembered that fully three-quarters of the vast total product has come from placer mines. It is the early placers that have been about worked out. The ledges which yielded the placer gold to the streams by erosion still reach down into the rocks from the present surface. These are not worked out. Of the total gold product of 1898, given at $15,906,478, but $1,841,473 came from surface placers, nearly $2,000,000 from ancient gravel deposits, worked as hydraulic and drift mines, and $12,488,321 from quartz mines. It is the lode mines of California which are now producing the gold, and yet the quartz gold deposits, continuously ribbing the State from its northern to its southern boundary, may be said to have been yet hardly prospected. It is particularly to quartz mining in California that the attention of mining investors is invited. Careful investigation will reveal to an intelligent and unprejudiced investigator that there are here opportunities for judicious investment, which those of no other large mining country in the world can rival in number and value. One of the leading mining engineers of this country recently stated, after visiting most of the mining regions of the world, that he considered California the best field for mining he had visited.

There are two classes of mining propositions to be encountered in California. Mere prospects, and but slightly developed mines, to be had for comparatively small initial investments, constitute one class. Of these there are thousands. Many of them have apparent merit, making them worthy of attention. The revived mining activity of the present has greatly stimulated prospecting, and the surface development of claims by owners of small capital, and the number is rapidly increasing. During the first eight months of the year 1899, from 300 to 500 mine location notices have been filed in some of the mining counties. Many of these claims and unimproved mines will be valuable mines in the future, and a few will be rich ones. They are spoken of here mainly to illustrate the possibilities of the State's mining future. They are rather favorite investments for Californians, but they are not the sort of propositions which eastern and foreign investors usually seek.

The other class of investments is made up of "going" mines, and of mines that have been developed to a demonstration of what there is in them, but which, for lack of capital or other reasons, are either idle or not worked in the most profitable way. The millions that eastern and foreign capitalists have been putting into this field lately have principally gone into such properties, and mainly with success. There is a great number of such investments open to capital, constituting as safe and profitable ones as can be found. Regarding such propositions, the question is often asked, Why are they offered outside of California if they are such good and promising properties? The answer is that the number of such properties is so great, and the field is so large, that there is not enough mining capital in California to occupy the field. California capital is going rapidly into California mines. Many millions are invested in California mines by Californians. Hundreds of new mining companies are incorporated in this State yearly. Several millions of San Francisco capital alone have been put into mining propositions in the last three years. But waiting opportunities are so many that no one State can supply the capital needed to embrace the opportunities offered.

In California, which is different from most other mining States, we have two classes of mines: first, those with small ledges of high-grade ore, and, second, those with large ledges of low-grade ore. So far, outside of Grass Valley, which has been a phenomenal district, the low-grade mines have been more universally good in depth than the others. With a large, low-grade mine, it is simply a manufacturing proposition. The question is, How much will it cost to extract the gold in the ore, and will working the ore on a wholesale plan give a profit? The opening and equipping of quartz mines in California has arrived at a point which makes necessary large capital, and a length of time that holds out little inducement to the average Californian, who generally looks for quick returns. For this and other reasons, we are obliged to look to the East and to Europe for necessary capital, and largely to people who are looking for twelve and fifteen per cent on the money invested, and who are willing to wait for a time sufficient to thoroughly explore and develop mines, before proceeding to equip them.

The development of large bodies of low-grade ore on a large scale is proving especially attractive to mining capitalists in recent years. Many of these propositions are mines abandoned years ago, when labor and supplies were much higher than to-day, and closed because they could not be made to pay. The history of the re-opening of old mines in California is so well known that it is hardly necessary to review it here. Probably the best example to be cited is that of the Utica Mine, at Angels Camp. In the report made, in 1867, by the Gold Commissioner of the Treasury Department, the Utica Mine was stated to have been worked to a depth of 100 feet, and then abandoned. This mine never did pay, until it was equipped with two sixty-stamp mills, and the ore was worked the whole width of the vein regardless of any pay streak in it. At one time, a few years ago, the Utica Company worked one stope over ninety feet in width, the ore being of high grade, averaging over $16 a ton. The average value of the ore worked in the Utica is, however, about $3.

The re-opening of old mines on the Mother Lode has, in recent years, demonstrated the depth and permanency of values in this lode, and this has greatly stimulated this re-opening of old mines at greater depth, and the development of new ones. In the Mother

Lode, which is probably to-day the best known mining region of the world, not excepting the Rand, no depth has yet been reached at which ore ceases. The Kennedy Mine, in Amador County, is now being worked on the 2,300-foot level, which is over 800 feet below the sea. Several mines are paying handsomely at what used to be considered great depths. The cheapening of production and the demonstration of deep values have thus brought about investments of large amounts of money in mines which have been, or are being, developed into great ones.

The cost of mining and milling in California, speaking now especially of the Mother Lode, which probably offers the greatest inducements to capital, is from $2.50 to $3 per ton, depending upon the size of the ore-body and the cost of timbering, which is the greatest item of cost in that region. Strange to say, the advancement which has been made in California in the cheapening of the cost of working gold ores, has been almost entirely mechanical—in other words, metallurgy has cut no figure. In milling low-grade ores, the old theory of endeavoring to amalgamate in the battery has been done away with to a great extent. Coarser screens are now being used, and mill men have increased the crushing capacity per day in consequence, from two and one-half and three tons to five and five and one-half tons. Mill men believe now in increased crushing capacity, depending more upon the outside plates and the concentrators. This is mentioned because many people have the idea that handling low-grade ores has been made more profitable mainly through the scientific advancement of metallurgy, whereas, we are to-day working ores, in a majority of the mines in California, in the same manner, from a metallurgical standpoint, as twenty-five years ago.

There are other fields for mining capital in this State. The deep gravel channels, on which much has been written by prominent engineers, have been barely prospected, for the reason that it requires capital to run long tunnels to reach the channels before pay gravel can be taken out. One marked improvement in prospecting these channels has been made in late years, and that is the use of ordinary oil-well boring machines, by which the channel can be located and prospected in a very satisfactory manner. It has been estimated that there are over 400 miles of these ancient, buried channels in California, containing many hundred millions in gold, and when the comparatively small proportion that has been even prospected, and the successes of many drift mines, noted elsewhere in this book, are considered, the vast possibilities of drift-mining here, and the extent of the field open to capital, may be realized.

While the rich surface placers worked in early days have ceased to yield large and quick returns, there are thousands of square miles of auriferous gravels, besides those of the buried channels referred to. In Northern California great areas of such placer ground may be freely worked by the hydraulic process, and throughout the Sierra Nevada region, which was the scene of the great hydraulic mining operations of past years, hundreds of mines will be profitably operated by this process in the future. Many placer deposits, which cannot, for various reasons, be worked in any other way, are available to gold dredges. This is a safe and inviting field, and this form of the industry is just now assuming large importance.

The economic minerals afford extensive opportunities, as can be realized by even a cursory study of the wonderful extent and variety of California's mineral resources. This is especially true of copper and petroleum, which are now receiving so much attention in this State. Yet, the extent of these resources is only partly known, and will appear much greater in the future. Northern California has been but slightly prospected, and the same may be said of Southern California.

Capital will find in California a less tendency to charge exhorbitant prices for small prospects than is generally the case in mining districts, and, in almost all cases, an investor is given ample time for proving a mine before purchase.

I do not assert that California is a richer mineral country than many other States of the Union, but that, given the same value per ton of ore, we can work it cheaper in California than in almost any other mining region of the world, partly because of climatic conditions so greatly in our favor. All California needs is an intelligent investigation by those seeking mining investments.

Quicksilver in California.

NE of California's chief mineral products, ever since early days, has been quicksilver. Production began before 1850, and since that date this State has produced all of the quicksilver mined in the United States, with the exception of a total of perhaps 300 flasks, produced in Oregon, Utah, and Texas. The California product, since 1850, has amounted to 1,770,000 flasks. Mainly because of the price of quicksilver, the production is not now as large as in some former years. In 1898, it amounted to 31,092 flasks, valued at $1,188,626. For many years quicksilver was second only to gold in the California record. Now it has been passed by copper and petroleum.

Ever since 1850, California has been one of the few great quicksilver fields of the world. Between 1850 and 1886, it produced more than any other district, including the world's oldest and richest one, that of Almaden, in Spain. Production in Almaden began several centuries before the Christian Era, in Idria and Huancavelica in the 16th century, and in California in 1850. Yet, in 1886, California's total product of 1,429,346 flasks was estimated to be nearly equal to that of Idria, in Austria, and Huancavelica, in Peru, while Almaden's total was placed at 3,965,812 flasks. This State was then producing about one-half the world's supply.

Cinnabar occurs in the coast range of the State, practically throughout its length, and several very rich and extensive deposits have been found. It has been discovered in most of the counties along or near the coast, and has been produced in several of them. Once there were thirty mines working, and about forty altogether have been producers. Now there are but ten producing mines of importance. In 1898, the product came almost entirely from the following sources: Santa Clara County—New Almaden Mine, 5,875 flasks; Napa County—Napa Consolidated, 6,850 flasks, Aetna, 3,450 flasks, Redington, 990 flasks; Lake County—Great Western, 1,128 flasks, Mirabel, 108 flasks, Abbott, 189 flasks; Trinity County—Altoona, 4,032 flasks; San Benito County—New Idria, 5,000 flasks;

Sonoma County—Great Eastern, 1,704 flasks. There were a few small producers besides. Nearly all of these mines have produced more in other years, and all could produce more.

OPEN STOPE IN THE ALMADEN QUICKSILVER MINE.

The great quicksilver mine of California, and one of the greatest in the world, has been the New Almaden, in Santa Clara County, near San Francisco Bay. Here was the first discovery of quicksilver ore in the United States in 1824, and later, the first production. It was first worked in a small and crude way in 1845, but production did not practically begin until 1850, when the yield was over 7,000 flasks.

Since then, its annual yield has risen to 47,000 flasks, in 1865, and for many years it varied between 25,000 and 30,000 flasks, in some years falling far below those figures. In this mine the ore-bodies are "stock-works," arranged along definite fissures, but having a vein-like character. Two main fissures have been developed, and in them great ore-bodies have been found.

In 1864, there were fifteen miles of shafts and drifts in this great property, and, in 1894, the shafts and drifts, exclusive of excavations for the extraction of ore, amounted to over fifty miles, extending over an area of one square mile. Since then, the workings have been considerably extended. The plant at this mine is, throughout, the result of the best technical knowledge and skill that this feature of the mining industry has developed. The extensive reduction works include eight furnaces, planned and operated according to the most improved methods of working quicksilver ores, and the plant is the most complete and perfect one of its kind in the world. For many years, this mine produced two-thirds of the yield in California. Its present comparatively small production is not due to the exhaustion of the mine, but to business conditions.

An illustration of what this mine has been in the past is afforded by a statement of the earnings and expenses of the Quicksilver Mining Company, from January 1, 1871, to December 31, 1885. In that time it produced 299,822 flasks of quicksilver, of the average value of $35.12, and of a total value of $10,529,851.70. The expenses were $6,939,637.26, leaving a profit balance of $4,189,735.64, of which $947,666.55 was expended in improvements to the plant, etc. There are six main shafts on the property, which include 8,500 acres of land.

Another great quicksilver mine, recently coming into renewed prominence, is the New Idria, in San Benito County. It has been worked since 1858, and in 1868 produced 12,180 flasks. This mine has been developed to an extent of 1,500 feet, and recently large investments in development and in new reduction works have greatly increased its producing capacity.

Several of the other important mines at present producing, as heretofore given, have extensive developments and reduction works. The Altoona, in Trinity County, has recently undergone extensive development, and is being operated on a large scale.

VIEW OF FURNACES, NEW ALMADEN QUICKSILVER MINE, SANTA CLARA COUNTY.

The Evans Hydraulic Gravel Elevator.

BY

Thomas J. Barbour.

 HYDRAULIC mining, in all countries, is frequently bothered by poor dumping facilities. To obviate same, many schemes have been originated so as to allow the tailings to be carried through a sufficient length of sluice box to save the gold, and, at the same time, far enough away from the face of the claim to provide ample surface.

One of the many devices, which have lately attained great prominence, is that of the Evans Hydraulic Gravel Elevator. This machine is the invention of George H. Evans, a mining engineer, who developed it during the course of his operations in New Zealand. On his arrival in California, the writer assisted Mr. Evans in the development of the machine to the success it has attained to-day. We believe there are more of this system in operation in the United States than any other hydraulic elevators constructed.

Its principal features are the three suctions. The patent was on an elevator with more than one suction inlet. It is now built, for most of the gravel claims, with three openings, one of which is called the main suction, the other two, auxiliaries. The auxiliaries are principally used to balance the intake, reducing the wear and tear of the machine. They also increase the efficiency of the elevator by allowing the proper proportions of air to enter when the water and material in the main opening might, from any cause, become choked. The auxiliaries can be extended with any size pipe to a distance beyond the elevator proper, and are frequently used for draining places in the bedrock below the line of the sluices connecting with the elevator. This is a great advantage and can be carried on without interfering with the sluicing of the material through the main opening on the elevator seat.

Another point about the Evans Elevator is that you can sink a shaft to bedrock and place the machine in position by simply lowering it away, using a 5 to 8-foot length of suction pipe on the main opening, carrying all the material to sluice boxes by water pressure.

The machine is made in various sizes, and the ones most in use are easily manipulated and so light that they will sink their own pits in very short time, and thereby save cost of pumping out and excavating a pit by hand and windlass, which was frequently the rule until the invention of this machine.

All the connections, such as main suction, auxiliary suctions and water supply, are connected with swivel joints, so that they will adapt themselves to any adjustment necessary for connections.

Mr. Evans placed three machines, a year ago, on the Feather River, California, and in twelve hours had them fitted up and in place, twenty-four feet below the surface of the water, and, on turning on water, discharged 18,000,000 gallons of water each twenty-four hours.

Another point, on ground where seepage water is very great, is the fact of being able to recover the elevator, if from any cause it should be buried by a fall. In this case, one of the auxiliaries can be connected up with a hose and the pit gradually pumped out, until the machine is recovered.

The Evans Elevator is made in four sizes. Nos. 1 and 2 are adapted for mule-back transportation. The Nos. 3 and 4 are in such large sections that this would not be possible. The smallest size, or No. 1, will utilize water up to the capacity of a three-inch nozzle; the No. 2 will handle water up to the capacity of a four-inch nozzle; the No. 3, up to the capacity of an eight-inch nozzle; and the No. 4, up to a capacity of a ten-inch nozzle. The various sizes of throats vary in size from 3 inches to 20 inches diameter, and one can get some idea of the immense power stored in a machine of this kind when an 18-inch diameter boulder can be carried up to a height of sixty feet, with a pressure of water of four hundred feet.

The Nos. 1 and 2 can be built of steel plate, thereby lessening the weight where transportation facilities are difficult.

The quantity of material that can be moved by various elevators will depend entirely upon the pressure of water and the height the material is to be lifted. It is absolutely safe to figure on handling all the material that can be carried to the suction opening to the elevator by one-half the quantity of water used for its operation; that is, supposing we had an elevator using 400 inches of water, we could then use a giant that would consume 200 inches of water, and if the ground was of such a character that four yards of gravel could be moved per inch of water, then the capacity of the elevator would be 800 cubic yards. This is the basis of figuring that the writer has arrived at after visiting fifteen to twenty elevating propositions, all working in the United States.

The quantity of material that can be moved by a miner's inch of water varies greatly in different sections of the United States. The Risdon Iron Works, who manufacture the Evans Elevator, issue a pamphlet in which is given the duty of a miner's inch of water under various conditions. By that we find that in California, at the time when hydraulic mining was objected to by the United States Government authorities, the duty of a miner's inch was discovered to vary from one cubic yard to five cubic yards per miner's inch. This largely depends on the character of the gravel and the grade of the sluices. In the northern part of the State of California, there is an elevator plant running on the Oro Fino Mine, which consists of three elevators. The maximum grade of the sluices is two and one-quarter inches to each twelve-foot box, and at that point three No. 2 style of Evans Elevators, each discharging 400 inches of water, are handling 300 yards of material each, per day of twenty-four hours.

At Round Hill, in New Zealand, Mr. Evans operated for several years with this machine. This same size elevator, using a grade of five inches to the box of twelve feet, moved 2,000 yards of material per day with 400 inches of water. This will show the great variation in the quantity of material that can be moved by an elevator.

To properly estimate on an elevator, it is necessary to acquire the following information, and the manufacturers of the Evans system always require it before making estimates as to the cost of a machine or the quantity of work performed:

> 1st — Quantity of water available.
> 2nd— Quantity of water for giant.
> 3rd— Quantity of water for elevator.
> 4th— Head of water in feet.
> 5th— Distance that elevater must lift.
> 6th— Distance from bedrock to top of bank.
> 7th — Largest size of gravel to elevate.
> 8th— Grade of bedrock.
> 9th— Grade of discharge flume.
> 10th— Length of pipe line and sizes.
> 11th— Sketch of ground, if possible.

Another point in reference to the elevator, which requires very careful attention, is that the pipes are made of sufficient size that the greater part of the head is not absorbed by friction.

The writer found a case a very short time ago, in making examination of a hydraulic proposition, where forty per cent of the head was absorbed by friction. A great many people are under the impression that, where they have very large quantities of water, they can afford a loss in head, owing to the fact that water costs them nothing; but if a careful examination is made of the circumstances, it is usually found that they have absorbed at least one-half the value of their water by economy in the size of pipe lines. In all elevating propositions, the higher the head the more work can be accomplished, and we usually try to have one hundred feet pressure for each twelve to fifteen feet lift; but work has been very successfully accomplished at running as high as twenty to twenty-five feet lift per one hundred feet head. On one of the elevators on the Feather River, with 300 feet head, very good work was accomplished, lifting sixty-five feet.

In the operation of elevator plants, it is well to make the discharge box three times the depth of the sluices. In this way, it will give the elevator a better chance to discharge, and not throw the material on top of the discharge column time after time.

The Evans Machine works best when set at an angle of 75 degrees, and care should be taken that the box is amply strong to support the striking plate, and, at the same time, resist the force of water due to the spouting velocity.

In all flumes should be placed wrought-iron riffles. The Evans Riffles, which are used in connection with this elevator, will offer less resistance in carrying off the material than the old-style wood blocks. They are easily placed in position, and can be cleaned up with great rapidity.

Under-currents should be placed in the flume as near as possible to the paddock, so as to allow fine gold to settle, and not be carried off by the action of large boulders running down the sluice ways. Where very fine gold is found in a deposit, we advise the placing of what is known as the New Zealand gold-saving tables. They are made on the principle of an under-current, having tables covered with cocoa matting, and the writer is of the opinion that they will save finer gold than any other that has ever come under his notice.

In making connections of the various parts of the Evans Elevator, it is best to have pipe equipped with wrought-iron or angle-iron flanges. They do not break, and should all be made inter-changeable, so that in lowering the elevator to bedrock, or changing it, quick connections can be made.

The pamphlet issued by the Risdon Iron Works is very interesting reading to any one interested in hydraulic elevators, and it illustrates various types of machines built by that company.

This system was used in the unwatering of the Comstock Lode, and, at the present writing, has reduced the water to a height of 250 feet below the Sutro Tunnel. The distance for which the manufacturers designed the machine is 500 feet below that point.

The writer visited the Comstock Lode during the past month, and found the machine working very satisfactorily, and it was at that time discharging about 4,000 gallons per minute, using 130 miner's inches (each one and one-half cubic feet) of pressure water. The water pressure was under a head of 2,000 feet, and the lift at that time was 250 feet. The highest efficiency attained since commencing to operate was about forty per cent.

A machine of this kind, where water pressure is available, and the expense of same not prohibitive, is a very cheap way of unwatering auriferous mines, particularly if the head is sufficient to allow the machine to work successfully.

The development of the Evans system on the Comstock Lode will be watched with interest by engineers throughout all parts of the country, and, up to date, the success attained is very gratifying to the manufacturers.

Brief History of Concentration and Description of the Johnston Concentrator.

BY

GEORGE JOHNSTON.

ITTLE attention was given in California to the concentration of tailings and the saving of sulphurets prior to 1863, before which date sluices, with riffles raised at intervals until the sluices were full, were used, when the partial concentration was shoveled into other sluices for further treatment, or was treated in rockers, the headings being saved for chlorination or pan amalgamation. The Cornish buddle and percussion tables succeeded the sluices, and were a decided advance. Then came the Prater and Hungerford Pan (the best in use at that date), improved and patented September, 1864, by Zenas Wheeler. On March 14, 1865, Guido Kustel patented a revolving belt or table, described as "A flexible band of leather, India rubber, or of metal or wooden plates, etc." On April 17, 1866, a patent was issued to Joshua Hendy for a further improvement on the Prater and Hungerford Pan. Henry Alderson, on November 27, 1866, obtained patent for a percussion table. On March 12, 1867, Mr. Hungerford received a patent for further improvement on the Prater and Hungerford Pan ; and on March 19, 1867, T. Varney made a further improvement on the same pan.

On July 9, 1867, George Johnston and E. G. Smith obtained patent No. 66,499 on a concentrating table, of which the following was the principal claim: "The revolving belt, or apron F, with its raised edges O, having a shaking or rocking motion from side to side, substantially as and for the purposes described."

This was the key-note of concentration, and, with very little change, it has been in use ever since, all other manufacturers of concentrators being compelled to adopt more or less of the above described invention.

This system was different from all its predecessors. The pyrites were settled on the belt by gravity and the shake of the machine, while the upward travel of the revolving belt carried the sulphurets from under the sands through the clear water to the sulphuret box, in a state of concentration which was not attainable prior to this time.

All metallic ores, when pulverized, contain a greater or less proportion of particles of ore or metal that will, even if pulverized in water, float on the surface of the water, and, the finer the substances are pulverized, the greater the proportion of floating particles. The floating particles appear to possess some peculiar quality which repels the water from their surfaces, especially when such particles are exposed, even momentarily, to the atmosphere, and, when such exposures take place, the water is repelled from a sufficient portion of their surface to cause such particles to float on the surface of the water. So minute are these particles of float, that even when under water, they will remain in suspension for hours. These floats are of more importance than is generally supposed, their value often being many times greater than the clean sulphurets from the same ore. In one case, where the sulphurets assayed from $250 to $300 per ton, the settlings assayed $1,350 to $1,400 per ton from the first settling box, and, from the second settling chamber, $500 per ton. Of course, the percentage saved in this way is small, but, as it requires no attention, and costs only the original price of the settling boxes, it is well worth attending to. It must be remembered, that these float sulphurets are saved with belt concentrators, for the reason that they become entangled and protected by the sulphurets as they get down on the rubber belt, and do not have to slide the length of the percussion table, with machines of that class. Almost any concentrator will save coarse sulphurets, but it requires a belt machine to save the very fine ones. All mill men know the tenacity with which the fine sulphurets adhere to sluices or belts.

There are various opinions regarding the sizing of ores before concentration. We think the pulp from the amalgamating plates should go direct to the concentrator, where the coarse sulphurets will be all saved, with more of the very fine floats than can be saved at any other stage. There then remain only the very fine sulphurets to be separated from the slimes, which should go to a canvas plant for further treatment. This is the usual course adopted in our best mills, and we do not see how it can be improved on. The canvas plant is made of No. 6 duck covering, tables eight feet wide by twelve feet long. There is a drop of four inches from table to table, and the table has a grade of one and one-half inches to a foot. Once an hour, what deposits on these tables is washed down with a hose, and the washings passed over a belt concentrator to be further concentrated. Even this does not save all the fine concentrates. What has passed over this concentrator should be again passed over sluice tables, and, where the sulphurets are of high grade, a third concentration may pay.

In January, 1893, three patents were issued to George Johnston for improvements on the belt concentrator, the first of which was an undulating motion imparted to the belt for the purpose of preventing sand corners. This was accomplished by hanging the shaking table from four non-parallel hangers, imparting to the table an undulating motion, similar to the pan or batea in panning. This places the machine so under the control of the operator that he can, by increasing the inclination of the hangers, throw the pulp to the center, as is the case with the rocker, or, by moving the hangers to a vertical position, pile the pulp against the edges of the belt.

The throw of the eccentrics was increased, which, with the undulating motion, admitted of the shake being reduced from 190 to 120 shakes per minute, thereby reducing wear and tear of the machine, and lessening the disturbance of the sulphurets which had settled on the belt and were on their way to the head of the table.

The number of patents issued prior to October, 1874, was only 54; from that time to July, 1887, 325, and from that date to January, 1898, about 65. Of all this number, there are not more than twenty kinds in practical use, the belt machines being more numerous than all the others.

A MASSIVE BRIDGE PIER OF UNIQUE CONSTRUCTION.

One of the most interesting feats in bridge engineering is the method in which the center pier for the new Webster-street Draw-bridge, crossing the Estuary of San Antonio, Oakland, California, was built; Cotton Bros. & Co. of said city being the designers and contractors.

The pier is cylindrical in form, thirty-two feet six inches in diameter at the top, with a base forty-five feet in diameter, and thirty-six feet six inches high, constructed of concrete, with an eighteen-inch granite coping on top. One hundred and seventy-seven piles were driven and cut off one foot above the bed of the stream, twenty-seven feet below low water and thirty-four feet below high water, there being seven feet between tides. A grillage four feet thick and forty-five feet in diameter was built of timber and floated into position over the piles. Four Howe trusses, one hundred feet long, were erected, the ends resting on two pontoons, each thirty-two by one hundred feet. Twelve chains were attached to grillage by eye-bolts and run to top of Howe trusses and attached to wooden beams acting as levers, with a twenty-ton screw jack under one end of each beam or lever, and blocking at the other end. The concrete was placed on grillage and built up three feet high and allowed to set for twelve hours, then lowered by means of the jacks and chains to within a foot of the top of concrete. This method was continued until the pier settled on top of the piles, and in settling on the piles the variation was only one-quarter of an inch. In order that the weight of the pier, in sinking, should not exceed the weight figured to be carried by the trusses, the pier was built hollow in the center. The actual portion of pier sunk by the above means was a ring six feet by ten feet high, then battering to four feet eleven inches at the top of pier, the center afterwards being filled in solid.

The weight of concrete sunk was thirteen hundred tons, and at no time did the weight on the chains exceed three hundred tons.

EARLY STAGE IN CONSTRUCTION OF WEBSTER-STREET DRAW-BRIDGE PIER.

PLACING THE TRUSS OVER THE CENTER PIER.

Heat Insulation; Cat-tail Fibre.

BY
Captain William Borrowe.

Heat is one of the most insidious of natural elements. Few things arrest its course. Electricity can be barred, the Roetgen ray "brings up" against many substances, but heat stops at no wall. It has to hesitate and struggle in some cases, but will get through in some degree, and, singular to say, the element of highest resistance, one of the most profuse in nature, is common air.

Air in motion is not a bad conductor of heat, but fixed air stands at the head of the list, not by name, but in fact, as we sometimes see spongy paper and other things so classed; but this simply means a body of fixed air in the pores of the material held in cells or passages so intricate as to prevent circulation.

All heat non-conductors, whether mineral or vegetable, have this cellular feature—infusorial earth, for example, which consists of minute shells, the abandoned houses of the diatoms, now filled with air. Of course, the air cells may consist of material itself more or less refractory; but the essential feature seems to be fixed air.

In the industrial world, we find the singular circumstance of, on one hand, a persistent effort to attain the highest conduction of heat, as in the case of steam-boilers, and when heat is to be applied to liquids or gases under pressure; and, on the other hand, an equally persistent effort to prevent conduction, or, as it may be called, to conserve heat. This latter brings out the various non-conductiug substances, such as asbestos, saturated fibre, fossil meal, magnesia, and others, including last, and notably, the cat-tail fibre (Typha-Latifolia) of the Pacific Coast.

This last, while seemingly as fine and combustible as flossy silk, is, structurally. tubes filled with air; not only this, these tubes are divided into sections by partitions or diaphragms, so that if one division is broken, the others remain intact. This cat-tail fibre, when mixed with diatomaceous earth to form a plastic compost, is an ideal non-conductor, and also has the other required characteristics of tenacity and stability.

This discovery came about, as most others of the kind have done, by accident. The author, in making experiments with the different fibres, found that this flossy substance, which one would expect to explode, if ignited, refused to burn when the fields of rushes were burned over, the "cat-tails" remaining unconsumed. A lighted match or taper thrown among the loose material would not ignite it, because of the fixed air locked in its cells. It was like mineral wool, but with only a fraction of its weight, and, when the idea of its uses was once grasped, it required only a short time to devise moulded, powdered, and plastic forms for insulating purposes to retain or to repel heat. The reed on which the cat-tails grow is prolific and endless in the fluvial lands of California, and of these lands there are enough.

The effects attained by this peculiar substance are not only well settled by general results in practice, but by crucial experiments in one case, at Chicago, in October, 1898, when Mr. Gillespie, engineer of the Rookery Building, made comparative tests of the cat-tail fibre covering and other well known compositions, as follows:

Quoting from Mr. Gillespie's report on these experiments, he used as follows:

"One section four-inch cat-tail, furnished by California Anti-Caloric Company, San Francisco.

"One section four-inch cork, furnished by Nonpareil Cork Manufacturing Company, Chicago.

"One section four-inch magnesia, furnished by Watson & Co., Chicago.

"One section four-inch magnesia, K. & M., furnished by Walch & Wyeth, Chicago.

"One section four-inch asbestos, furnished by H. W. Jones & Co., Chicago."

He continues:

"Each of these sections was three feet long.

"The test was made in the following manner: The steam pipe used was four inches in diameter, and connected at one end with the main steam heater, sixteen inches in diameter; the other end to a high-speed engine. While the test was being made, this engine was put in commission, that there might be a constant flow of steam through this pipe. The sections of covering were placed on this pipe by a professional pipe coverer, in the following order: First, cork; second, asbestos; third, magnesia; fourth, magnesia, K. & M.; fifth, cat-tail fibre.

"In the center of each section a piece of cork, two inches thick and eight inches square, was circled out to fit about one-third the circumference of the pipe covering sections, and thoroughly cemented to the covering and made perfectly air-tight. The thermometer was inserted through the cork by making a hole in the cork sufficiently large to insert the thermometer. The mercury bulb was allowed to rest on the top of the covering. The hole made for the thermometer was then sealed, that there might be no circulation of air. The thermometer was then placed on the top of the bare steam pipe and thoroughly insulated, for the purpose of ascertaining the correct temperature on the outside of the steam pipe. A thermometer was then placed within one foot from the covering, and at the same height as the steam pipe from the floor. The steam pressure from the boilers, and the reading from each thermometer, were taken in a very careful manner every ten minutes for one hour and twenty minutes, with the following results:

Temperature resulting.   Fah. scale.

Steam.	Room.	Cork.	Asbestos.	Magnesia.	Magnesia, K. & M.	Cat-tail fibre.
332	133.4	182.6	178.5	172.1	178.8	165.5 "

From these results, Mr. Gillespie prepares the following table, showing the comparative resisting power of the various substances experimented with:

Cat-tail (anti-caloric) covering	1,000
Asbestos	800
Magnesia	813
Cork	639
Magnesia, K. & M.	541

CONTENTS

INDEX